Symmetry Breaking & Symmetry in Cosmology

" The Fundamentals of The Cosmological Timeline "

Edited by Paul F. Kisak

Contents

Chapter 1

Symmetry breaking

In physics, **symmetry breaking** is a phenomenon in which (infinitesimally) small fluctuations acting on a system crossing a critical point decide the system's fate, by determining which branch of a bifurcation is taken. To an outside observer unaware of the fluctuations (or "noise"), the choice will appear arbitrary. This process is called symmetry "breaking", because such transitions usually bring the system from a symmetric but disorderly state into one or more definite states. Symmetry breaking is supposed to play a major role in pattern formation.

In 1972, Nobel laureate P.W. Anderson used the idea of symmetry breaking to show some of the drawbacks of reductionism in his paper titled "More is different" in *Science*.[1]

Symmetry breaking can be distinguished into two types, explicit symmetry breaking and spontaneous symmetry breaking, characterized by whether the equations of motion fail to be invariant or the ground state fails to be invariant.

1.1 Explicit symmetry breaking

Main article: Explicit symmetry breaking

In explicit symmetry breaking, the equations of motion describing a system are variant under the broken symmetry.

1.2 Spontaneous symmetry breaking

Main article: Spontaneous symmetry breaking

In spontaneous symmetry breaking, the equations of motion of the system are invariant, but the system is not because the background (spacetime) of the system, its vacuum, is non-invariant. Such a symmetry breaking is parametrized by an order parameter. A special case of this type of symmetry breaking is dynamical symmetry breaking.

1.3 Examples

Symmetry breaking can cover any of the following scenarios:[2]

- The breaking of an exact symmetry of the underlying laws of physics by the random formation of some structure;

- A situation in physics in which a minimal energy state has less symmetry than the system itself;

- Situations where the actual state of the system does not reflect the underlying symmetries of the dynamics because the manifestly symmetric state is unstable (stability is gained at the cost of local asymmetry);

- Situations where the equations of a theory may have certain symmetries, though their solutions may not (the symmetries are "hidden").

One of the first cases of broken symmetry discussed in the physics literature is related to the form taken by a uniformly rotating body of incompressible fluid in gravitational and hydrostatic equilibrium. Jacobi[3] and soon later Liouville,[4] in 1834, discussed the fact that a tri-axial ellipsoid was an equilibrium solution for this problem when the kinetic energy compared to the gravitational energy of the rotating body exceeded a certain critical value. The axial symmetry presented by the McLaurin spheroids is broken at this bifurcation point. Furthermore, above this bifurcation point, and for constant angular momentum, the solutions that minimize the kinetic energy are the *non*-axially symmetric Jacobi ellipsoids instead of the Maclaurin spheroids.

1.4   See also

- Anomalous symmetry breaking

- Higgs mechanism

- QCD vacuum

- Goldstone boson

- 1964 PRL symmetry breaking papers

- J. J. Sakurai Prize for Theoretical Particle Physics

1.5   References

[1] Anderson, P.W. (1972). "More is Different" (PDF). *Science* **177** (4047): 393–396. Bibcode:1972Sci...177..393A. doi:10.1126/science. PMID 17796623.

[2] http://www.angelfire.com/stars5/astroinfo/gloss/s.html

[3] Jacobi, C.G.J. (1834). "Über die figur des gleichgewichts". *Annalen der Physik und Chemie* (33): 229–238.

[4] Liouville, J. (1834). "Sur la figure d'une masse fluide homogène, en équilibre et douée d'un mouvement de rotation". *Journal de l'École Polytechnique* (14): 289–296.

Chapter 2

Explicit symmetry breaking

In theoretical physics, **explicit symmetry breaking** is the breaking of a symmetry of a theory by terms in its defining equations of motion (most typically, to the Lagrangian or the Hamiltonian) that do not respect the symmetry. Usually this term is used in situations where these symmetry-breaking terms are small, so that the symmetry is approximately respected by the theory. An example is the spectral line splitting in the Zeeman effect, due to a magnetic interaction perturbation in the Hamiltonian of the atoms involved.

Explicit symmetry breaking differs from spontaneous symmetry breaking. In the latter, the defining equations respect the symmetry but the ground state (vacuum) of the theory breaks it.[1]

2.1   See also

- Symmetry breaking

2.2   References

[1] Castellani, E. (2003) "On the meaning of Symmetry Breaking" in Brading, K. and Castellani, E. (eds) Symmetries in Physics: New Reflections, Cambridge: Cambridge University Press

Chapter 3

Spontaneous symmetry breaking

Spontaneous symmetry breaking[1][2][3] is a mode of realization of symmetry breaking in a physical system, where the underlying laws are invariant under a symmetry transformation, but the system as a whole changes under such transformations, in contrast to explicit symmetry breaking. It is a spontaneous process by which a system in a symmetrical state ends up in an asymmetrical state. It thus describes systems where the equations of motion or the Lagrangian obey certain symmetries, but the lowest-energy solutions do not exhibit that symmetry.

Consider a symmetrical upward dome with a trough circling the bottom. If a ball is put at the very peak of the dome, the system is symmetrical with respect to a rotation around the center axis. But the ball may *spontaneously break* this symmetry by rolling down the dome into the trough, a point of lowest energy. Afterward, the ball has come to a rest at some fixed point on the perimeter. The dome and the ball retain their individual symmetry, but the system does not.[4]

Most simple phases of matter and phase transitions, like crystals, magnets, and conventional superconductors can be simply understood from the viewpoint of spontaneous symmetry breaking. Notable exceptions include topological phases of matter like the fractional quantum Hall effect.

3.1   Spontaneous symmetry breaking in physics

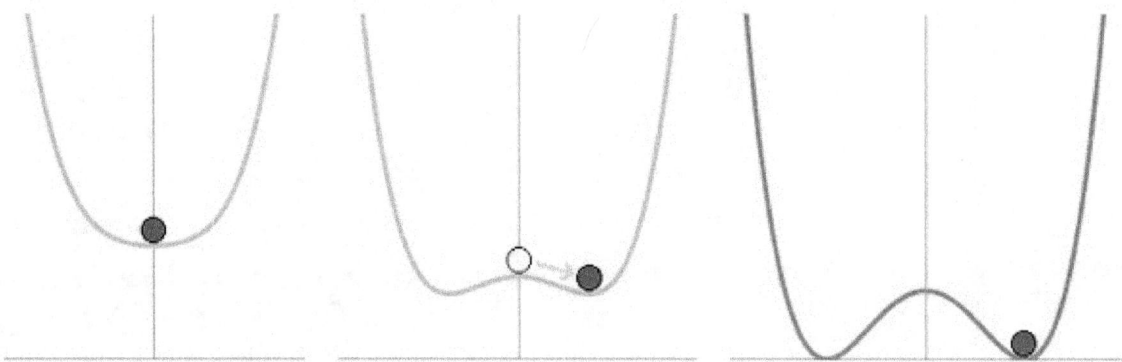

Spontaneous symmetry breaking simplified: – *At high energy levels (left) the ball settles in the center, and the result is symmetrical. At lower energy levels (right), the overall "rules" remain symmetrical, but the "Mexican hat" potential comes into effect: "local" symmetry is inevitably broken since eventually the ball must roll one way (at random) and not another.*

3.1.1   Particle physics

In particle physics the force carrier particles are normally specified by field equations with gauge symmetry; their equations predict that certain measurements will be the same at any point in the field. For instance, field equations might predict that the mass of two quarks is constant. Solving the equations to find the mass of each quark might give two solutions. In one solution, quark A is heavier than quark B. In the second solution, quark B is heavier than quark A *by the same amount*. The symmetry of the equations is not reflected by the individual solutions, but it is reflected by the range of solutions. An actual measurement reflects only one solution, representing a breakdown in the symmetry of the underlying theory. "Hidden" is perhaps a better term than "broken" because the symmetry is always there in these equations. This phenomenon is called *spontaneous* symmetry breaking because *nothing* (that we know) breaks the symmetry in the equations.[5]:194–195

Chiral symmetry

Main article: Chiral symmetry breaking

Chiral symmetry breaking is an example of spontaneous symmetry breaking affecting the chiral symmetry of the strong interactions in particle physics. It is a property of quantum chromodynamics, the quantum field theory describing these interactions, and is responsible for the bulk of the mass (over 99%) of the nucleons, and thus of all common matter, as it converts very light bound quarks into 100 times heavier constituents of baryons. The approximate Nambu–Goldstone bosons in this spontaneous symmetry breaking process are the pions, whose mass is an order of magnitude lighter than the mass of the nucleons. It served as the prototype and significant ingredient of the Higgs mechanism underlying the electroweak symmetry breaking.

Higgs mechanism

Main articles: Brout–Englert–Higgs mechanism and Yukawa interaction

The strong, weak, and electromagnetic forces can all be understood as arising from gauge symmetries. The Higgs mechanism, the spontaneous symmetry breaking of gauge symmetries, is an important component in understanding the superconductivity of metals and the origin of particle masses in the standard model of particle physics. One important consequence of the distinction between true symmetries and *gauge symmetries*, is that the spontaneous breaking of a gauge symmetry does not give rise to characteristic massless Nambu–Goldstone modes, but only massive modes, like the plasma mode in a superconductor, or the Higgs mode observed in particle physics.

In the standard model of particle physics, spontaneous symmetry breaking of the SU(2) × U(1) gauge symmetry associated with the electro-weak force generates masses for several particles, and separates the electromagnetic and weak forces. The W and Z bosons are the elementary particles that mediate the weak interaction, while the photon mediates the electromagnetic interaction. At energies much greater than 100 GeV all these particles behave in a similar manner. The Weinberg–Salam theory predicts that, at lower energies, this symmetry is broken so that the photon and the massive W and Z bosons emerge.[6] In addition, fermions develop mass consistently.

Without spontaneous symmetry breaking, the Standard Model of elementary particle interactions requires the existence of a number of particles. However, some particles (the W and Z bosons) would then be predicted to be massless, when, in reality, they are observed to have mass. To overcome this, spontaneous symmetry breaking is augmented by the Higgs mechanism to give these particles mass. It also suggests the presence of a new particle, the Higgs boson, reported as possibly identifiable with a boson detected in 2012. (If the Higgs boson were not confirmed to have been found, it would mean that the simplest implementation of the Higgs mechanism and spontaneous symmetry breaking *as they are currently formulated* require modification.)

Superconductivity of metals is a condensed-matter analog of the Higgs phenomena, in which a condensate of Cooper pairs of electrons spontaneously breaks the U(1) gauge "symmetry" associated with light and electromagnetism.

3.1.2   Condensed matter physics

Most phases of matter can be understood through the lens of spontaneous symmetry breaking. For example, crystals are periodic arrays of atoms that are not invariant under all translations (only under a small subset of translations by a lattice vector). Magnets have north and south poles that are oriented in a specific direction, breaking rotational symmetry. In addition to these examples, there are a whole host of other symmetry-breaking phases of matter including nematic phases of liquid crystals, charge- and spin-density waves, superfluids and many others.

There are several known examples of matter that cannot be described by spontaneous symmetry breaking, including: topologically ordered phases of matter like fractional quantum Hall liquids, and spin-liquids. These states do not break any symmetry, but are distinct phases of matter. Unlike the case of spontaneous symmetry breaking, there is not a general framework for describing such states.

Continuous symmetry

The ferromagnet is the canonical system which spontaneously breaks the continuous symmetry of the spins below the Curie temperature and at $h = 0$, where h is the external magnetic field. Below the Curie temperature the energy of the system is invariant under inversion of the magnetization $m(\mathbf{x})$ such that $m(\mathbf{x}) = -m(-\mathbf{x})$. The symmetry is spontaneously broken as $h \to 0$ when the Hamiltonian becomes invariant under the inversion transformation, but the expectation value is not invariant.

Spontaneously, symmetry broken phases of matter are characterized by an order parameter that describes the quantity which breaks the symmetry under consideration. For example, in a magnet, the order parameter is the local magnetization.

Spontaneously breaking of a continuous symmetry is inevitably accompanied by gapless (meaning that these modes do not cost any energy to excite) Nambu–Goldstone modes associated with slow long-wavelength fluctuations of the order parameter. For example, vibrational modes in a crystal, known as phonons, are associated with slow density fluctuations of the crystal's atoms. The associated Goldstone mode for magnets are oscillating waves of spin known as spin-waves. For symmetry-breaking states, whose order parameter is not a conserved quantity, Nambu–Goldstone modes are typically massless and propagate at a constant velocity.

An important theorem, due to Mermin and Wagner, states that, at finite temperature, thermally activated fluctuations of Nambu–Goldstone modes destroy the long-range order, and prevent spontaneous symmetry breaking in one- and two-dimensional systems. Similarly, quantum fluctuations of the order parameter prevent most types of continuous symmetry breaking in one-dimensional systems even at zero temperature (an important exception is ferromagnets, whose order parameter, magnetization, is an exactly conserved quantity and does not have any quantum fluctuations).

Other long-range interacting systems such as cylindrical curved surfaces interacting via the Coulomb potential or Yukawa potential has been shown to break translational and rotational symmetries.[7] It was shown, in the presence of a symmetric Hamiltonian, and in the limit of infinite volume, the system spontaneously adopts a chiral configuration, i.e. breaks mirror plane symmetry.

3.1.3   Dynamical symmetry breaking

Dynamical symmetry breaking (DSB) is a special form of spontaneous symmetry breaking where the ground state of the system has reduced symmetry properties compared to its theoretical description (Lagrangian).

Dynamical breaking of a global symmetry is a spontaneous symmetry breaking, that happens not at the (classical) tree level (i.e. at the level of the bare action), but due to quantum corrections (i.e. at the level of the effective action).

Dynamical breaking of a gauge symmetry is subtler. In the conventional spontaneous gauge symmetry breaking, there exists an unstable Higgs particle in the theory, which drives the vacuum to a symmetry-broken phase (see e.g. Electroweak interaction). In dynamical gauge symmetry breaking, however, no unstable Higgs particle operates in the theory, but the bound states of the system itself provide the unstable fields that render the phase transition. For example, Bardeen, Hill, and Lindner published a paper which attempts to replace the conventional Higgs mechanism in the standard model, by a DSB that is driven by a bound state of top-antitop quarks (such models, where a composite particle plays the role of the Higgs boson, are often referred to as "Composite Higgs models").[8] Dynamical breaking of gauge symmetries is often due to creation of a fermionic condensate; for example the quark condensate, which is connected to the dynamical breaking of chiral symmetry in quantum chromodynamics. Conventional superconductivity is the paradigmatic example from the condensed matter side, where phonon-mediated attractions lead electrons to become bound in pairs and then condense, thereby breaking the electromagnetic gauge symmetry.

3.2 Generalisation and technical usage

For spontaneous symmetry breaking to occur, there must be a system in which there are several equally likely outcomes. The system as a whole is therefore symmetric with respect to these outcomes. (If we consider any two outcomes, the probability is the same. This contrasts sharply to explicit symmetry breaking.) However, if the system is sampled (i.e. if the system is actually used or interacted with in any way), a specific outcome must occur. Though the system as a whole is symmetric, it is never encountered with this symmetry, but only in one specific asymmetric state. Hence, the symmetry is said to be spontaneously broken in that theory. Nevertheless, the fact that each outcome is equally likely is a reflection of the underlying symmetry, which is thus often dubbed "hidden symmetry", and has crucial formal consequences. (See the article on the Goldstone boson).

When a theory is symmetric with respect to a symmetry group, but requires that one element of the group be distinct, then spontaneous symmetry breaking has occurred. The theory must not dictate *which* member is distinct, only that *one is*. From this point on, the theory can be treated as if this element actually is distinct, with the proviso that any results found in this way must be resymmetrized, by taking the average of each of the elements of the group being the distinct one.

The crucial concept in physics theories is the order parameter. If there is a field (often a background field) which acquires an expectation value (not necessarily a *vacuum* expectation value) which is not invariant under the symmetry in question, we say that the system is in the ordered phase, and the symmetry is spontaneously broken. This is because other subsystems interact with the order parameter, which specifies a "frame of reference" to be measured against. In that case, the vacuum state does not obey the initial symmetry (which would keep it invariant, in the linearly realized **Wigner mode** in which it would be a singlet), and, instead changes under the (hidden) symmetry, now implemented in the (nonlinear) **Nambu–Goldstone mode**. Normally, in the absence of the Higgs mechanism, massless Goldstone bosons arise.

The symmetry group can be discrete, such as the space group of a crystal, or continuous (e.g., a Lie group), such as the rotational symmetry of space. However, if the system contains only a single spatial dimension, then only discrete symmetries may be broken in a vacuum state of the full quantum theory, although a classical solution may break a continuous symmetry.

3.3 A pedagogical example: the Mexican hat potential

In the simplest idealized relativistic model, the spontaneously broken symmetry is summarized through an illustrative scalar field theory. The relevant Lagrangian, which essentially dictates how a system behaves, can be split up into kinetic and potential terms,

It is in this potential term $V(\Phi)$ that the symmetry breaking is triggered. An example of a potential, due to Jeffrey Goldstone[9] is illustrated in the graph at the right.

This potential has an infinite number of possible minima (vacuum states) given by

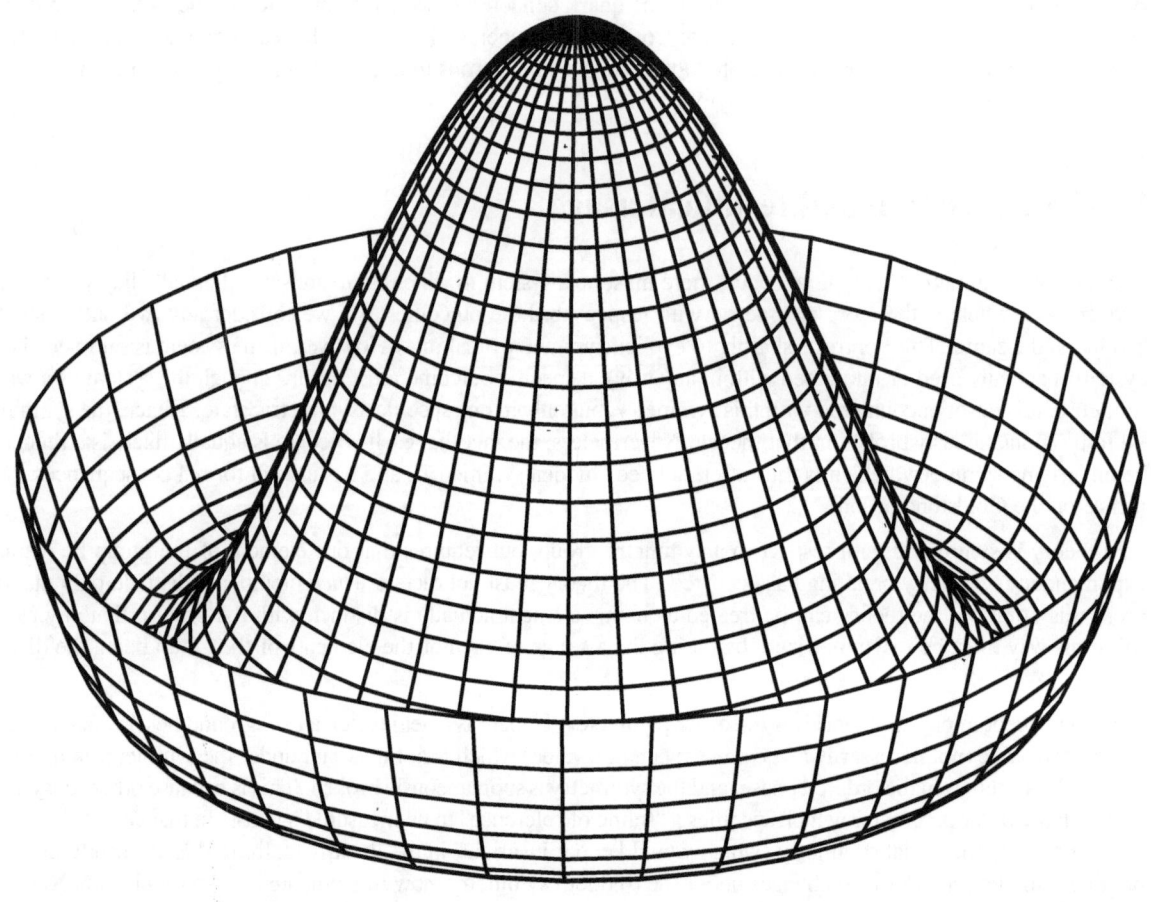

Graph of Goldstone's "Mexican hat" potential function V *versus* φ.

for any real θ between 0 and 2π. The system also has an unstable vacuum state corresponding to $\Phi = 0$. This state has a U(1) symmetry. However, once the system falls into a specific stable vacuum state (amounting to a choice of θ), this symmetry will appear to be lost, or "spontaneously broken".

In fact, any other choice of θ would have exactly the same energy, implying the existence of a massless Nambu–Goldstone boson, the mode running around the circle at the minimum of this potential, and indicating there is some memory of the original symmetry in the Lagrangian.

3.4   Other examples

- For ferromagnetic materials, the underlying laws are invariant under spatial rotations. Here, the order parameter is the magnetization, which measures the magnetic dipole density. Above the Curie temperature, the order parameter is zero, which is spatially invariant, and there is no symmetry breaking. Below the Curie temperature, however, the magnetization acquires a constant nonvanishing value, which points in a certain direction (in the idealized situation where we have full equilibrium; otherwise, translational symmetry gets broken as well). The residual rotational symmetries which leave the orientation of this vector invariant remain unbroken, unlike the other rotations which do not and are thus spontaneously broken.

- The laws describing a solid are invariant under the full Euclidean group, but the solid itself spontaneously breaks this group down to a space group. The displacement and the orientation are the order parameters.

- General relativity has a Lorentz symmetry, but in FRW cosmological models, the mean 4-velocity field defined by averaging over the velocities of the galaxies (the galaxies act like gas particles at cosmological scales) acts as an order parameter breaking this symmetry. Similar comments can be made about the cosmic microwave background.

- For the electroweak model, as explained earlier, a component of the Higgs field provides the order parameter breaking the electroweak gauge symmetry to the electromagnetic gauge symmetry. Like the ferromagnetic example, there is a phase transition at the electroweak temperature. The same comment about us not tending to notice broken symmetries suggests why it took so long for us to discover electroweak unification.

- In superconductors, there is a condensed-matter collective field ψ, which acts as the order parameter breaking the electromagnetic gauge symmetry.

- Take a thin cylindrical plastic rod and push both ends together. Before buckling, the system is symmetric under rotation, and so visibly cylindrically symmetric. But after buckling, it looks different, and asymmetric. Nevertheless, features of the cylindrical symmetry are still there: ignoring friction, it would take no force to freely spin the rod around, displacing the ground state in time, and amounting to an oscillation of vanishing frequency, unlike the radial oscillations in the direction of the buckle. This spinning mode is effectively the requisite Nambu–Goldstone boson.

- Consider a uniform layer of fluid over an infinite horizontal plane. This system has all the symmetries of the Euclidean plane. But now heat the bottom surface uniformly so that it becomes much hotter than the upper surface. When the temperature gradient becomes large enough, convection cells will form, breaking the Euclidean symmetry.

- Consider a bead on a circular hoop that is rotated about a vertical diameter. As the rotational velocity is increased gradually from rest, the bead will initially stay at its initial equilibrium point at the bottom of the hoop (intuitively stable, lowest gravitational potential). At a certain critical rotational velocity, this point will become unstable and the bead will jump to one of two other newly created equilibria, equidistant from the center. Initially, the system is symmetric with respect to the diameter, yet after passing the critical velocity, the bead ends up in one of the two new equilibrium points, thus breaking the symmetry.

3.5 Nobel Prize

On October 7, 2008, the Royal Swedish Academy of Sciences awarded the 2008 Nobel Prize in Physics to three scientists for their work in subatomic physics symmetry breaking. Yoichiro Nambu, of the University of Chicago, won half of the prize for the discovery of the mechanism of spontaneous broken symmetry in the context of the strong interactions, specifically chiral symmetry breaking. Physicists Makoto Kobayashi and Toshihide Maskawa shared the other half of the prize for discovering the origin of the explicit breaking of CP symmetry in the weak interactions.[10] This origin is ultimately reliant on the Higgs mechanism, but, so far understood as a "just so" feature of Higgs couplings, not a spontaneously broken symmetry phenomenon.

3.6 See also

- Autocatalytic reactions and order creation

- Catastrophe theory

- Chiral symmetry breaking

- CP-violation

- Explicit symmetry breaking

- Gauge gravitation theory

- Goldstone boson

- Grand unified theory

- Higgs mechanism

- Higgs boson

- Higgs field (classical)

- Irreversibility

- Magnetic catalysis of chiral symmetry breaking

- Mermin–Wagner theorem

- Quantum fluctuation

- Sakurai Prize for Theoretical Particle Physics

- Second-order phase transition

- Symmetry breaking

- Tachyon condensation

- Tachyonic field

- Wheeler–Feynman absorber theory

- 1964 PRL symmetry breaking papers

3.7   Notes

- ^ Note that (as in fundamental Higgs driven spontaneous gauge symmetry breaking) the term "symmetry breaking" is a misnomer when applied to gauge symmetries.

3.8   References

[1] *Dynamical Symmetry Breaking in Quantum Field Theories*. By Vladimir A. Miranskij. Pg 15.

[2] Patterns of Symmetry Breaking. Edited by Henryk Arodz, Jacek Dziarmaga, Wojciech Hubert Zurek. Pg 141.

[3] Bubbles, Voids and Bumps in Time: The New Cosmology. Edited by James Cornell. Pg 125.

[4] Gerald M. Edelman, Bright Air, Brilliant Fire: On the Matter of the Mind (New York: BasicBooks, 1992) 203.

[5] Steven Weinberg (20 April 2011). *Dreams of a Final Theory: The Scientist's Search for the Ultimate Laws of Nature*. Knopf Doubleday Publishing Group. ISBN 978-0-307-78786-6.

[6] A Brief History of Time, Stephen Hawking, Bantam; 10th anniversary edition (September 1, 1998). pp. 73–74.

[7] Kohlstedt, K.L.; Vernizzi, G.; Solis, F.J.; Olvera de la Cruz, M. (2007). "Spontaneous Chirality via Long-range Electrostatic Forces". *Physical Review Letters* **99**: 030602. arXiv:0704.3435. Bibcode:2007PhRvL..99c0602K. doi:10.1103/PhysRevLett.99.03060

[8] William A. Bardeen; Christopher T. Hill; Manfred Lindner (1990). "Minimal dynamical symmetry breaking of the standard model". *Physical Review D* **41** (5): 1647–1660. Bibcode:1990PhRvD..41.1647B. doi:10.1103/PhysRevD.41.1647.

[9] Goldstone, J. (1961). "Field theories with " Superconductor " solutions". *Il Nuovo Cimento* **19**: 154–164. doi:10.1007/BF02812722.

[10] The Nobel Foundation. "The Nobel Prize in Physics 2008". *nobelprize.org*. Retrieved January 15, 2008.

3.9 External links

- Spontaneous symmetry breaking

- Physical Review Letters – 50th Anniversary Milestone Papers

- In CERN Courier, Steven Weinberg reflects on spontaneous symmetry breaking

- Englert–Brout–Higgs–Guralnik–Hagen–Kibble Mechanism on Scholarpedia

- History of Englert–Brout–Higgs–Guralnik–Hagen–Kibble Mechanism on Scholarpedia

- The History of the Guralnik, Hagen and Kibble development of the Theory of Spontaneous Symmetry Breaking and Gauge Particles

- International Journal of Modern Physics A: The History of the Guralnik, Hagen and Kibble development of the Theory of Spontaneous Symmetry Breaking and Gauge Particles

- Guralnik, G S; Hagen, C R and Kibble, T W B (1967). Broken Symmetries and the Goldstone Theorem. Advances in Physics, vol. 2 Interscience Publishers, New York. pp. 567–708 ISBN 0-470-17057-3

- Spontaneous Symmetry Breaking in Gauge Theories: a Historical Survey

Chapter 4

Age of the universe

This article is about scientific estimates of the age of the universe. For religious and other non-scientific estimates, see Dating creation.

In physical cosmology, the **age of the universe** is the time elapsed since the Big Bang. The current measurement of the age of the universe is 13.799±0.021 billion years ((13.799±0.021)×10^9 years) within the Lambda-CDM concordance model.[1][2][3] The uncertainty of 21 million years has been obtained by the agreement of a number of scientific research projects, such as microwave background radiation measurements by the Planck satellite, the Wilkinson Microwave Anisotropy Probe and other probes. Measurements of the cosmic background radiation give the cooling time of the universe since the Big Bang,[4] and measurements of the expansion rate of the universe can be used to calculate its approximate age by extrapolating backwards in time.

4.1 Explanation

The Lambda-CDM concordance model describes the evolution of the universe from a very uniform, hot, dense primordial state to its present state over a span of about 13.8 billion years[5] of cosmological time. This model is well understood theoretically and strongly supported by recent high-precision astronomical observations such as WMAP. In contrast, theories of the origin of the primordial state remain very speculative. If one extrapolates the Lambda-CDM model backward from the earliest well-understood state, it quickly (within a small fraction of a second) reaches a singularity called the "Big Bang singularity". This singularity is not understood as having a physical significance in the usual sense, but it is convenient to quote times measured "since the Big Bang" even though they do not correspond to a physically measurable time. For example, "10^{-6} seconds after the Big Bang" is a well-defined era in the universe's evolution. If one referred to the same era as "13.8 billion years minus 10^{-6} seconds ago", the precision of the meaning would be lost because the minuscule latter time interval is swamped by uncertainty in the former.

Though the universe might in theory have a longer history, the International Astronomical Union[6] presently use "age of the universe" to mean the duration of the Lambda-CDM expansion, or equivalently the elapsed time since the Big Bang in the current observable universe.

4.2 Observational limits

Since the universe must be at least as old as the oldest thing in it, there are a number of observations which put a lower limit on the age of the universe; these include the temperature of the coolest white dwarfs, which gradually cool as they age, and the dimmest turnoff point of main sequence stars in clusters (lower-mass stars spend a greater amount of time on the main sequence, so the lowest-mass stars that have evolved off of the main sequence set a minimum age).

4.3   Cosmological parameters

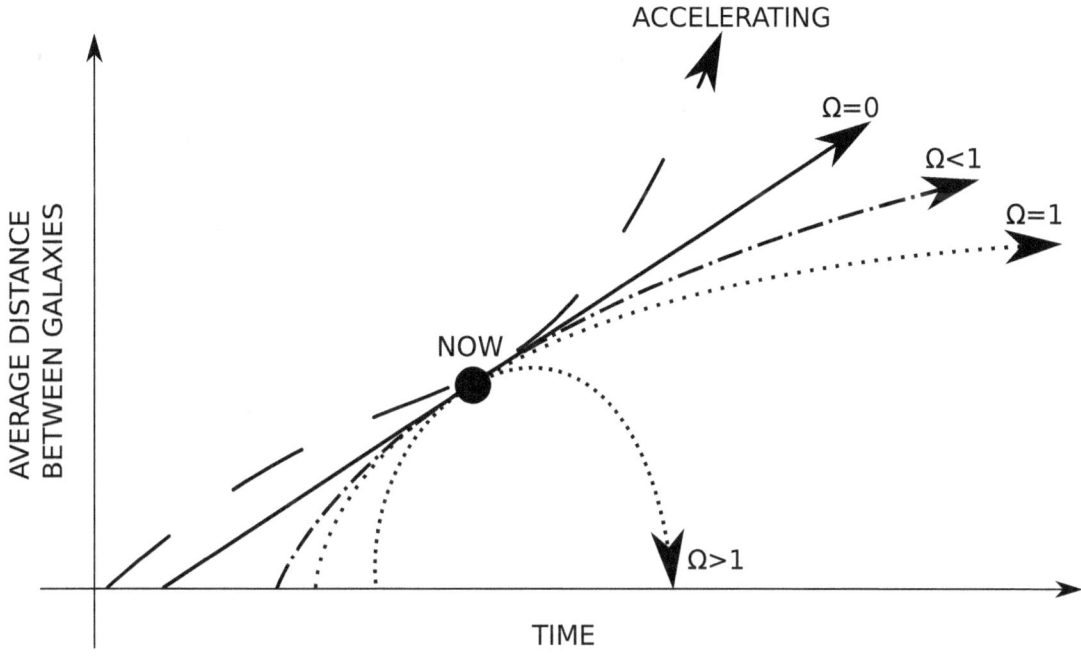

The age of the universe can be determined by measuring the Hubble constant today and extrapolating back in time with the observed value of density parameters (Ω). Before the discovery of dark energy, it was believed that the universe was matter-dominated, and so Ω on this graph corresponds to Ω_{m}. Note that the accelerating universe has the greatest age, while the Big Crunch universe has the least age.

The problem of determining the age of the universe is closely tied to the problem of determining the values of the cosmological parameters. Today this is largely carried out in the context of the ΛCDM model, where the universe is assumed to contain normal (baryonic) matter, cold dark matter, radiation (including both photons and neutrinos), and a cosmological constant. The fractional contribution of each to the current energy density of the universe is given by the density parameters Ωm, Ωr, and $\Omega \Lambda$. The full ΛCDM model is described by a number of other parameters, but for the purpose of computing its age these three, along with the Hubble parameter H_0, are the most important.

If one has accurate measurements of these parameters, then the age of the universe can be determined by using the Friedmann equation. This equation relates the rate of change in the scale factor $a(t)$ to the matter content of the universe. Turning this relation around, we can calculate the change in time per change in scale factor and thus calculate the total age of the universe by integrating this formula. The age t_0 is then given by an expression of the form

$$t_0 = \frac{1}{H_0} F(\Omega_r, \Omega_m, \Omega_\Lambda, \dots)$$

where H_0 is the Hubble parameter and the function F depends only on the fractional contribution to the universe's energy content that comes from various components. The first observation that one can make from this formula is that it is the Hubble parameter that controls that age of the universe, with a correction arising from the matter and energy content. So a rough estimate of the age of the universe comes from the Hubble time, the inverse of the Hubble parameter. With a value for H_0 around 68 km/s/Mpc, the Hubble time evaluates to $1/H_0 = 14.4$ billion years.[7]

To get a more accurate number, the correction factor F must be computed. In general this must be done numerically, and the results for a range of cosmological parameter values are shown in the figure. For the Planck values $(\Omega m, \Omega \Lambda) = (0.3086, 0.6914)$, shown by the box in the upper left corner of the figure, this correction factor is about $F = 0.956$. For a flat universe without any cosmological constant, shown by the star in the lower right corner, $F = {}^2/_3$ is much smaller and thus the universe is younger for a fixed value of the Hubble parameter. To make this figure, Ωr is held constant (roughly

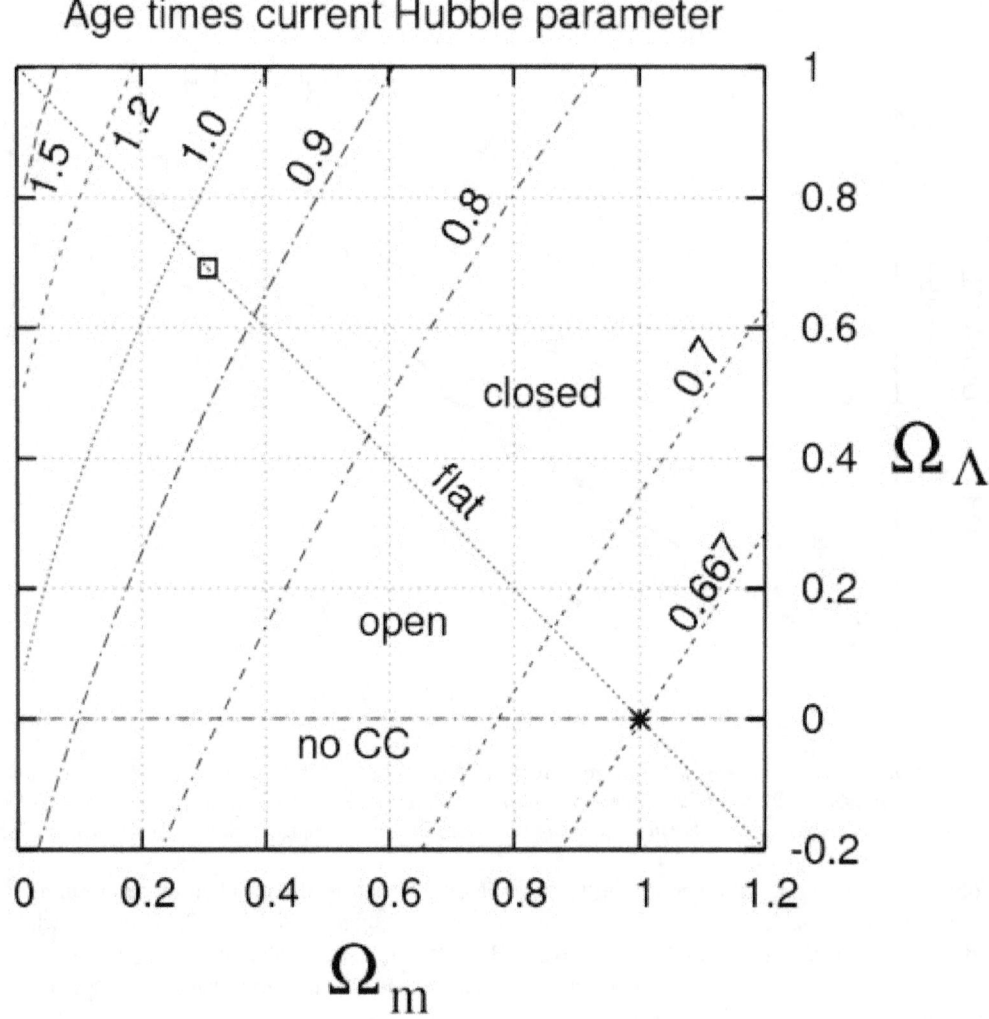

The value of the age correction factor, F, is shown as a function of two cosmological parameters: the current fractional matter density Ω_m and cosmological constant density $\Omega\Lambda$. The best-fit values of these parameters are shown by the box in the upper left; the matter-dominated universe is shown by the star in the lower right.

equivalent to holding the CMB temperature constant) and the curvature density parameter is fixed by the value of the other three.

Apart from the Planck satellite, the Wilkinson Microwave Anisotropy Probe (WMAP) was instrumental in establishing an accurate age of the universe, though other measurements must be folded in to gain an accurate number. CMB measurements are very good at constraining the matter content Ωm[8] and curvature parameter Ωk.[9] It is not as sensitive to $\Omega\Lambda$ directly,[9] partly because the cosmological constant becomes important only at low redshift. The most accurate determinations of the Hubble parameter H_0 come from Type Ia supernovae. Combining these measurements leads to the generally accepted value for the age of the universe quoted above.

The cosmological constant makes the universe "older" for fixed values of the other parameters. This is significant, since before the cosmological constant became generally accepted, the Big Bang model had difficulty explaining why globular clusters in the Milky Way appeared to be far older than the age of the universe as calculated from the Hubble parameter and a matter-only universe.[10][11] Introducing the cosmological constant allows the universe to be older than these clusters, as well as explaining other features that the matter-only cosmological model could not.[12]

4.4 WMAP

NASA's Wilkinson Microwave Anisotropy Probe (WMAP) project's nine-year data release in 2012 estimated the age of the universe to be $(13.772\pm0.059)\times10^9$ years (13.772 billion years, with an uncertainty of plus or minus 59 million years).[4]

However, this age is based on the assumption that the project's underlying model is correct; other methods of estimating the age of the universe could give different ages. Assuming an extra background of relativistic particles, for example, can enlarge the error bars of the WMAP constraint by one order of magnitude.[13]

This measurement is made by using the location of the first acoustic peak in the microwave background power spectrum to determine the size of the decoupling surface (size of the universe at the time of recombination). The light travel time to this surface (depending on the geometry used) yields a reliable age for the universe. Assuming the validity of the models used to determine this age, the residual accuracy yields a margin of error near one percent.[14]

4.5 Planck

In 2015, the European Space Agency's Planck spacecraft team estimated the age of the universe to be 13.813 ± 0.038 billion years,[1] slightly higher but within the uncertainties of the earlier number derived from the WMAP data. By combining the Planck data with external data, the best combined estimate of the age of the universe is $(13.799\pm0.021)\times10^9$ years old.[1][2]

4.6 Assumption of strong priors

Calculating the age of the universe is accurate only if the assumptions built into the models being used to estimate it are also accurate. This is referred to as strong priors and essentially involves stripping the potential errors in other parts of the model to render the accuracy of actual observational data directly into the concluded result. Although this is not a valid procedure in all contexts (as noted in the accompanying caveat: "based on the fact we have assumed the underlying model we used is correct"), the age given is thus accurate to the specified error (since this error represents the error in the instrument used to gather the raw data input into the model).

The age of the universe based on the best fit to Planck 2015 data alone is 13.813 ± 0.038 billion years (the estimate of 13.799 ± 0.021 billion years uses Gaussian priors based on earlier estimates from other studies to determine the combined uncertainty). This number represents an accurate "direct" measurement of the age of the universe (other methods typically involve Hubble's law and the age of the oldest stars in globular clusters, etc.). It is possible to use different methods for determining the same parameter (in this case – the age of the universe) and arrive at different answers with no overlap in the "errors". To best avoid the problem, it is common to show two sets of uncertainties; one related to the actual measurement and the other related to the systematic errors of the model being used.

An important component to the analysis of data used to determine the age of the universe (e.g. from Planck) therefore is to use a Bayesian statistical analysis, which normalizes the results based upon the priors (i.e. the model).[14] This quantifies any uncertainty in the accuracy of a measurement due to a particular model used.[15][16]

4.7 History

Main article: Cosmic age problem

In the 18th century, the concept that the age of the Earth was millions, if not billions, of years began to appear. However, most scientists throughout the 19th century and into the first decades of the 20th century presumed that the universe itself was Steady State and eternal, with maybe stars coming and going but no changes occurring at the largest scale known at the time.

The first scientific theories indicating that the age of the universe might be finite were the studies of thermodynamics, formalized in the mid-19th century. The concept of entropy dictates that if the universe (or any other closed system) were infinitely old, then everything inside would be at the same temperature, and thus there would be no stars and no life. No scientific explanation for this contradiction was put forth at the time.

In 1915 Albert Einstein published the theory of general relativity[17] and in 1917 constructed the first cosmological model based on his theory. In order to remain consistent with a steady state universe, Einstein added what was later called a cosmological constant to his equations. However, already in 1922, also using Einstein's theory, Alexander Friedmann, and independently five years later Georges Lemaître, showed that the universe cannot be static and must be either expanding or contracting. Einstein's model of a static universe was in addition proved unstable by Arthur Eddington.

The first direct observational hint that the universe has a finite age came from the observations of 'recession velocities', mostly by Vesto Slipher, combined with distances to the 'nebulae' (galaxies) by Edwin Hubble in a work published in 1929.[18] Earlier in the 20th century, Hubble and others resolved individual stars within certain nebulae, thus determining that they were galaxies, similar to, but external to, our Milky Way Galaxy. In addition, these galaxies were very large and very far away. Spectra taken of these distant galaxies showed a red shift in their spectral lines presumably caused by the Doppler effect, thus indicating that these galaxies were moving away from the Earth. In addition, the farther away these galaxies seemed to be (the dimmer they appeared to us) the greater was their redshift, and thus the faster they seemed to be moving away. This was the first direct evidence that the universe is not static but expanding. The first estimate of the age of the universe came from the calculation of when all of the objects must have started speeding out from the same point. Hubble's initial value for the universe's age was very low, as the galaxies were assumed to be much closer than later observations found them to be.

The first reasonably accurate measurement of the rate of expansion of the universe, a numerical value now known as the Hubble constant, was made in 1958 by astronomer Allan Sandage.[19] His measured value for the Hubble constant came very close to the value range generally accepted today.

However Sandage, like Einstein, did not believe his own results at the time of discovery. His value for the age of the universe was too short to reconcile with the 25-billion-year age estimated at that time for the oldest known stars. Sandage and other astronomers repeated these measurements numerous times, attempting to reduce the Hubble constant and thus increase the resulting age for the universe. Sandage even proposed new theories of cosmogony to explain this discrepancy. This issue was finally resolved by improvements in the theoretical models used for estimating the ages of stars. As of 2013, using the latest models for stellar evolution, the estimated age of the oldest known star is 14.46±0.8 billion years.[20]

The discovery of microwave cosmic background radiation announced in 1965[21] finally brought an effective end to the remaining scientific uncertainty over the expanding universe. The recently launched space probes WMAP, launched in 2001, and Planck, launched in 2009, produced data that determines the Hubble constant and the age of the universe independent of galaxy distances, removing the largest source of error.[14]

More recently, in February 2015, an alternative view to extend the Big Bang model was presented that suggests the Universe had no beginning or singularity and that the age of the Universe may be infinite.[22][23][24]

4.8   See also

- Age of the Earth

- Anthropic principle

- Cosmic age problem

- Cosmic Calendar (age of universe scaled to a single year)

- Cosmology

- Dark Ages Radio Explorer (DARE)

- Hubble Deep Field

- Illustris project

- Metric expansion of space

- Multiverse

- Observable universe

- Red shift observations in astronomy

- Static universe

- *The First Three Minutes* (1977 book by Steven Weinberg).

4.9 References

[1] Planck Collaboration (2015). "Planck 2015 results. XIII. Cosmological parameters (See Table 4 on page 31 of PDF).". arXiv:1502.01589.

[2] C. R. Lawrence, JPL, for the Planck Collaboration, Astrophysics Subcommittee, NASA HQ (18 March 2015) "Planck 2015 Results" (See page 29 of pdf) http://science.nasa.gov/media/medialibrary/2015/04/08/CRL_APS_2015-03-18_compressed2.pdf

[3] http://www.teachastronomy.com/astropedia/article/Precision-Cosmology

[4] Bennett, C.L.; et al. (2013). "Nine-Year Wilkinson Microwave Anisotropy Probe (WMAP) Observations: Final Maps and Results". arXiv:1212.5225 [astro-ph.CO].

[5] "Cosmic Detectives". European Space Agency. 2 April 2013. Retrieved 2013-04-15.

[6] Chang, K. (9 March 2008). "Gauging Age of Universe Becomes More Precise". *The New York Times*.

[7] Liddle, A. R. (2003). *An Introduction to Modern Cosmology* (2nd ed.). Wiley. p. 57. ISBN 0-470-84835-9.

[8] Hu, W. "Animation: Matter Content Sensitivity. The matter-radiation ratio is raised while keeping all other parameters fixed.". University of Chicago. Archived from the original on 23 February 2008. Retrieved 2008-02-23.

[9] Hu, W. "Animation: Angular diameter distance scaling with curvature and lambda". University of Chicago. Archived from the original on 23 February 2008. Retrieved 2008-02-23.

[10] "Globular Star Clusters". SEDS. 1 July 2011. Archived from the original on 24 February 2008. Retrieved 2013-07-19.

[11] Iskander, E. (11 January 2006). "Independent age estimates". University of British Columbia. Archived from the original on 6 March 2008. Retrieved 2008-02-23.

[12] Ostriker, J. P.; Steinhardt, P. J. (1995). "Cosmic Concordance". arXiv:astro-ph/9505066.

[13] de Bernardis, F.; Melchiorri, A.; Verde, L.; Jimenez, R. (2008). "The Cosmic Neutrino Background and the Age of the Universe". *Journal of Cosmology and Astroparticle Physics* **2008** (3): 20. arXiv:0707.4170. Bibcode:2008JCAP...03..020D. doi:10.1088/1475-7516/2008/03/020.

[14] Spergel, D. N.; et al. (2003). "First-Year Wilkinson Microwave Anisotropy Probe (WMAP) Observations: Determination of Cosmological Parameters". *The Astrophysical Journal Supplement Series* **148** (1): 175–194. arXiv:astro-ph/0302209. Bibcode:2003ApJS..148..175S. doi:10.1086/377226.

[15] Loredo, T. J. (1992). "The Promise of Bayesian Inference for Astrophysics" (PDF). In Feigelson, E. D.; Babu, G. J. *Statistical Challenges in Modern Astronomy*. Springer-Verlag. pp. 275–297. Bibcode:1992scma.conf..275L. doi:10.1007/978-1-4613-9290-3_31. ISBN 978-1-4613-9292-7.

[16] Colistete, R.; Fabris, J. C.; Concalves, S. V. B. (2005). "Bayesian Statistics and Parameter Constraints on the Generalized Chaplygin Gas Model Using SNe ia Data". *International Journal of Modern Physics D* **14** (5): 775–796. arXiv:astro-ph/0409245. Bibcode:2005IJMPD..14..775C. doi:10.1142/S0218271805006729.

[17] Einstein, A. (1915). "Zur allgemeinen Relativitätstheorie". *Sitzungsberichte der Königlich Preußischen Akademie der Wissenschaften* (in German): 778–786. Bibcode:1915SPAW.......778E.

[18] Hubble, E. (1929). "A relation between distance and radial velocity among extra-galactic nebulae". *Proceedings of the National Academy of Sciences* **15** (3): 168–173. Bibcode:1929PNAS...15..168H. doi:10.1073/pnas.15.3.168. PMC 522427. PMID 16577160.

[19] Sandage, A. R. (1958). "Current Problems in the Extragalactic Distance Scale". *The Astrophysical Journal* **127** (3): 513–526. Bibcode:1958ApJ...127..513S. doi:10.1086/146483.

[20] Bond, H. E.; Nelan, E. P.; Vandenberg, D. A.; Schaefer, G. H.; Harmer, D. (2013). "HD 140283: A Star in the Solar Neighborhood that Formed Shortly After the Big Bang". *The Astrophysical Journal* **765** (12): L12. arXiv:1302.3180. Bibcode:2013ApJ...765L..12B. doi:10.1088/2041-8205/765/1/L12.

[21] Penzias, A. A.; Wilson, R .W. (1965). "A Measurement of Excess Antenna Temperature at 4080 Mc/s". *The Astrophysical Journal* **142**: 419–421. Bibcode:1965ApJ...142..419P. doi:10.1086/148307.

[22] Ghose, Ti a (26 February 2015). "Bi g Bang, De flated? Universe Ma y Hav e Had No Beginning". *Live Science.* February 2015.

[23] Das, Saurya; Bhaduri, Rajat K. (18 November 2014). "Dark matter and dark energy from Bose-Einstein condensate"

[24] Ali, Ahmed Faraq (4 February 2015). "Cosmology from quantum potential". *Physics Letters B* **741**: 276–279. doi Retrieved 28 February 2015.

4.10   External links

- Ned Wright's Cosmology Tutorial

- Wright, Edward L. (2 July 2005). "Age of the Universe".

- Wayne Hu's cosmological parameter animations

- Ostriker; Steinhardt (1995). "Cosmic Concordance". arXiv:astro-ph/9505066.

- SEDS page on "Globular Star Clusters"

- Douglas Scott "Independent Age Estimates"

- KryssTal "The Scale of the Universe" Space and Time scaled for the beginner.

- iCosmos: Cosmology Calculator (With Graph Generation)

- The Expanding Universe (American Institute of Physics)

Chapter 5

Chronology of the universe

See also: Timeline of the formation of the Universe

The **chronology of the universe** describes the history and future of the universe according to Big Bang cosmology, the

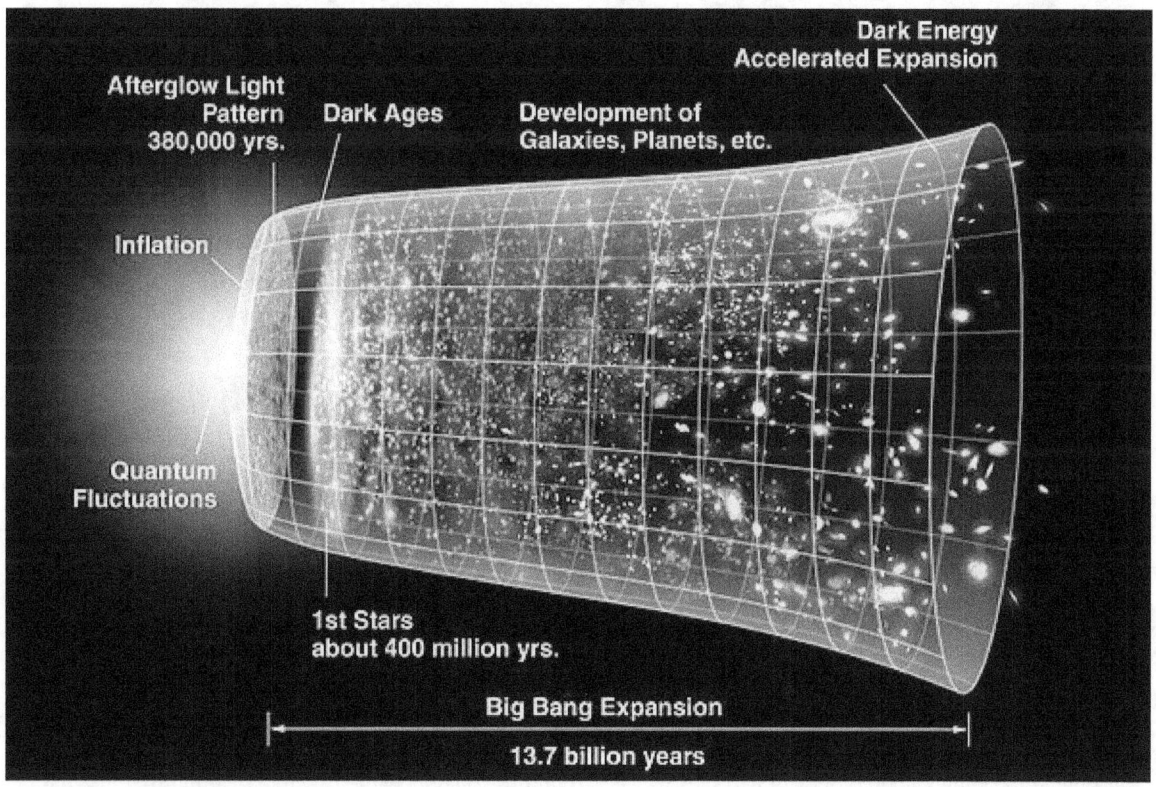

Diagram of evolution of the (observable part) of the universe from the Big Bang (left) - to the present.

prevailing scientific model of how the universe developed over time from the Planck epoch, using the cosmological time parameter of comoving coordinates. The model of the universe's expansion is known as the Big Bang. As of 2015, this expansion is estimated to have begun 13.799 ± 0.021 billion years ago.[1] It is convenient to divide the evolution of the universe so far into three phases.

5.1   Summary

In the first phase, the very earliest universe was so hot, or energetic, that initially no matter particles existed or could exist perhaps only fleetingly. According to prevailing scientific theories, at this time the distinct forces we see around us today were joined in one unified force. Space-time itself expanded during an inflationary epoch due to the immensity of the energies involved. Gradually the immense energies cooled – still to a temperature inconceivably hot compared to any we see around us now, but sufficiently to allow forces to gradually undergo symmetry breaking, a kind of repeated condensation from one status quo to another, leading finally to the separation of the strong force from the electroweak force and the first particles.

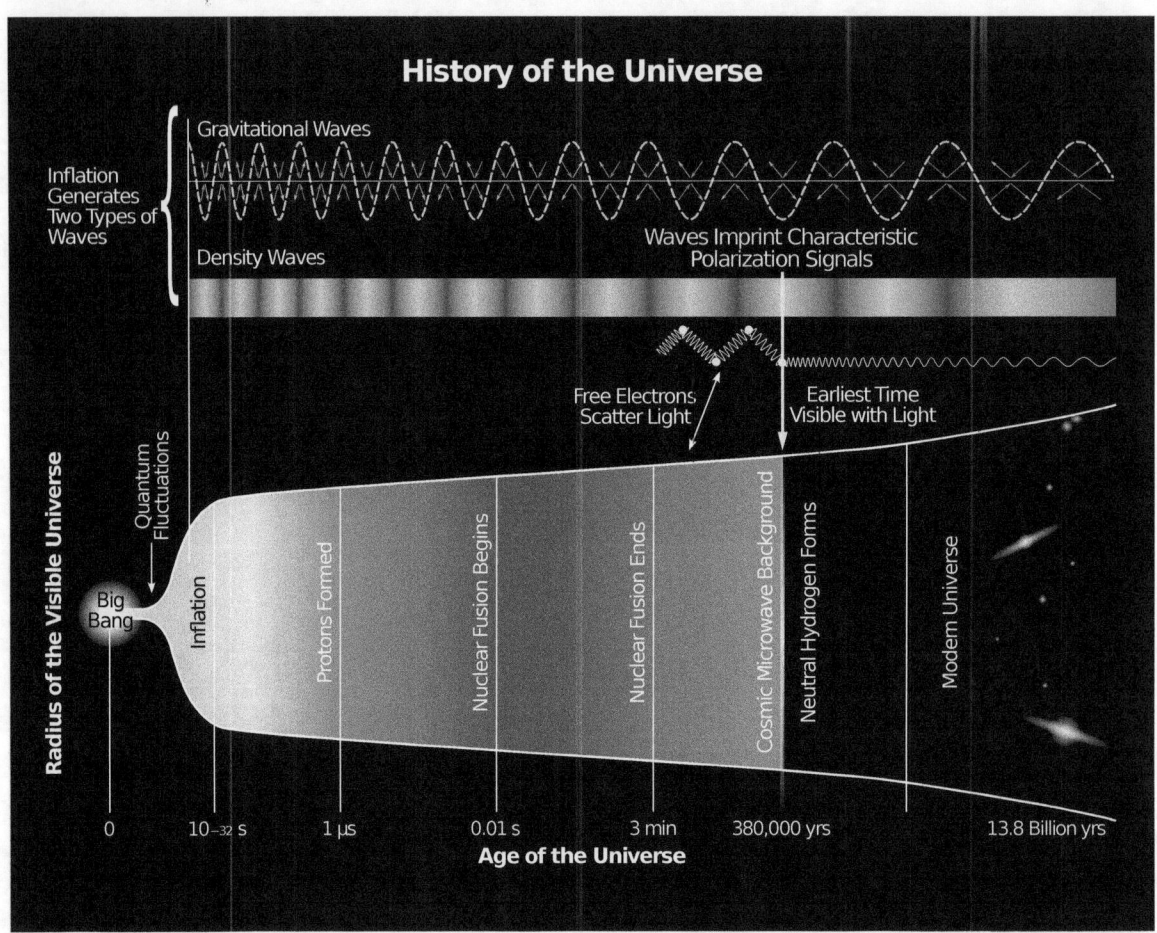

History of the Universe - gravitational waves are hypothesized to arise from cosmic inflation, a faster-than-light expansion just after the Big Bang (17 March 2014).[2][3][4]

In the second phase, the resulting quark–gluon plasma universe then cooled further, the current fundamental forces we know take their present forms through further symmetry breaking – notably the breaking of electroweak symmetry – and the full range of complex and composite particles we see around us today became possible, leading to a gravitationally dominated universe, the first neutral atoms (~ 80% hydrogen), and the cosmic microwave background radiation we can detect today. Modern high energy particle physics theories are satisfactory at these energy levels, and so physicists believe they have a good understanding of this and subsequent development of the fundamental universe around us. Because of these changes, space had also become largely transparent to light and other electromagnetic energy, rather than "foggy", by the end of this phase.

The third phase started after a short dark age with a universe whose fundamental particles and forces were as we know them, and witnessed the emergence of large scale stable structures, such as the earliest stars, quasars, galaxies, clusters of galaxies and superclusters, and the development of these to create the kind of universe we see today. Some researchers call the development of all this physical structure over billions of years "cosmic evolution". Other, more interdisciplinary, researchers refer to "cosmic evolution" as the entire scenario of growing complexity from big bang to humankind, thereby incorporating biology and culture into a unified view of all complex systems in the universe to date.[5]

Beyond the present day, scientists anticipate that the Earth will cease to be able to support life in about a billion years, and will be enveloped by a greatly-expanded Sun in about 5 billion years. On a far longer timescale, the Stelliferous Era will end as stars eventually die and fewer are born to replace them, leading to a darkening universe. Various theories suggest a number of subsequent possibilities. If particles such as protons are unstable then eventually matter may evaporate into low level energy in a kind of entropy related heat death. Alternatively the universe may collapse in a big crunch, although current data shows the rate of expansion is still increasing. If this is correct then it may end in a "big freeze" as matter and energy become very thinly spread and cool down. Alternative suggestions include a false vacuum catastrophe or a Big Rip as possible ends to the universe.

5.2 Very early universe

All ideas concerning the very early universe (cosmogony) are speculative. No accelerator experiments have yet probed energies of sufficient magnitude to provide any experimental insight into the behavior of matter at the energy levels that prevailed during this period. Proposed scenarios differ radically. Some examples are the Hartle–Hawking initial state, string landscape, brane inflation, string gas cosmology, and the ekpyrotic universe. Some of these are mutually compatible, while others are not.

5.2.1 Planck epoch

0 to 10^{-43} second after the Big Bang

Main article: Planck epoch

The Planck epoch is an era in traditional (non-inflationary) big bang cosmology wherein the temperature was so high that the four fundamental forces—electromagnetism, gravitation, weak nuclear interaction, and strong nuclear interaction— were one fundamental force. Little is understood about physics at this temperature; different hypotheses propose different scenarios. Traditional big bang cosmology predicts a gravitational singularity before this time, but this theory relies on general relativity and is expected to break down due to quantum effects.

In inflationary cosmology, times before the end of inflation (roughly 10^{-32} second after the Big Bang) do not follow the traditional big bang timeline.

5.2.2 Grand unification epoch

Between 10^{-43} second and 10^{-36} second after the Big Bang[6]

Main article: Grand unification epoch

As the universe expanded and cooled, it crossed transition temperatures at which forces separate from each other. These are phase transitions much like condensation and freezing. The grand unification epoch began when gravitation separated from the other forces of nature, which are collectively known as gauge forces. The non-gravitational physics in this epoch would be described by a so-called grand unified theory (GUT). The grand unification epoch ended when the GUT forces further separate into the strong and electroweak forces.

5.2.3 Electroweak epoch

Between 10^{-36} second (or the end of inflation) and 10^{-32} second after the Big Bang[6]

Main article: Electroweak epoch

According to traditional big bang cosmology, the Electroweak epoch began 10^{-36} second after the Big Bang, when the temperature of the universe was low enough (10^{28} K) to separate the strong force from the electroweak force (the name for the unified forces of electromagnetism and the weak interaction). In inflationary cosmology, the electroweak epoch ends when the inflationary epoch begins, at roughly 10^{-32} second.

Inflationary epoch

Unknown duration, ending $10^{-32}(?)$ second after the Big Bang

Main article: Inflationary epoch

Cosmic inflation was an era of accelerating expansion produced by a hypothesized field called the inflaton, which would have properties similar to the Higgs field and dark energy. While decelerating expansion would magnify deviations from homogeneity, making the universe more chaotic, accelerating expansion would make the universe more homogeneous. A sufficiently long period of inflationary expansion in the past could explain the high degree of homogeneity that is observed in the universe today at large scales, even if the state of the universe before inflation was highly disordered.

Inflation ended when the inflaton field decayed into ordinary particles in a process called "reheating", at which point ordinary Big Bang expansion began. The time of reheating is usually quoted as a time "after the Big Bang". This refers to the time that would have passed in traditional (non-inflationary) cosmology between the Big Bang singularity and the universe dropping to the same temperature that was produced by reheating, even though, in inflationary cosmology, the traditional Big Bang did not occur.

According to the simplest inflationary models, inflation ended at a temperature corresponding to roughly 10^{-32} second after the Big Bang. As explained above, this does not imply that the inflationary era lasted less than 10^{-32} second. In fact, in order to explain the observed homogeneity of the universe, the duration must be longer than 10^{-32} second. In inflationary cosmology, the earliest meaningful time "after the Big Bang" is the time of the end of inflation.

On March 17, 2014, astrophysicists of the BICEP2 collaboration announced the detection of inflationary gravitational waves in the B-mode power spectrum which was interpreted as clear experimental evidence for the theory of inflation.[2][3][4][7][8][9] However, on June 19, 2014, lowered confidence in confirming the cosmic inflation findings was reported [8][10][11] and finally, on February 2, 2015, a joint analysis of data from BICEP2/Keck and Planck satellite concluded that the statistical "significance [of the data] is too low to be interpreted as a detection of primordial B-modes" and can be attributed mainly to polarized dust in the Milky Way.[12][13][14][15]

Baryogenesis

Main article: Baryogenesis

There is currently insufficient observational evidence to explain why the universe contains far more baryons than antibaryons. A candidate explanation for this phenomenon must allow the Sakharov conditions to be satisfied at some time after the end of cosmological inflation. While particle physics suggests asymmetries under which these conditions are met, these asymmetries are too small empirically to account for the observed baryon-antibaryon asymmetry of the universe.

5.3   Early universe

After cosmic inflation ends, the universe is filled with a quark–gluon plasma. From this point onwards the physics of the early universe is better understood, and less speculative.

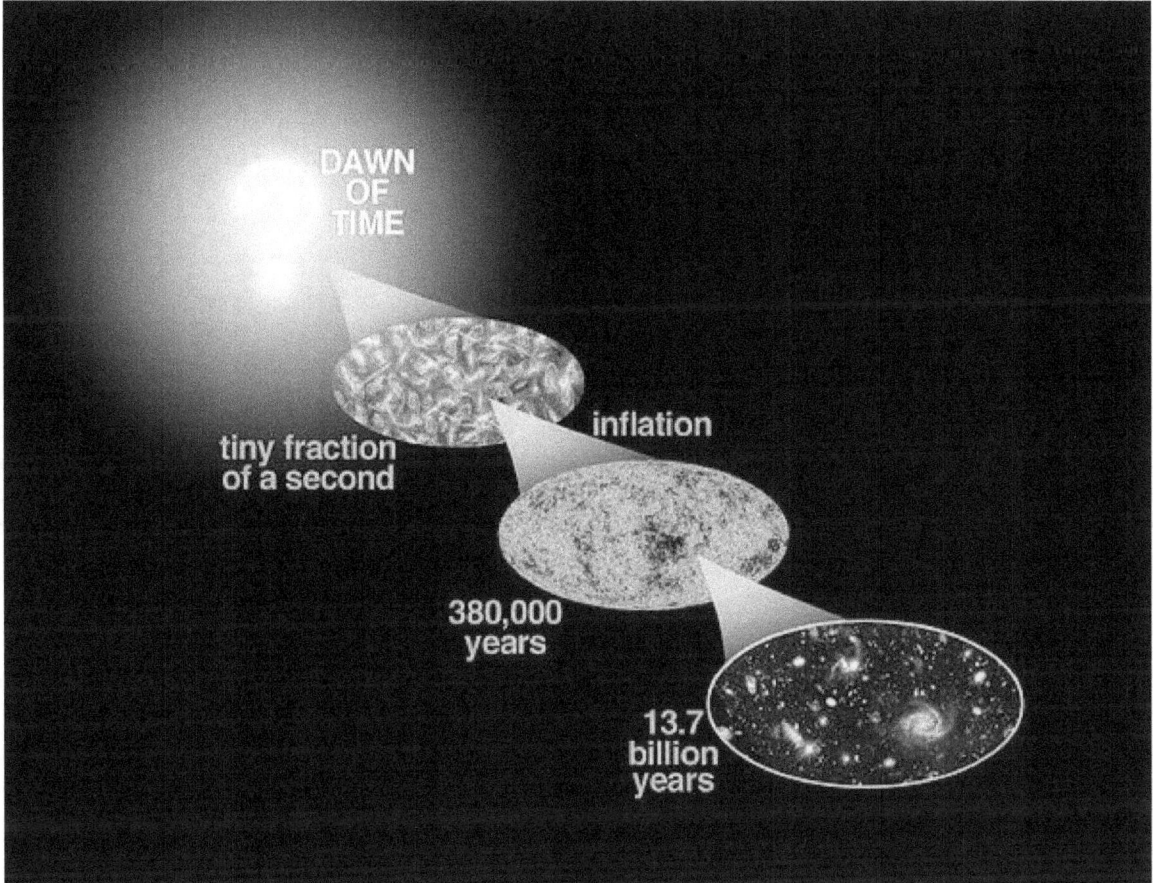

Cosmic History

5.3.1 Supersymmetry breaking (speculative)

Main article: Supersymmetry breaking

If supersymmetry is a property of our universe, then it must be broken at an energy that is no lower than 1 TeV, the electroweak symmetry scale. The masses of particles and their superpartners would then no longer be equal, which could explain why no superpartners of known particles have ever been observed.

5.3.2 Electroweak symmetry breaking and the quark epoch

Between 10^{-12} second and 10^{-6} second after the Big Bang

Main articles: Electroweak symmetry breaking and Quark epoch

As the universe's temperature falls below a certain very high energy level, it is believed that the Higgs field spontaneously acquires a vacuum expectation value, which breaks electroweak gauge symmetry. This has two related effects:

1. The weak force and electromagnetic force, and their respective bosons (the W and Z bosons and photon) manifest differently in the present universe, with different ranges;

2. Via the Higgs mechanism, all elementary particles interacting with the Higgs field become massive, having been massless at higher energy levels.

At the end of this epoch, the fundamental interactions of gravitation, electromagnetism, the strong interaction and the weak interaction have now taken their present forms, and fundamental particles have mass, but the temperature of the universe is still too high to allow quarks to bind together to form hadrons.

5.3.3   Hadron epoch

Between 10^{-6} second and 1 second after the Big Bang

Main article: Hadron epoch

The quark–gluon plasma that composes the universe cools until hadrons, including baryons such as protons and neutrons, can form. At approximately 1 second after the Big Bang neutrinos decouple and begin traveling freely through space. This cosmic neutrino background, while unlikely to ever be observed in detail since the neutrino energies are very low, is analogous to the cosmic microwave background that was emitted much later. (See above regarding the quark–gluon plasma, under the String Theory epoch.) However, there is strong indirect evidence that the cosmic neutrino background exists, both from Big Bang nucleosynthesis predictions of the helium abundance, and from anisotropies in the cosmic microwave background.

5.3.4   Lepton epoch

Between 1 second and 10 seconds after the Big Bang

Main article: Lepton epoch

The majority of hadrons and anti-hadrons annihilate each other at the end of the hadron epoch, leaving leptons and anti-leptons dominating the mass of the universe. Approximately 10 seconds after the Big Bang the temperature of the universe falls to the point at which new lepton/anti-lepton pairs are no longer created and most leptons and anti-leptons are eliminated in annihilation reactions, leaving a small residue of leptons.[16]

5.3.5   Photon epoch

Between 10 seconds and 380,000 years after the Big Bang

Main article: Photon epoch

After most leptons and anti-leptons are annihilated at the end of the lepton epoch the energy of the universe is dominated by photons. These photons are still interacting frequently with charged protons, electrons and (eventually) nuclei, and continue to do so for the next 380,000 years.

Nucleosynthesis

Between 3 minutes and 20 minutes after the Big Bang[17]

Main article: Big Bang nucleosynthesis

During the photon epoch the temperature of the universe falls to the point where atomic nuclei can begin to form. Protons (hydrogen ions) and neutrons begin to combine into atomic nuclei in the process of nuclear fusion. Free neutrons combine with protons to form deuterium. Deuterium rapidly fuses into helium-4. Nucleosynthesis only lasts for about seventeen minutes, since the temperature and density of the universe has fallen to the point where nuclear fusion cannot continue. By this time, all neutrons have been incorporated into helium nuclei. This leaves about three times more hydrogen than helium-4 (by mass) and only trace quantities of other light nuclei.

Matter domination

70,000 years after the Big Bang

At this time, the densities of non-relativistic matter (atomic nuclei) and relativistic radiation (photons) are equal. The Jeans length, which determines the smallest structures that can form (due to competition between gravitational attraction and pressure effects), begins to fall and perturbations, instead of being wiped out by free-streaming radiation, can begin to grow in amplitude.

According to ΛCDM, at this stage, cold dark matter dominates, paving the way for gravitational collapse to amplify the tiny inhomogeneities left by cosmic inflation, making dense regions denser and rarefied regions more rarefied. However, because present theories as to the nature of dark matter are inconclusive, there is as yet no consensus as to its origin at earlier times, as currently exist for baryonic matter.

Recombination

ca. 377,000 years after the Big Bang

Main article: Recombination (cosmology)
 Hydrogen and helium *atoms* begin to form as the density of the universe falls. This is thought to have occurred about

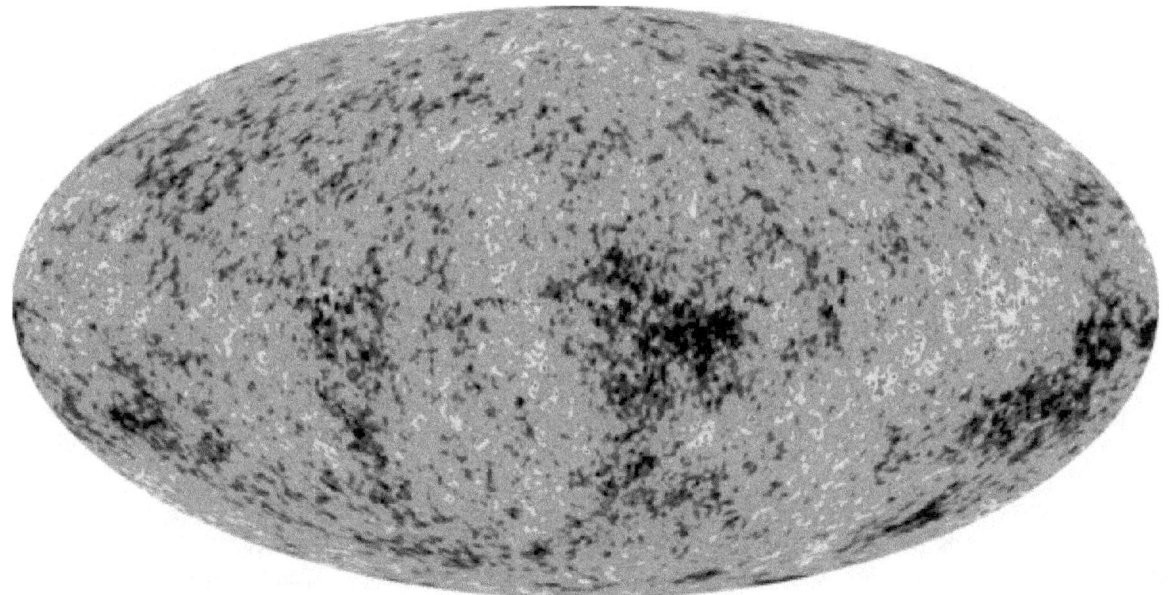

9 year WMAP data (2012) shows the cosmic microwave background radiation variations throughout the universe from our perspective, though the actual variations are much smoother than the diagram suggests.[18][19]

377,000 years after the Big Bang.[20] Hydrogen and helium are at the beginning ionized, i.e., no electrons are bound to the nuclei, which (containing positively charged protons) are therefore electrically charged (+1 and +2 respectively). As the universe cools down, the electrons get captured by the ions, forming electrically neutral atoms. This process is relatively fast (and faster for the helium than for the hydrogen), and is known as recombination.[21] At the end of recombination, most of the protons in the universe are bound up in neutral atoms. Therefore, the photons' mean free path becomes effectively infinite and the photons can now travel freely (see Thomson scattering): the universe has become transparent. This cosmic event is usually referred to as *decoupling*.

The photons present at the time of decoupling are the same photons that we see in the cosmic microwave background (CMB) radiation, after being greatly cooled by the expansion of the universe. Around the same time, existing pressure waves within the electron-baryon plasma — known as baryon acoustic oscillations — became embedded in the distribution of matter as it condensed, giving rise to a very slight preference in distribution of large scale objects. Therefore the cosmic microwave background is a picture of the universe at the end of this epoch including the tiny fluctuations generated during inflation (see diagram), and the spread of objects such as galaxies in the universe is an indication of the scale and size of the universe as it developed over time.[22]

Habitable epoch

See also: Abiogenesis

The chemistry of life may have begun shortly after the Big Bang, 13.8 billion years ago, during a habitable epoch when the Universe was only 10-17 million years old.[23][24][25]

Dark Ages

See also: Hydrogen line

Before decoupling occurred, most of the photons in the universe were interacting with electrons and protons in the photon–baryon fluid. The universe was opaque or "foggy" as a result. There was light but not light we can now observe through telescopes. The baryonic matter in the universe consisted of ionized plasma, and it only became neutral when it gained free electrons during "recombination", thereby releasing the photons creating the CMB. When the photons were released (or decoupled) the universe became transparent. At this point the only radiation emitted was the 21 cm spin line of neutral hydrogen. There is currently an observational effort underway to detect this faint radiation, as it is in principle an even more powerful tool than the cosmic microwave background for studying the early universe. The Dark Ages are currently thought to have lasted between 150 million to 800 million years after the Big Bang. The October 2010 discovery of UDFy-38135539, the first observed galaxy to have existed during the following reionization epoch, gives us a window into these times. The galaxy earliest in this period observed and thus also the most distant galaxy ever observed is currently on the record of Leiden University's Richard J. Bouwens and Garth D. Illingsworth from UC Observatories/Lick Observatory. They found the galaxy UDFj-39546284 to be at a time some 480 million years after the Big Bang or about halfway through the Cosmic Dark Ages at a distance of about 13.2 billion light-years. More recently, the UDFj-39546284 galaxy was found to be around "380 million years" after the Big Bang and at a distance of 13.37 billion light-years.[26]

5.4   Structure formation

See also: Large-scale structure of the cosmos and Structure formation
 Structure formation in the big bang model proceeds hierarchically, with smaller structures forming before larger ones. The first structures to form are quasars, which are thought to be bright, early active galaxies, and population III stars. Before this epoch, the evolution of the universe could be understood through linear cosmological perturbation theory: that is, all structures could be understood as small deviations from a perfect homogeneous universe. This is computationally relatively easy to study. At this point non-linear structures begin to form, and the computational problem becomes much more difficult, involving, for example, N-body simulations with billions of particles.

5.4.1   Reionization

150 million to 1 billion years after the Big Bang

See also: Reionization and 21 centimeter radiation

The first stars and quasars form from gravitational collapse. The intense radiation they emit reionizes the surrounding universe. From this point on, most of the universe is composed of plasma.

The Hubble Ultra Deep Fields often showcase galaxies from an ancient era that tell us what the early Stelliferous Age was like.

5.4.2 Formation of stars

See also: Star formation

The first stars, most likely Population III stars, form and start the process of turning the light elements that were formed in the Big Bang (hydrogen, helium and lithium) into heavier elements. However, as yet there have been no observed Population III stars, and understanding of them is currently based on computational models of their formation and evolution. Fortunately observations of the Cosmic Microwave Background radiation can be used to date when star formation began in earnest. Analysis of such observations made by the European Space Agency's Planck telescope, as reported by BBC News in early February, 2015, concludes that the first generation of stars lit up 560 million years after the Big Bang. [27] [28]

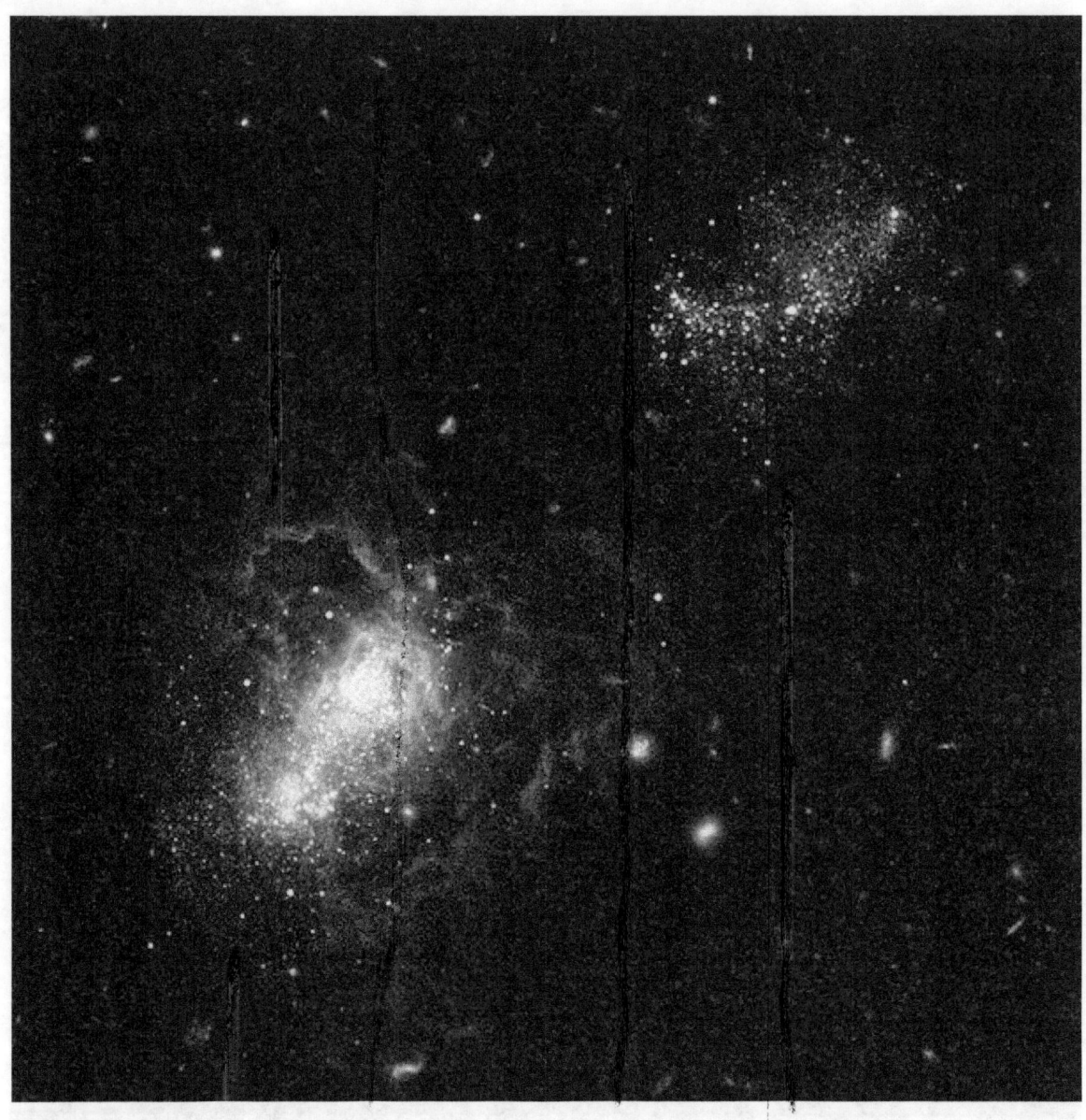

Another Hubble image shows an infant galaxy forming nearby, which means this happened very recently on the cosmological timescale. This shows that new galaxy formation in the universe is still occurring.

5.4.3 Formation of galaxies

See also: Galaxy formation and evolution

Large volumes of matter collapse to form a galaxy. Population II stars are formed early on in this process, with Population I stars formed later.

Johannes Schedler's project has identified a quasar CFHQS 1641+3755 at 12.7 billion light-years away,[29] when the universe was just 7% of its present age.

On July 11, 2007, using the 10-metre Keck II telescope on Mauna Kea, Richard Ellis of the California Institute of Technology at Pasadena and his team found six star forming galaxies about 13.2 billion light years away and therefore created when the universe was only 500 million years old.[30] Only about 10 of these extremely early objects are currently known.[31] More recent observations have shown these ages to be shorter than previously indicated. The most distant galaxy observed as of October 2013 has been reported to be 13.1 billion light years away.[32]

The Hubble Ultra Deep Field shows a number of small galaxies merging to form larger ones, at 13 billion light years, when the universe was only 5% its current age.[33] This age estimate is now believed to be slightly shorter.[32]

Based upon the emerging science of nucleocosmochronology, the Galactic thin disk of the Milky Way is estimated to have been formed 8.8 ± 1.7 billion years ago.[34]

5.4.4 Formation of groups, clusters and superclusters

See also: Large-scale structure of the cosmos

Gravitational attraction pulls galaxies towards each other to form groups, clusters and superclusters.

5.4.5 Formation of the Solar System

9 billion years after the Big Bang

Main article: Formation and evolution of the Solar System

The Solar System began forming about 4.6 billion years ago, or about 9 billion years after the Big Bang. A fragment of a molecular cloud made mostly of hydrogen and traces of other elements began to collapse, forming a large sphere in the center which would become the Sun, as well as a surrounding disk. The surrounding accretion disk would coalesce into a multitude of smaller objects that would become planets, asteroids, and comets. The Sun is a late-generation star, and the Solar System incorporates matter created by previous generations of stars.

5.4.6 Today

13.8 billion years after the Big Bang

The Big Bang is estimated to have occurred about 13.8 billion years ago.[35] Since the expansion of the universe appears to be accelerating, its large-scale structure is likely to be the largest structure that will ever form in the universe. The present accelerated expansion prevents any more inflationary structures entering the horizon and prevents new gravitationally bound structures from forming.

5.5 Ultimate fate of the universe

Main article: Ultimate fate of the universe

As with interpretations of what happened in the very early universe, advances in fundamental physics are required before it will be possible to know the ultimate fate of the universe with any certainty. Below are some of the main possibilities.

5.5.1 Fate of the Solar System: 1 to 5 billion years

Main articles: Formation and evolution of the Solar System § Future, Stability of the Solar System, Future of the Earth § Solar evolution and Red giant § The Sun as a red giant
 Over a timescale of a billion years or more, the Earth and Solar System are unstable. Earth's existing biosphere is expected to vanish in about a billion years, as the Sun's heat production gradually increases to the point that liquid water and life are unlikely;[36] the Earth's magnetic fields, axial tilt and atmosphere are subject to long-term change; and the Solar System

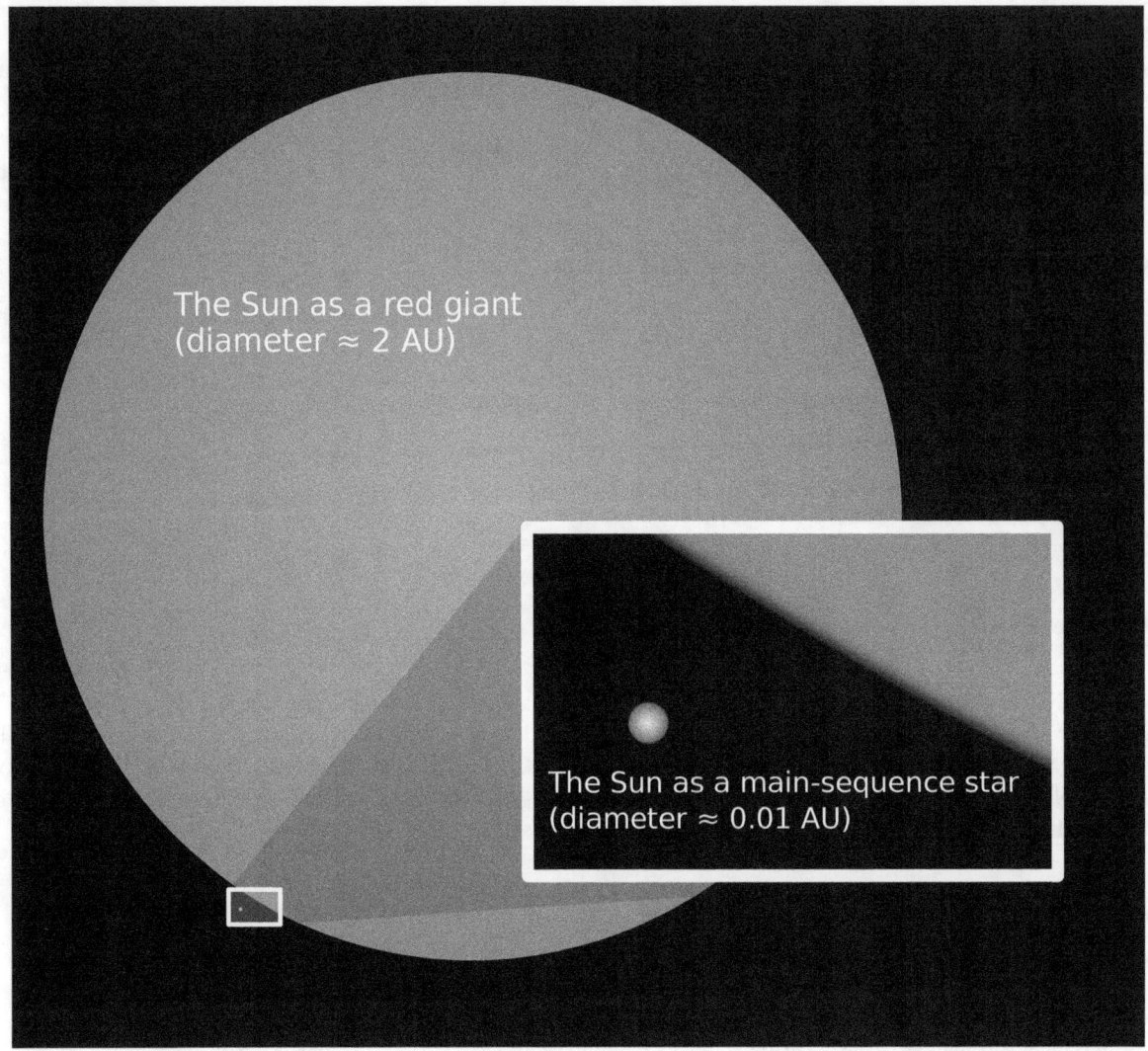

Relative size of our Sun as it is now (inset) compared to its estimated future size as a red giant

itself is chaotic over million- and billion-year timescales.[37] Eventually in around 5.4 billion years from now, the core of the Sun will become hot enough to trigger hydrogen fusion in its surrounding shell.[36] This will cause the outer layers of the star to expand greatly, and the star will enter a phase of its life in which it is called a red giant.[38][39] Within 7.5 billion years, the Sun will have expanded to a radius of 1.2 AU—256 times its current size, and studies announced in 2008 show that due to tidal interaction between Sun and Earth, Earth would actually fall back into a lower orbit, and get engulfed and incorporated inside the Sun before the Sun reaches its largest size, despite the Sun losing about 38% of its mass.[40] The Sun itself will continue to exist for many billions of years, passing through a number of phases, and eventually ending up as a long-lived white dwarf. Eventually, after billions more years, the Sun will finally cease to shine altogether, becoming a black dwarf.[41]

5.5.2   Big Rip: ≥20 billion years from now

See also: Big Rip

This scenario is possible only if the energy density of dark energy actually increases without limit over time. Such dark energy is called phantom energy and is unlike any known kind of energy. In this case, the expansion rate of the universe will increase without limit. Gravitationally bound systems, such as clusters of galaxies, galaxies, and ultimately the Solar System will be torn apart. Eventually the expansion will be so rapid as to overcome the electromagnetic forces holding molecules and atoms together. Finally even atomic nuclei will be torn apart and the universe as we know it will end in an unusual kind of gravitational singularity. At the time of this singularity, the expansion rate of the universe will reach infinity, so that any and all forces (no matter how strong) that hold composite objects together (no matter how closely) will be overcome by this expansion, literally tearing everything apart.

5.5.3 Big Crunch: ≥100 billion years from now

See also: Big Crunch

If the energy density of dark energy were negative or the universe were closed, then it would be possible that the expansion of the universe would reverse and the universe would contract towards a hot, dense state. This is a required element of oscillatory universe scenarios, such as the cyclic model, although a Big Crunch does not necessarily imply an oscillatory universe. Current observations suggest that this model of the universe is unlikely to be correct, and the expansion will continue or even accelerate.

5.5.4 Big Freeze: ≥10^5 billion years from now

Main articles: Future of an expanding universe and Heat death of the universe

This scenario is generally considered to be the most likely, as it occurs if the universe continues expanding as it has been. Over a time scale on the order of 10^{14} years or less, existing stars burn out, stars cease to be created, and the universe goes dark.[42], §IID. Over a much longer time scale in the eras following this, the galaxy evaporates as the stellar remnants comprising it escape into space, and black holes evaporate via Hawking radiation.[42], §III, §IVG. In some grand unified theories, proton decay after at least 10^{34} years will convert the remaining interstellar gas and stellar remnants into leptons (such as positrons and electrons) and photons. Some positrons and electrons will then recombine into photons.[42], §IV, §VF. In this case, the universe has reached a high-entropy state consisting of a bath of particles and low-energy radiation. It is not known however whether it eventually achieves thermodynamic equilibrium.[42], §VIB, VID.

5.5.5 Heat Death: 10^{1000} years from now

See also: Heat death of the universe

The heat death is a possible final state of the universe, estimated at after 10^{1000} years, in which it has "run down" to a state of no thermodynamic free energy to sustain motion or life. In physical terms, it has reached maximum entropy (because of this, the term "entropy" has often been confused with heat death, to the point of entropy being labelled as the "force killing the universe"). The hypothesis of a universal heat death stems from the 1850s ideas of William Thomson (Lord Kelvin)[43] who extrapolated the theory of heat views of mechanical energy loss in nature, as embodied in the first two laws of thermodynamics, to universal operation.

5.5.6 Vacuum metastability event

See also: False vacuum

If our universe is in a very long-lived false vacuum, it is possible that a small region of the universe will tunnel into a lower energy state (see Bubble nucleation). If this happens, all structures within will be destroyed instantaneously and the region will expand at near light speed, bringing destruction without any forewarning.

5.6 See also

- Cosmic Calendar (age of universe scaled to a single year)

- Cyclic model

- Dark-energy-dominated era

- Dyson's eternal intelligence

- Entropy (arrow of time)

- Graphical timeline from Big Bang to Heat Death

- Graphical timeline of the Big Bang

- Graphical timeline of the Stelliferous Era

- Illustris project

- Matter-dominated era

- Radiation-dominated era

- Timeline of the far future

- Ultimate fate of the universe

5.7 References

[1] Planck Collaboration (2015). "Planck 2015 results. XIII. Cosmological parameters (See Table 4 on page 31 of pfd).". arXiv:1502.01589. Bibcode:2015arXiv150201589P.

[2] Staff (17 March 2014). "BICEP2 2014 Results Release". *National Science Foundation*. Retrieved 18 March 2014.

[3] Clavin, Whitney (17 March 2014). "NASA Technology Views Birth of the Universe". *NASA*. Retrieved 17 March 2014.

[4] Overbye, Dennis (17 March 2014). "Detection of Waves in Space Buttresses Landmark Theory of Big Bang". *The New York Times*. Retrieved 17 March 2014.

[5] Chaisson, E., (2001). *Cosmic Evolution: The Rise of Complexity in Nature*, Harvard University Press, ISBN 0-674-00987-8; see also Cosmic Evolution

[6] Ryden B: "Introduction to Cosmology", pg. 196 Addison-Wesley 2003

[7] Overbye, Dennis (March 24, 2014). "Ripples From the Big Bang". *New York Times*. Retrieved March 24, 2014.

[8] Ade, P.A.R. (BICEP2 Collaboration); et al. (June 19, 2014). "Detection of B-Mode Polarization at Degree Angular Scales by BICEP2" (PDF). *Physical Review Letters* **112**: 241101. arXiv:1403.3985. Bibcode:2014PhRvL.112x1101A. doi:10.1103/PhysRevLett.112.241101.PMID 24996078. Retrieved June 20, 2014.

[9] http://www.math.columbia.edu/~{}woit/wordpress/?p=6865

[10] Overbye, Dennis (June 19, 2014). "Astronomers Hedge on Big Bang Detection Claim". *New York Times*. Retrieved June 20, 2014.

[11] Amos, Jonathan (June 19, 2014). "Cosmic inflation: Confidence lowered for Big Bang signal". *BBC News*. Retrieved June 20, 2014.

[12] BICEP2/Keck, Planck Collaborations (2015). "A Joint Analysis of BICEP2/Keck Array and Planck Data (Provisionally accepted by PRL)". *arXiv*. arXiv:1502.00612v1. Retrieved 13 February 2015.

[13] Clavin, Whitney (30 January 2015). "Gravitational Waves from Early Universe Remain Elusive". *NASA*. Retrieved 30 January 2015.

[14] Overbye, Dennis (30 January 2015). "Speck of Interstellar Dust Obscures Glimpse of Big Bang". *New York Times*. Retrieved 31 January 2015.

[15] "Gravitational waves from early universe remain elusive". *Science Daily*. 31 January 2015. Retrieved 3 February 2015.

[16] The Timescale of Creation

[17] Detailed timeline of Big Bang nucleosynthesis processes

[18] Gannon, Megan (December 21, 2012). "New 'Baby Picture' of Universe Unveiled". Space.com. Retrieved December 21, 2012.

[19] Bennett, C.L.; Larson, L.; Weiland, J.L.; Jarosk, N.; Hinshaw, N.; Odegard, N.; Smith, K.M.; Hill, R.S.; Gold, B.; Halpern, M.; Komatsu, E.; Nolta, M.R.; Page, L.; Spergel, D.N.; Wollack, E.; Dunkley, J.; Kogut, A.; Limon, M.; Meyer, S.S.; Tucker, G.S.; Wright, E.L. (December 20, 2012). "Nine-Year Wilkinson Microwave Anisotropy Probe (WMAP) Observations: Final Maps and Results". *The Astrophysical Journal Supplement Series* **208**: 20. arXiv:1212.5225. Bibcode:2013ApJS..208...20B. doi:10.1088/0067-0049/208/2/20. Retrieved December 22, 2012.

[20] Hinshaw, G.; et al. (2009). "Five-Year Wilkinson Microwave Anisotropy Probe (WMAP) Observations: Data Processing, Sky Maps, and Basic Results" (PDF). *Astrophysical Journal Supplement* **180** (2): 225–245. arXiv:0803.0732. Bibcode:2009ApJS.. 180..225H.doi:10.1088/0067-0049/180/2/225.

[21] Mukhanov, V: "Physical foundations of Cosmology", pg. 120, Cambridge 2005

[22] Amos, Jonathan (2012-11-13). "Quasars illustrate dark energy's roller coaster ride". *BBC News*. Retrieved 13 November 2012.

[23] Loeb, Abraham (October 2014). "The Habitable Epoch of the Early Universe". *International Journal of Astrobiology* **13** (04): 337–339. arXiv:1312.0613. Bibcode:2014IJAsB..13..337L. doi:10.1017/S1473550414000196. Retrieved 15 December 2014.

[24] Loeb, Abraham (2 December 2013). "The Habitable Epoch of the Early Universe" (PDF). *Arxiv*. arXiv:1312.0613v3. Retrieved 15 December 2014.

[25] Dreifus, Claudia (2 December 2014). "Much-Discussed Views That Go Way Back - Avi Loeb Ponders the Early Universe, Nature and Life". *New York Times*. Retrieved 3 December 2014.

[26] Wall, Mike (December 12, 2012). "Ancient Galaxy May Be Most Distant Ever Seen". Space.com. Retrieved December 12, 2012.

[27] *Ferreting Out The First Stars*; physorg.com

[28]

[29] APOD: 2007 September 6 - Time Tunnel

[30] "New Scientist" 14 July 2007

[31] HET Helps Astronomers Learn Secrets of One of Universe's Most Distant Objects

[32] Scientists confirm most distant galaxy ever

[33] APOD: 2004 March 9 – The Hubble Ultra Deep Field

[34] Eduardo F. del Peloso a1a, Licio da Silva a1, Gustavo F. Porto de Mello and Lilia I. Arany-Prado (2005), "The age of the Galactic thin disk from Th/Eu nucleocosmochronology: extended sample" (Proceedings of the International Astronomical Union (2005), 1: 485-486 Cambridge University Press)

[35] "Cosmic Detectives". The European Space Agency (ESA). 2013-04-02. Retrieved 2013-04-15.

[36] K. P. Schroder, Robert Connon Smith (2008). "Distant future of the Sun and Earth revisited". *Monthly Notices of the Royal Astronomical Society* **386** (1): 155–163. arXiv:0801.4031. Bibcode:2008MNRAS.386..155S. doi:10.1111/j.1365-2966.2008.13022.x.

[37] J. Laskar (1994). "Large-scale chaos in the solar system". *Astronom y and Astrophysics* **287**: L9–L12. Bibcode:1994A&A ...28

CHAPTER 5. CHRONOLOGY OF THE UNIVERSE

[38] Zeilik & Gregory 1998, p. 320–321.

[39] "Introduction to Cataclysmic Variables (CVs)". *NASA Goddard Space Center*. 2006. Retrieved 2006-12-29.

[40] Palmer, Jason (22 February 2008). "Hope dims that Earth will survive Sun's death". *New Scientist*.

[41] G. Fontaine, P. Brassard, P. Bergeron (2001). "The Potential of White Dwarf Cosmochronology". *Publications of the Astronomical Society of the Pacific* **113** (782): 409–435. Bibcode:2001PASP..113..409F. doi:10.1086/319535. Retrieved 2008-05-11.

[42] A dying universe: the long-term fate and evolution of astrophysical objects, Fred C. Adams and Gregory Laughlin, *Reviews of Modern Physics* **69**, #2 (April 1997), pp. 337–372. Bibcode: 1997RvMP...69..337A. doi:10.1103/RevModPhys.69.337.

[43] Thomson, William. (1851). "On the Dynamical Theory of Heat, with numerical results deduced from Mr Joule's equivalent of a Thermal Unit, and M. Regnault's Observations on Steam." Excerpts. [§§1-14 & §§99-100], *Transactions of the Royal Society of Edinburgh*, March, 1851; and *Philosophical Magazine* IV. 1852, [from *Mathematical and Physical Papers*, vol. i, art. XLVIII, pp. 174]

5.8 External links

- PBS Online (2000). From the Big Bang to the End of the Universe – The Mysteries of Deep Space Timeline. Retrieved March 24, 2005.

- Schulman, Eric (1997). The History of the Universe in 200 Words or Less. Retrieved March 24, 2005.

- Space Telescope Science Institute Office of Public Outreach (2005). Home of the Hubble Space Telescope. Retrieved March 24, 2005.

- Fermilab graphics (see "Energy time line from the Big Bang to the present" and "History of the Universe Poster")

- Exploring Time from Planck time to the lifespan of the Universe

- Cosmic Evolution is a multi-media web site that explores the cosmic-evolutionary scenario from big bang to humankind.

- Astronomers' first detailed hint of what was going on less than a trillionth of a second after time began

- The Universe Adventure

- Cosmology FAQ, Professor Edward L. Wright, UCLA

- Sean Carroll on the arrow of time (Part 1), *The origin of the universe and the arrow of time*, Sean Carroll, video, CHAST 2009, Templeton, Faculty of science, University of Sydney, November 2009, TED.com

- A Universe From Nothing, video, Lawrence Krauss, AAI 2009, YouTube.com

- Once Upon A Universe - Story of the Universe told in 13 chapters. Science communication site supported by STFC.

- Cosmic Evolution through Time - an interactive timeline explains the main events in the history of our Universe

Chapter 6

Cosmic Calendar

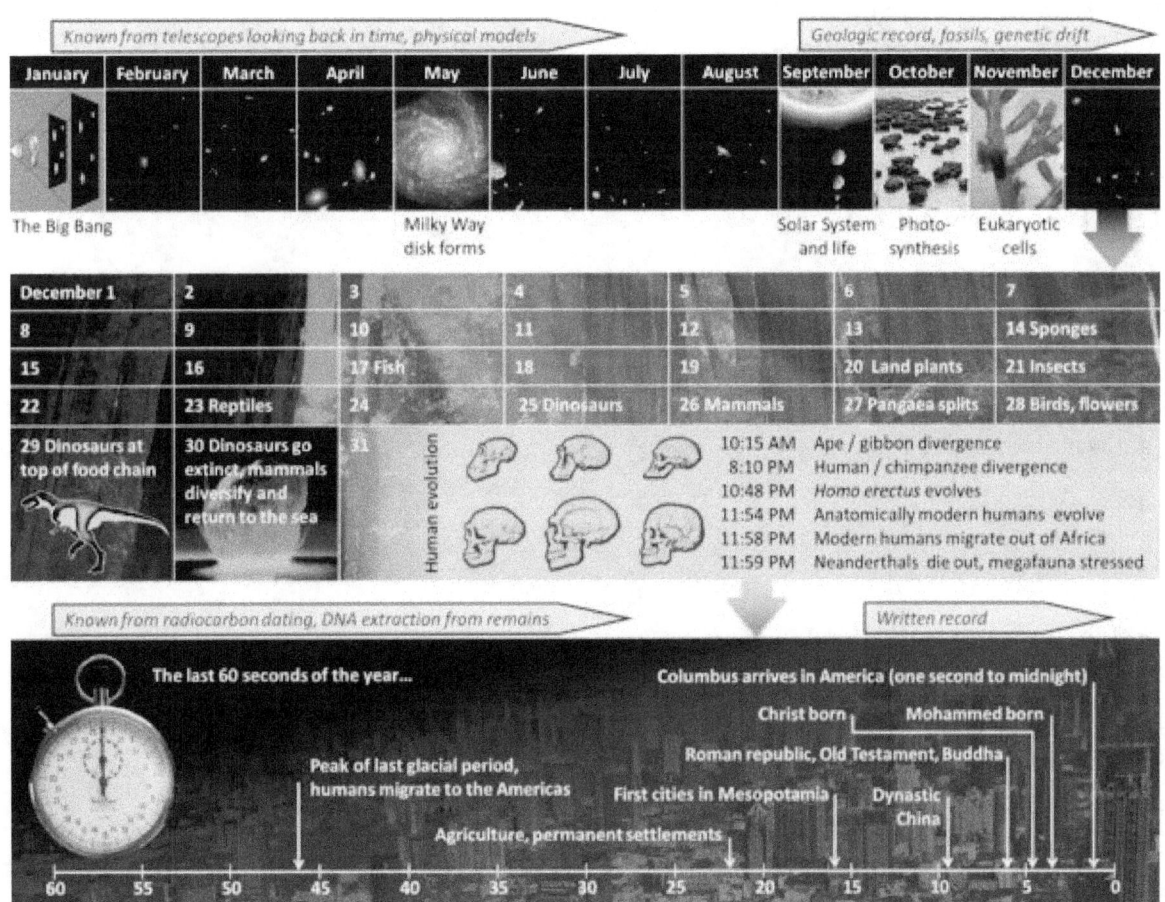

The 13.8 billion year history of the universe mapped onto a single year, as popularized by Carl Sagan. At this scale the Big Bang takes place on January 1 at midnight, the current time is December 31 at midnight, and the longest human life is a blink of an eye (about 1/4 of a second).

The **Cosmic Calendar** is a method to visualize the vast history of the universe in which its 13.8 billion year lifetime is condensed down into a single year. In this visualization, the Big Bang took place at the beginning of January 1 at midnight, and the current moment is mapped onto the end of December 31 at midnight.[1] At this scale, there are 438 years per second, 1.58 million years per hour, and 37.8 million years per day. This concept was popularized by Carl Sagan in his book *The Dragons of Eden* and on his television series *Cosmos*.[2] In the 2014 sequel series, *Cosmos: A Spacetime Odyssey*, host Neil deGrasse Tyson presents the same concept of a Cosmic Calendar, but using the revised age of the universe of 13.8 billion years as an improvement on Sagan's 1980 figure of 15 billion years. Sagan goes on to extend the comparison in terms of surface area, explaining that if the Cosmic Calendar is scaled to the size of a football field, then "all of human history would occupy an area the size of [his] hand".[3]

6.1   The Cosmic Year

6.1.1   Big Bang

Date in year calculated from formula

T(days) = 365.25 days * (1- T_bya/13.8)

6.1.2   Evolution of life on Earth

6.1.3   Human evolution

6.1.4   History begins

6.1.5   The current second

6.2   See also

- Big History

- Detailed logarithmic timeline

- List of timelines

- Timeline of ancient history

- Timeline of early modern history

- Timeline of evolution

- Timeline of the far future

- Timeline of human evolution

- Timeline of human prehistory

- Timeline of modern history

- Timeline of natural history

- Timeline of plant evolution

- Timeline of the Big Bang

- Timeline of the Middle Ages

6.3 References

[1] Therese Puyau Blanchard (1995). "The Universe At Your Fingertips Activity: Cosmic Calendar". Astronomical Society of the Pacific. Retrieved 2007-12-15.

[2] Cosmos, episode 1 (1980)

[3]

[4] "First Galaxies Born Sooner After Big Bang Than Thought". *Space.com*. Retrieved 2015-11-07.

[5] Borenstein, Seth (19 October 2015). "Hints of life on what was thought to be desolate early Earth". *Excite* (Yonkers, NY: Mindspark Interactive Network). Associated Press. Retrieved 2015-10-20.

[6] Bell, Elizabeth A.; Boehnike, Patrick; Harrison, T. Mark; et al. (19 October 2015). "Potentially biogenic carbon preserved in a 4.1 billion-year-old zircon" (PDF). *Proc. Natl. Acad. Sci. U.S.A.* (Washington, D.C.: National Academy of Sciences) **112**: 14518–21. doi:10.1073/pnas.1517557112. ISSN 1091-6490. PMC 4664351. PMID 26483481. Retrieved 2015-10-20. Early edition, published online before print.

[7] Yoko Ohtomo, Takeshi Kakegawa, Akizumi Ishida, Toshiro Nagase, Minik T. Rosing (8 December 2013). "Evidence for biogenic graphite in early Archaean Isua metasedimentary rocks". *Nature Geoscience*. doi:10.1038/ngeo2025. Retrieved 9 December 2013.

[8] Borenstein, Seth (13 November 2013). "Oldest fossil found: Meet your microbial mom". *AP News*. Retrieved 15 November 2013.

[9] Noffke, Nora; Christian, Daniel; Wacey, David; Hazen, Robert M. (8 November 2013). "Microbially Induced Sedimentary Structures Recording an Ancient Ecosystem in the ca. 3.48 Billion-Year-Old Dresser Formation, Pilbara, Western Australia". *Astrobiology (journal)* **13** (12): 1103–24. Bibcode:2013AsBio..13.1103N. doi:10.1089/ast.2013.1030. PMC 3870916. PMID 24205812. Retrieved 15 November 2013.

[10] Erwin, Douglas H. (9 November 2015). "Early metazoan life: divergence, environment and ecology". *Phil. Trans. R. Soc. B* **370** (20150036). doi:10.1098/rstb.2015.0036. Retrieved 7 January 2016.

[11] Cosmos: A Spacetime Odyssey (@35min)

6.4 External links

- More information on the image used for this article.

- The Cosmic Calendar in a Google Calendar format

- The Cosmic Calendar relayed in real time.

Chapter 7

Planck epoch

In physical cosmology, the **Planck epoch** or **Planck era** is the earliest period of time in the history of the universe, from zero to approximately 10^{-43} seconds (Planck time). While there is no proven theory that correctly describes the universe at this period, it is postulated that quantum effects of gravity dominated physical interactions due to the small scale of the universe. During this period, approximately 13.79 billion years ago, gravitation is believed to have been as strong as the other fundamental forces, and all the forces may have been unified. Inconceivably hot and dense, the state of the universe during the Planck epoch was unstable. As it expanded and cooled, the familiar manifestations of the fundamental forces arose through a process known as symmetry breaking.

Modern cosmology now suggests that the Planck epoch may have inaugurated a period of unification, known as the grand unification epoch, and that symmetry breaking then quickly led to the era of cosmic inflation, the Inflationary epoch, during which the universe greatly expanded in scale over a very short period of time.[1]

7.1 Theoretical ideas

As there presently exists no widely accepted framework for how to combine quantum mechanics with relativistic gravity, science is not currently able to make predictions about events occurring over intervals shorter than the Planck time or distances shorter than one Planck length, the distance light travels in one Planck time—about 1.616×10^{-35} meters. Without an understanding of quantum gravity, a theory unifying quantum mechanics and relativistic gravity, the physics of the Planck epoch are unclear, and the exact manner in which the fundamental forces were unified, and how they came to be separate entities, is still poorly understood. Three of the four forces have been successfully integrated in a common framework, but gravity remains problematic. If quantum effects are ignored, the universe starts from a singularity with an infinite density. This conclusion could change when quantum gravity is taken into account. String theory and loop quantum gravity are leading candidates for a theory of unification, which have yielded meaningful insights already, but work in noncommutative geometry and other fields also holds promise for our understanding of the very beginning.

7.2 Experiments exploring this time

Experimental data casting light on this cosmological epoch has been scant or non-existent until now, but results from the WMAP and Planck probes have allowed scientists to test hypotheses about the universe's first trillionth of a second (although the cosmic microwave background radiation observed by these probes originated when the universe was already several hundred thousand years old). Although this interval is still orders of magnitude longer than the Planck time, other experiments promise to push back our 'cosmic clock' further to reveal quite a bit more about the very first moments of our universe's history, hopefully giving us some insight into the Planck epoch itself. Data from particle accelerators provides meaningful insight into the early universe as well. Experiments with the Relativistic Heavy Ion Collider have allowed physicists to determine that the quark–gluon plasma (an early phase of matter) behaved more like a liquid than a gas, and the Large Hadron Collider at CERN will probe still earlier phases of matter, but no accelerator (current or planned) will be capable of probing the Planck scale directly.

7.3 See also

- Big Bang

- Chronology of the universe

- Planck particle

- Planck scale

- Quantum gravity

- Unified field theory

7.4 Footnotes

[1] Edward W. Kolb; Michael S. Turner (1994). *The Early Universe*. Basic Books. p. 447. ISBN 978-0-201-62674-2. Retrieved 10 April 2010.

7.5 References

- *A Brief History of the Universe*.

- *The Planck Epoch*.

- *Genesis I: The Planck Epoch*.

- *Evolution of the Universe through the Planck Epoch*.

7.6 External links

- The Planck Era from U of Tennessee Astrophysics pages

- The Planck Era from U of Oregon Cosmology pages

- The Planck Era by Sten Odenwald from Astronomy Cafe

- The Plank Epoch by professor James Schombert 39O

- The Planck Era - definition from U of Ottawa's Astronomy Knowledge Base

Chapter 8

Supersymmetry breaking

In particle physics, **supersymmetry breaking** is the process to obtain a seemingly non-supersymmetric physics from a supersymmetric theory which is a necessary step to reconcile supersymmetry with actual experiments. It is an example of spontaneous symmetry breaking. In supergravity, this results in a slightly modified counterpart of the Higgs mechanism where the gravitinos become massive.

Supersymmetry breaking occurs at supersymmetry breaking scale. The superpartners, whose mass would otherwise be equal to the mass of the regular particles in the absence of the SUSY breaking, become much heavier.

8.1   See also

- Soft SUSY breaking

- Timeline of the Big Bang

- Chronology of the universe

- Big Bang

-

Chapter 9

Spacetime

For other uses of this term, see Spacetime (disambiguation).

In physics, **spacetime** is any mathematical model that combines space and time into a single interwoven continuum. The spacetime of our universe is "usually" interpreted from a Euclidean space perspective, which regards space as consisting of three dimensions, and time as consisting of one dimension, the "fourth dimension". By combining space and time into a single manifold called Minkowski space, physicists have significantly simplified a large number of physical theories, as well as described in a more uniform way the workings of the universe at both the supergalactic and subatomic levels.

9.1   Explanation

In non-relativistic classical mechanics, the use of Euclidean space instead of spacetime is appropriate, because time is treated as universal with a constant rate of passage that is independent of the state of motion of an observer. In relativistic contexts, time cannot be separated from the three dimensions of space, because the observed rate at which time passes for an object depends on the object's velocity relative to the observer and also on the strength of gravitational fields, which can slow the passage of time for an object as seen by an observer outside the field.

In cosmology, the concept of spacetime combines space and time to a single abstract universe. Mathematically it is a manifold consisting of "events" which are described by some type of coordinate system. Typically **three spatial dimensions** (length, width, height), and one **temporal dimension** (time) are required. Dimensions are independent components of a coordinate grid needed to locate a point in a certain defined "space". For example, on the globe the latitude and longitude are two independent coordinates which together uniquely determine a location. In spacetime, a coordinate grid that spans the 3+1 dimensions locates events (rather than just points in space), i.e., time is added as another dimension to the coordinate grid. This way the coordinates specify *where* and *when* events occur. However, the unified nature of spacetime and the freedom of coordinate choice it allows imply that to express the temporal coordinate in one coordinate system requires both temporal and spatial coordinates in another coordinate system. Unlike in normal spatial coordinates, there are still restrictions for how measurements can be made spatially and temporally (see Spacetime intervals). These restrictions correspond roughly to a particular mathematical model which differs from Euclidean space in its manifest symmetry.

Until the beginning of the 20th century, time was believed to be independent of motion, progressing at a fixed rate in all reference frames; however, following its prediction by special relativity, later experiments confirmed that time slows at higher speeds of the reference frame relative to another reference frame. Such slowing, called time dilation, is explained in special relativity theory. Many experiments have confirmed time dilation, such as the relativistic decay of muons from cosmic ray showers and the slowing of atomic clocks aboard a Space Shuttle relative to synchronized Earth-bound inertial clocks.[1] The duration of time can therefore vary according to events and reference frames.

When dimensions are understood as mere components of the grid system, rather than physical attributes of space, it is easier to understand the alternate dimensional views as being simply the result of coordinate transformations.

The term *spacetime* has taken on a generalized meaning beyond treating spacetime events with the normal 3+1 dimensions. It is really the combination of space and time. Other proposed spacetime theories include additional dimensions—normally spatial but there exist some speculative theories that include additional temporal dimensions and even some that include dimensions that are neither temporal nor spatial (e.g., superspace). How many dimensions are needed to describe the universe is still an open question. Speculative theories such as string theory predict 10 or 26 dimensions (with M-theory predicting 11 dimensions: 10 spatial and 1 temporal), but the existence of more than four dimensions would only appear to make a difference at the subatomic level.[2]

9.2   Spacetime in literature

Incas regarded space and time as a single concept, referred to as **pacha** (Quechua: *pacha*, Aymara: *pacha*).[3][4] The peoples of the Andes maintain a similar understanding.[5]

The idea of a unified spacetime is stated by Edgar Allan Poe in his essay on cosmology titled *Eureka* (1848) that "Space and duration are one". In 1895, in his novel *The Time Machine*, H. G. Wells wrote, "There is no difference between time and any of the three dimensions of space except that our consciousness moves along it", and that "any real body must have extension in four directions: it must have Length, Breadth, Thickness, and Duration".

Marcel Proust, in his novel *Swann's Way* (published 1913), describes the village church of his childhood's Combray as "a building which occupied, so to speak, four dimensions of space—the name of the fourth being Time".

9.2.1   Mathematical concept

In Encyclopedie under the term *dimension* Jean le Rond d'Alembert speculated that duration (time) might be considered a fourth dimension if the idea was not too novel.[6]

Another early venture was by Joseph Louis Lagrange in his *Theory of Analytic Functions* (1797, 1813). He said, "One may view mechanics as a geometry of four dimensions, and mechanical analysis as an extension of geometric analysis".[7]

The ancient idea of the cosmos gradually was described mathematically with differential equations, differential geometry, and abstract algebra. These mathematical articulations blossomed in the nineteenth century as electrical technology stimulated men like Michael Faraday and James Clerk Maxwell to describe the reciprocal relations of electric and magnetic fields. Daniel Siegel phrased Maxwell's role in relativity as follows:

> [...] the idea of the propagation of forces at the velocity of light through the electromagnetic field as described by Maxwell's equations—rather than instantaneously at a distance—formed the necessary basis for relativity theory.[8]

Maxwell used vortex models in his papers on On Physical Lines of Force, but ultimately gave up on any substance but the electromagnetic field. Pierre Duhem wrote:

> [Maxwell] was not able to create the theory that he envisaged except by giving up the use of any model, and by extending by means of analogy the abstract system of electrodynamics to displacement currents.[9]

In Siegel's estimation, "this very abstract view of the electromagnetic fields, involving no visualizable picture of what is going on out there in the field, is Maxwell's legacy."[10] Describing the behaviour of electric fields and magnetic fields led Maxwell to view the combination as an electromagnetic field. These fields have a value at every point of spacetime. It is the intermingling of electric and magnetic manifestations, described by Maxwell's equations, that give spacetime its structure. In particular, the rate of motion of an observer determines the electric and magnetic profiles of the electromagnetic field. The propagation of the field is determined by the electromagnetic wave equation, which requires spacetime for description.

Spacetime was described as an affine space with quadratic form in Minkowski space of 1908.[11] In his 1914 textbook *The Theory of Relativity*, Ludwik Silberstein used biquaternions to represent events in Minkowski space. He also exhibited the Lorentz transformations between observers of differing velocities as biquaternion mappings. Biquaternions were described in 1853 by W. R. Hamilton, so while the physical interpretation was new, the mathematics was well known in English literature, making relativity an instance of applied mathematics.

The first inkling of general relativity in spacetime was articulated by W. K. Clifford. Description of the effect of gravitation on space and time was found to be most easily visualized as a "warp" or stretching in the geometrical fabric of space and time, in a smooth and continuous way that changed smoothly from point-to-point along the spacetime fabric. In 1947 James Jeans provided a concise summary of the development of spacetime theory in his book *The Growth of Physical Science*.[12]

9.3 Basic concepts

The basic elements of spacetime are events. In any given spacetime, an event is a unique position at a unique time. Because events are spacetime points, an example of an event in classical relativistic physics is (x, y, z, t), the location of an elementary (point-like) particle at a particular time. A spacetime itself can be viewed as the union of all events in the same way that a line is the union of all of its points, formally organized into a manifold, a space which can be described at small scales using coordinate systems.

A spacetime is independent of any observer.[13] However, in describing physical phenomena (which occur at certain moments of time in a given region of space), each observer chooses a convenient metrical coordinate system. Events are specified by four real numbers in any such coordinate system. The trajectories of elementary (point-like) particles through space and time are thus a continuum of events called the world line of the particle. Extended or composite objects (consisting of many elementary particles) are thus a union of many world lines twisted together by virtue of their interactions through spacetime into a "world-braid".

However, in physics, it is common to treat an extended object as a "particle" or "field" with its own unique (e.g., center of mass) position at any given time, so that the world line of a particle or light beam is the path that this particle or beam takes in the spacetime and represents the history of the particle or beam. The world line of the orbit of the Earth (in such a description) is depicted in two spatial dimensions x and y (the plane of the Earth's orbit) and a time dimension orthogonal to x and y. The orbit of the Earth is an ellipse in space alone, but its world line is a helix in spacetime.[14]

The unification of space and time is exemplified by the common practice of selecting a metric (the measure that specifies the interval between two events in spacetime) such that all four dimensions are measured in terms of units of distance: representing an event as $(x_0, x_1, x_2, x_3) = (ct, x, y, z)$ (in the Lorentz metric) or $(x_1, x_2, x_3, x_4) = (x, y, z, ict)$ (in the original Minkowski metric) where c is the speed of light.[15] The metrical descriptions of Minkowski Space and spacelike, lightlike, and timelike intervals given below follow this convention, as do the conventional formulations of the Lorentz transformation.

9.3.1 Spacetime intervals in flat space

In a Euclidean space, the separation between two points is measured by the distance between the two points. The distance is purely spatial, and is always positive. In spacetime, the displacement four-vector ΔR is given by the space displacement vector Δr and the time difference Δt between the events. The *spacetime interval*, also called *invariant interval*, between the two events, s^2,[16] is defined as:

$$s^2 = \Delta r^2 - c^2 \Delta t^2 \text{ (spacetime interval)},$$

where c is the speed of light. The choice of signs for s^2 above follows the space-like convention (−+++).[17] Spacetime intervals may be classified into three distinct types, based on whether the temporal separation ($c^2 \Delta t^2$) or the spatial separation (Δr^2) of the two events is greater: time-like, light-like or space-like.

Certain types of world lines are called geodesics of the spacetime – straight lines in the case of Minkowski space and their closest equivalent in the curved spacetime of general relativity. In the case of purely time-like paths, geodesics are (locally) the paths of greatest separation (spacetime interval) as measured along the path between two events, whereas in Euclidean space and Riemannian manifolds, geodesics are paths of shortest distance between two points.[18][19] The concept of geodesics becomes central in general relativity, since geodesic motion may be thought of as "pure motion" (inertial motion) in spacetime, that is, free from any external influences.

Time-like interval

$$c^2 \Delta t^2 > \Delta r^2$$
$$s^2 < 0$$

For two events separated by a time-like interval, enough time passes between them that there could be a cause–effect relationship between the two events. For a particle traveling through space at less than the speed of light, any two events which occur to or by the particle must be separated by a time-like interval. Event pairs with time-like separation define a negative spacetime interval ($s^2 < 0$) and may be said to occur in each other's future or past. There exists a reference frame such that the two events are observed to occur in the same spatial location, but there is no reference frame in which the two events can occur at the same time.

The measure of a time-like spacetime interval is described by the proper time interval, $\Delta\tau$:

$$\Delta\tau = \sqrt{\Delta t^2 - \frac{\Delta r^2}{c^2}} \text{ (proper time interval)}.$$

The proper time interval would be measured by an observer with a clock traveling between the two events in an inertial reference frame, when the observer's path intersects each event as that event occurs. (The proper time interval defines a real number, since the interior of the square root is positive.)

Light-like interval

$$c^2 \Delta t^2 = \Delta r^2$$
$$s^2 = 0$$

In a light-like interval, the spatial distance between two events is exactly balanced by the time between the two events. The events define a spacetime interval of zero ($s^2 = 0$). Light-like intervals are also known as "null" intervals.

Events which occur to or are initiated by a photon along its path (i.e., while traveling at c , the speed of light) all have light-like separation. Given one event, all those events which follow at light-like intervals define the propagation of a light cone, and all the events which preceded from a light-like interval define a second (graphically inverted, which is to say "*pastward*") light cone.

Space-like interval

$$c^2 \Delta t^2 < \Delta r^2$$
$$s^2 > 0$$

When a space-like interval separates two events, not enough time passes between their occurrences for there to exist a causal relationship crossing the spatial distance between the two events at the speed of light or slower. Generally, the events are considered not to occur in each other's future or past. There exists a reference frame such that the two events are observed to occur at the same time, but there is no reference frame in which the two events can occur in the same spatial location.

For these space-like event pairs with a positive spacetime interval ($s^2 > 0$), the measurement of space-like separation is the proper distance, $\Delta\sigma$:

$$\Delta\sigma = \sqrt{s^2} = \sqrt{\Delta r^2 - c^2 \Delta t^2} \text{ (proper distance)}.$$

Like the proper time of time-like intervals, the proper distance of space-like spacetime intervals is a real number value.

9.3.2 Interval as area

The interval has been presented as the area of an oriented rectangle formed by two events and isotropic lines through them. Time-like or space-like separations correspond to oppositely oriented rectangles, one type considered to have rectangles of negative area. The case of two events separated by light corresponds to the rectangle degenerating to the segment between the events and zero area.[20] The transformations leaving interval-length invariant are the area-preserving squeeze mappings.

The parameters traditionally used rely on quadrature of the hyperbola, which is the natural logarithm. This transcendental function is essential in mathematical analysis as its inverse unites circular functions and hyperbolic functions: The exponential function, e^t, t a real number, used in the hyperbola (e^t, e^{-t}), generates hyperbolic sectors and the hyperbolic angle parameter. The functions cosh and sinh, used with rapidity as hyperbolic angle, provide the common representation of squeeze in the form $\begin{pmatrix} \cosh\phi & \sinh\phi \\ \sinh\phi & \cosh\phi \end{pmatrix}$, or as the split-complex unit $e^{j\phi} = \cosh\phi + j\,\sinh\phi$.

9.4 Mathematics of spacetimes

For physical reasons, a spacetime continuum is mathematically defined as a four-dimensional, smooth, connected Lorentzian manifold (M, g). This means the smooth Lorentz metric g has signature $(3, 1)$. The metric determines the geometry of spacetime, as well as determining the geodesics of particles and light beams. About each point (event) on this manifold, coordinate charts are used to represent observers in reference frames. Usually, Cartesian coordinates (x, y, z, t) are used. Moreover, for simplicity's sake, units of measurement are usually chosen such that the speed of light c is equal to 1.

A reference frame (observer) can be identified with one of these coordinate charts; any such observer can describe any event p. Another reference frame may be identified by a second coordinate chart about p. Two observers (one in each reference frame) may describe the same event p but obtain different descriptions.

Usually, many overlapping coordinate charts are needed to cover a manifold. Given two coordinate charts, one containing p (representing an observer) and another containing q (representing another observer), the intersection of the charts represents the region of spacetime in which both observers can measure physical quantities and hence compare results. The relation between the two sets of measurements is given by a non-singular coordinate transformation on this intersection. The idea of coordinate charts as local observers who can perform measurements in their vicinity also makes good physical sense, as this is how one actually collects physical data—locally.

For example, two observers, one of whom is on Earth, but the other one who is on a fast rocket to Jupiter, may observe a comet crashing into Jupiter (this is the event p). In general, they will disagree about the exact location and timing of this impact, i.e., they will have different 4-tuples (x, y, z, t) (as they are using different coordinate systems). Although their kinematic descriptions will differ, dynamical (physical) laws, such as momentum conservation and the first law of thermodynamics, will still hold. In fact, relativity theory requires more than this in the sense that it stipulates these (and all other physical) laws must take the same form in all coordinate systems. This introduces tensors into relativity, by which all physical quantities are represented.

Geodesics are said to be time-like, null, or space-like if the tangent vector to one point of the geodesic is of this nature. Paths of particles and light beams in spacetime are represented by time-like and null (light-like) geodesics, respectively.

9.4.1 Topology

Main article: Spacetime topology

The assumptions contained in the definition of a spacetime are usually justified by the following considerations.

The connectedness assumption serves two main purposes. First, different observers making measurements (represented by coordinate charts) should be able to compare their observations on the non-empty intersection of the charts. If the connectedness assumption were dropped, this would not be possible. Second, for a manifold, the properties of connectedness and path-connectedness are equivalent, and one requires the existence of paths (in particular, geodesics) in the spacetime to represent the motion of particles and radiation.

Every spacetime is paracompact. This property, allied with the smoothness of the spacetime, gives rise to a smooth linear connection, an important structure in general relativity. Some important theorems on constructing spacetimes from compact and non-compact manifolds include the following:

- A compact manifold can be turned into a spacetime if, and only if, its Euler characteristic is 0. (Proof idea: the existence of a Lorentzian metric is shown to be equivalent to the existence of a nonvanishing vector field.)

- Any non-compact 4-manifold can be turned into a spacetime.[21]

9.4.2 Spacetime symmetries

Main article: Spacetime symmetries

Often in relativity, spacetimes that have some form of symmetry are studied. As well as helping to classify spacetimes, these symmetries usually serve as a simplifying assumption in specialized work. Some of the most popular ones include:

- Axisymmetric spacetimes

- Spherically symmetric spacetimes

- Static spacetimes

- Stationary spacetimes

9.4.3 Causal structure

Main article: Causal structure
See also: Causality (physics) and Causality

The causal structure of a spacetime describes causal relationships between pairs of points in the spacetime based on the existence of certain types of curves joining the points.

9.5 Spacetime in special relativity

Main article: Minkowski space

The geometry of spacetime in special relativity is described by the Minkowski metric on R^4. This spacetime is called Minkowski space. The Minkowski metric is usually denoted by η and can be written as a four-by-four matrix:

$$\eta_{ab} = \mathrm{diag}(1, -1, -1, -1)$$

where the Landau–Lifshitz time-like convention is being used. A basic assumption of relativity is that coordinate transformations must leave spacetime intervals invariant. Intervals are invariant under Lorentz transformations. This invariance property leads to the use of four-vectors (and other tensors) in describing physics.

Strictly speaking, one can also consider events in Newtonian physics as a single spacetime. This is Galilean–Newtonian relativity, and the coordinate systems are related by Galilean transformations. However, since these preserve spatial and temporal distances independently, such a spacetime can be decomposed into spatial coordinates plus temporal coordinates, which is not possible in the general case.

9.6 Spacetime in general relativity

In general relativity, it is assumed that spacetime is curved by the presence of matter (energy), this curvature being represented by the Riemann tensor. In special relativity, the Riemann tensor is identically zero, and so this concept of "non-curvedness" is sometimes expressed by the statement *Minkowski spacetime is flat.*

The earlier discussed notions of time-like, light-like and space-like intervals in special relativity can similarly be used to classify one-dimensional curves through curved spacetime. A time-like curve can be understood as one where the interval between any two infinitesimally close events on the curve is time-like, and likewise for light-like and space-like curves. Technically the three types of curves are usually defined in terms of whether the tangent vector at each point on the curve is time-like, light-like or space-like. The world line of a slower-than-light object will always be a time-like curve, the world line of a massless particle such as a photon will be a light-like curve, and a space-like curve could be the world line of a hypothetical tachyon. In the local neighborhood of any event, time-like curves that pass through the event will remain inside that event's past and future light cones, light-like curves that pass through the event will be on the surface of the light cones, and space-like curves that pass through the event will be outside the light cones. One can also define the notion of a three-dimensional "spacelike hypersurface", a continuous three-dimensional "slice" through the four-dimensional property with the property that every curve that is contained entirely within this hypersurface is a space-like curve.[22]

Many spacetime continua have physical interpretations which most physicists would consider bizarre or unsettling. For example, a compact spacetime has closed timelike curves, which violate our usual ideas of causality (that is, future events could affect past ones). For this reason, mathematical physicists usually consider only restricted subsets of all the possible spacetimes. One way to do this is to study "realistic" solutions of the equations of general relativity. Another way is to add some additional "physically reasonable" but still fairly general geometric restrictions and try to prove interesting things about the resulting spacetimes. The latter approach has led to some important results, most notably the Penrose–Hawking singularity theorems.

9.7 Quantized spacetime

Main article: Quantum spacetime

In general relativity, spacetime is assumed to be smooth and continuous—and not just in the mathematical sense. In the theory of quantum mechanics, there is an inherent discreteness present in physics. In attempting to reconcile these two theories, it is sometimes postulated that spacetime should be quantized at the very smallest scales. Current theory is focused on the nature of spacetime at the Planck scale. Causal sets, loop quantum gravity, string theory, causal dynamical triangulation, and black hole thermodynamics all predict a quantized spacetime with agreement on the order of magnitude. Loop quantum gravity makes precise predictions about the geometry of spacetime at the Planck scale.

Spin networks provide a language to describe quantum geometry of space. Spin foam does the same job on spacetime. A spin network is a one-dimensional graph, together with labels on its vertices and edges which encodes aspects of a spatial geometry.

9.8 See also

- Anthropic_principle § Applications of the principle §§ Spacetime

- Basic introduction to the mathematics of curved spacetime

- Four-vector

- Frame-dragging

- Global spacetime structure

- Hole argument

- List of mathematical topics in relativity

- Local spacetime structure

- Lorentz invariance

- Manifold

- Mathematics of general relativity

- Metric space

- Philosophy of space and time

- Relativity of simultaneity

- Strip photography

- World manifold

9.9 References

[1] Ashby, Neil (2003). "Relativity in the Global Positioning System" (PDF). *Living Reviews in Relativity* **6**: 16. Bibcode:2003LRR.....6....1A. doi:10.12942/lrr-2003-1.

[2] Kopeikin, Sergei; Efroimsky, Michael; Kaplan, George (2011). *Relativistic Celestial Mechanics of the Solar System*. John Wiley & Sons. p. 157. ISBN 3527634576., Extract of page 157

[3] Atuq Eusebio Manga Qespi, Instituto de lingüística y Cultura Amerindia de la Universidad de Valencia. *Pacha: un concepto andino de espacio y tiempo*. Revísta española de Antropología Americana, 24, p. 155–189. Edit. Complutense, Madrid. 1994

[4] Paul Richard Steele, Catherine J. Allen, *Handbook of Inca mythology*, p. 86, (ISBN 1-57607-354-8)

[5] Shirley Ardener, University of Oxford, *Women and space: ground rules and social maps*, p. 36 (ISBN 0-85496-728-1)

[6] Jean d'Alembert (1754) Dimension from ARTFL Encyclopedie project

[7] R.C. Archibald (1914) *Time as a fourth dimension Bulletin of the American Mathematical Society 20:409.*

[8] Daniel M. Siegel (2014) "Maxwell's contributions to electricity and magnetism", chapter 10 in *James Clerk Maxwell: Perspectives on his Life and Work*, Raymond Flood, Mark McCartney, Andrew Whitaker, editors, Oxford University Press ISBN 978-0-19-966437-5

[9] Pierre Duhem (1954) *The Aim and Structure of Physical Theory*, page 98, Princeton University Press

[10] Siegel 2014 p 191

[11] Minkowski, Hermann (1909), "Raum und Zeit", *Physikalische Zeitschrift* **10**: 75–88

- Various English translations on Wikisource: Space and Time.

[12] James Jeans (1947) The Growth of Physical Science, "Space-time", pp. 205–301, link from Internet Archive

[13] Matolcsi, Tamás (1994). *Spacetime Without Reference Frames*. Budapest: Akadémiai Kiadó.

[14] Ellis, G. F. R.; Williams, Ruth M. (2000). *Flat and curved space–times* (2nd ed.). Oxford University Press. p. 9. ISBN 0-19-850657-0.

[15] Petkov, Vesselin (2010). *Minkowski Spacetime: A Hundred Years Later*. Springer. p. 70. ISBN 90-481-3474-9., Section 3.4, p. 70

[16] Note that the term *spacetime interval* is applied by several authors to the quantity s^2 and not to s. The reason that the quantity s^2 is used and not s is that s^2 can be positive, zero or negative, and is a more generally convenient and useful quantity than the Minkowski norm with a timelike/null/spacelike distinguisher: the pair $(\sqrt{|s^2|}, \mathrm{sgn}(s^2))$. Despite the notation, it should not be regarded as the square of a number, but as a symbol. The cost for this convenience is that this "interval" is quadratic in linear separation along a straight line.

[17] More generally the spacetime interval in flat space can be written as $s^2 = g_{\alpha\beta}\Delta x^\alpha \Delta x^\beta$ with metric tensor g independent of spacetime position.

[18] This characterization is not universal: both the arcs between two points of a great circle on a sphere are geodesics.

[19] Berry, Michael V. (1989). *Principles of Cosmology and Gravitation*. CRC Press. p. 58. ISBN 0-85274-037-9., Extract of page 58, caption of Fig. 25

[20] I. M. Yaglom (1979) *A Simple Non-Euclidean Geometry and its Physical Basis*, page 178, Springer, ISBN 0387-90332-1, MR 520230

[21] Geroch, Robert; Horowitz, Gary T. (1979). "Chapter 5. Global structure of spacetimes". In Hawking, S.W.; Israel, W. *General Relativity An Einstein Centenary Survey*. Cambridge University Press. p. 219. ISBN 0521299284.

[22] See "Quantum Spacetime and the Problem of Time in Quantum Gravity" by Leszek M. Sokolowski, where on this page he writes "Each of these hypersurfaces is spacelike, in the sense that every curve, which entirely lies on one of such hypersurfaces, is a spacelike curve." More commonly a space-like hypersurface is defined technically as a surface such that the normal vector at every point is time-like, but the definition above may be somewhat more intuitive.

9.10 Further Reading

- Albert Einstein on Space-Time 13th edition Encyclopedia Britannica Historical: Albert Einstein's 1926 article

- Ehrenfest, Paul (1920) "How do the fundamental laws of physics make manifest that Space has 3 dimensions?" *Annalen der Physik 366*: 440.

- George F. Ellis and Ruth M. Williams (1992) *Flat and curved space–times*. Oxford Univ. Press. ISBN 0-19-851164-7

- Encyclopedia of Space-time and gravitation Scholarpedia Expert articles

9.11 External links

- http://universaltheory.org

- Barrow, John D.; Tipler, Frank J. (1988). *The Anthropic Cosmological Principle*. Oxford University Press. ISBN 978-0-19-282147-8. LCCN 87028148.

- Isenberg, J. A. (1981). "Wheeler–Einstein–Mac h spacetimes". *Phys. Rev. D* **24** (2): 251–256. Bibcode doi:10.1103/PhysRevD.24.251.

- Kant, Immanuel (1929) "Thoughts on the true estimation of living forces" in J. Handyside, trans., *Kant's Inaugural Dissertation and Early Writings on Space*. Univ. of Chicago Press.

- Lorentz, H. A., Einstein, Albert, Minkowski, Hermann, and Weyl, Hermann (1952) *The Principle of Relativity: A Collection of Original Memoirs*. Dover.

- Lucas, John Randolph (1973) *A Treatise on Time and Space*. London: Methuen.

- Penrose, Roger (2004). *The Road to Reality*. Oxford: Oxford University Press. ISBN 0-679-45443-8. Chpts. 17–18.

- Poe, Edgar A. (1848). *Eureka; An Essay on the Material and Spiritual Universe*. Hesperus Press Limited. ISBN 1-84391-009-8.

- Robb, A. A. (1936). *Geometry of Time and Space*. University Press.

- Erwin Schrödinger (1950) *Space–time structure*. Cambridge Univ. Press.

- Schutz, J. W. (1997). *Independent axioms for Minkowski Space–time*. Addison-Wesley Longman. ISBN 0-582-31760-6.

- Tangherlini, F. R. (1963). "Schwarzschild Field in n Dimensions and the Dimensionality of Space Problem". *Nuovo Cimento* **14** (27): 636.

- Taylor, E. F.; Wheeler, John A. (1963). *Spacetime Physics*. W. H. Freeman. ISBN 0-7167-2327-1.

- Wells, H.G. (2004). *The Time Machine*. New York: Pocket Books. ISBN 0-671-57554-6. (pp. 5–6)

- Stanford Encyclopedia of Philosophy: "Space and Time: Inertial Frames" by Robert DiSalle.

Chapter 10

Spacetime symmetries

For the notation, see Ricci calculus.

Spacetime symmetries are features of spacetime that can be described as exhibiting some form of symmetry. The role of symmetry in physics is important in simplifying solutions to many problems, spacetime symmetries are used in the study of exact solutions of Einstein's field equations of general relativity.

10.1   Physical motivation

Physical problems are often investigated and solved by noticing features which have some form of symmetry. For example, in the Schwarzschild solution, the role of spherical symmetry is important in deriving the Schwarzschild solution and deducing the physical consequences of this symmetry (such as the non-existence of gravitational radiation in a spherically pulsating star). In cosmological problems, symmetry finds a role to play in the cosmological principle which restricts the type of universes that are consistent with large-scale observations (e.g. the Friedmann-Lemaître-Robertson-Walker (FLRW) metric). Symmetries usually require some form of preserving property, the most important of which in general relativity include the following:

- preserving geodesics of the spacetime

- preserving the metric tensor

- preserving the curvature tensor

These and other symmetries will be discussed in more detail later. This preservation feature can be used to motivate a useful definition of symmetries.

10.2   Mathematical definition

A rigorous definition of symmetries in general relativity has been given by Hall (2004). In this approach, the idea is to use (smooth) vector fields whose local flow diffeomorphisms preserve some property of the spacetime. This preserving property of the diffeomorphisms is made precise as follows. A smooth vector field X on a spacetime M is said to *preserve* a smooth tensor T on M (or T is **invariant** under X) if, for each smooth local flow diffeomorphism ϕt associated with X, the tensors T and $\phi t^*(T)$ are equal on the domain of ϕt. This statement is equivalent to the more usable condition that the Lie derivative of the tensor under the vector field vanishes:

$$\mathcal{L}_X T = 0$$

on *M*. This has the consequence that, given any two points p and q on *M*, the coordinates of T in a coordinate system around p are equal to the coordinates of T in a coordinate system around q. A *symmetry on the spacetime* is a smooth vector field whose local flow diffeomorphisms preserve some (usually geometrical) feature of the spacetime. The (geometrical) feature may refer to specific tensors (such as the metric, or the energy-momentum tensor) or to other aspects of the spacetime such as its geodesic structure. The vector fields are sometimes referred to as *collineations*, *symmetry vector fields* or just *symmetries*. The set of all symmetry vector fields on *M* forms a Lie algebra under the Lie bracket operation as can be seen from the identity:

$$\mathcal{L}_{[X,Y]}T = \mathcal{L}_X(\mathcal{L}_Y T) - \mathcal{L}_Y(\mathcal{L}_X T)$$

the term on the right usually being written, with an abuse of notation, as $[\mathcal{L}_X, \mathcal{L}_Y]T$.

10.3   Killing symmetry

Main article: Killing vector field

A Killing vector field is one of the most important types of symmetries and is defined to be a smooth vector field that preserves the metric tensor:

$$\mathcal{L}_X g_{ab} = 0$$

This is usually written in the expanded form as:

$$X_{a;b} + X_{b;a} = 0$$

Killing vector fields find extensive applications (including in classical mechanics) and are related to conservation laws.

10.4   Homothetic symmetry

Main article: Homothetic vector field

A homothetic vector field is one which satisfies:

$$\mathcal{L}_X g_{ab} = 2c g_{ab}$$

where c is a real constant. Homothetic vector fields find application in the study of singularities in general relativity.

10.5   Affine symmetry

Main article: Affine vector field

An affine vector field is one that satisfies:

$$(\mathcal{L}_X g_{ab})_{;c} = 0$$

An affine vector field preserves geodesics and preserves the affine parameter.

The above three vector field types are special cases of projective vector fields which preserve geodesics without necessarily preserving the affine parameter.

10.6 Conformal symmetry

Main article: Conformal vector field

A conformal vector field is one which satisfies:

$$\mathcal{L}_X g_{ab} = \phi g_{ab}$$

where ϕ is a smooth real-valued function on M .

10.7 Curvature symmetry

Main article: Curvature collineation

A curvature collineation is a vector field which preserves the Riemann tensor:

$$\mathcal{L}_X R^a{}_{bcd} = 0$$

where $R^a bcd$ are the components of the Riemann tensor. The set of all smooth curvature collineations forms a Lie algebra under the Lie bracket operation (if the smoothness condition is dropped, the set of all curvature collineations need not form a Lie algebra). The Lie algebra is denoted by $CC(M)$ and may be infinite-dimensional. Every affine vector field is a curvature collineation.

10.8 Matter symmetry

Main article: Matter collineation

A less well-known form of symmetry concerns vector fields that preserve the energy-momentum tensor. These are variously referred to as matter collineations or matter symmetries and are defined by:

$$\mathcal{L}_X T_{ab} = 0$$

where *Tab* are the energy-momentum tensor components. The intimate relation between geometry and physics may be highlighted here, as the vector field X is regarded as preserving certain physical quantities along the flow lines of X, this being true for any two observers. In connection with this, it may be shown that *every Killing vector field is a matter collineation* (by the Einstein field equations, with or without cosmological constant). Thus, given a solution of the EFE, *a vector field that preserves the metric necessarily preserves the corresponding energy-momentum tensor.* When the energy-momentum tensor represents a perfect fluid, every Killing vector field preserves the energy density, pressure and the fluid flow vector field. When the energy-momentum tensor represents an electromagnetic field, a Killing vector field does *not necessarily* preserve the electric and magnetic fields.

10.9 Local and global symmetries

Main articles: Local symmetry and Global symmetry

10.10 Applications

As mentioned at the start of this article, the main application of these symmetries occur in general relativity, where solutions of Einstein's equations may be classified by imposing some certain symmetries on the spacetime.

10.10.1 Spacetime classifications

Classifying solutions of the EFE constitutes a large part of general relativity research. Various approaches to classifying spacetimes, including using the Segre classification of the energy-momentum tensor or the Petrov classification of the Weyl tensor have been studied extensively by many researchers, most notably Stephani et al. (2003). They also classify spacetimes using symmetry vector fields (especially Killing and homothetic symmetries). For example, Killing vector fields may be used to classify spacetimes, as there is a limit to the number of global, smooth Killing vector fields that a spacetime may possess (the maximum being 10 for 4-dimensional spacetimes). Generally speaking, the higher the dimension of the algebra of symmetry vector fields on a spacetime, the more symmetry the spacetime admits. For example, the Schwarzschild solution has a Killing algebra of dimension 4 (3 spatial rotational vector fields and a time translation), whereas the Friedmann-Lemaître-Robertson-Walker (FLRW) metric (excluding the Einstein static subcase) has a Killing algebra of dimension 6 (3 translations and 3 rotations). The Einstein static metric has a Killing algebra of dimension 7 (the previous 6 plus a time translation).

The assumption of a spacetime admitting a certain symmetry vector field can place restrictions on the spacetime.

10.11 See also

- Field (physics)

- Killing tensor

- Lie groups

- Noether's theorem

- Ricci decomposition

- Symmetry in physics

- Symmetry in quantum mechanics

- Derivations of the Lorentz transformations

10.12 References

- Hall, Graham (2004). *Symmetries and Curvature Structure in General Relativity (World Scientific Lecture Notes in Physics)*. Singapore: World Scientific Pub. Co. ISBN 981-02-1051-5. See *Section 10.1* for a definition of symmetries.

- Stephani, Hans; Kramer, Dietrich; MacCallum, Malcolm; Hoenselaers, Cornelius & Herlt, Eduard (2003). *Exact Solutions of Einstein's Field Equations*. Cambridge: Cambridge University Press. ISBN 0-521-46136-7.

- Schutz, Bernard (1980). *Geometrical Methods of Mathematical Physics*. Cambridge: Cambridge University Press. ISBN 0-521-29887-3. See *Chapter 3* for properties of the Lie derivative and *Section 3.10* for a definition of invariance.

Chapter 11

Phase transition

A **phase transition** is the transformation of a thermodynamic system from one phase or state of matter to another one by heat transfer. The term is most commonly used to describe transitions between solid, liquid and gaseous states of matter, and, in rare cases, plasma. A phase of a thermodynamic system and the states of matter have uniform physical properties. During a phase transition of a given medium certain properties of the medium change, often discontinuously, as a result of the change of some external condition, such as temperature, pressure, or others. For example, a liquid may become gas upon heating to the boiling point, resulting in an abrupt change in volume. The measurement of the external conditions at which the transformation occurs is termed the phase transition. Phase transitions are common in nature and used today in many technologies.

11.1   Types of phase transition

Examples of phase transitions include:

- The transitions between the solid, liquid, and gaseous phases of a single component, due to the effects of temperature and/or pressure:

    - (see also vapor pressure and phase diagram)

- A eutectic transformation, in which a two component single phase liquid is cooled and transforms into two solid phases. The same process, but beginning with a solid instead of a liquid is called a eutectoid transformation.

- A peritectic transformation, in which a two component single phase solid is heated and transforms into a solid phase and a liquid phase.

- A spinodal decomposition, in which a single phase is cooled and separates into two different compositions of that same phase.

- Transition to a mesophase between solid and liquid, such as one of the "liquid crystal" phases.

- The transition between the ferromagnetic and paramagnetic phases of magnetic materials at the Curie point.

- The transition between differently ordered, commensurate or incommensurate, magnetic structures, such as in cerium antimonide.

- The martensitic transformation which occurs as one of the many phase transformations in carbon steel and stands as a model for displacive phase transformations.

- Changes in the crystallographic structure such as between ferrite and austenite of iron.

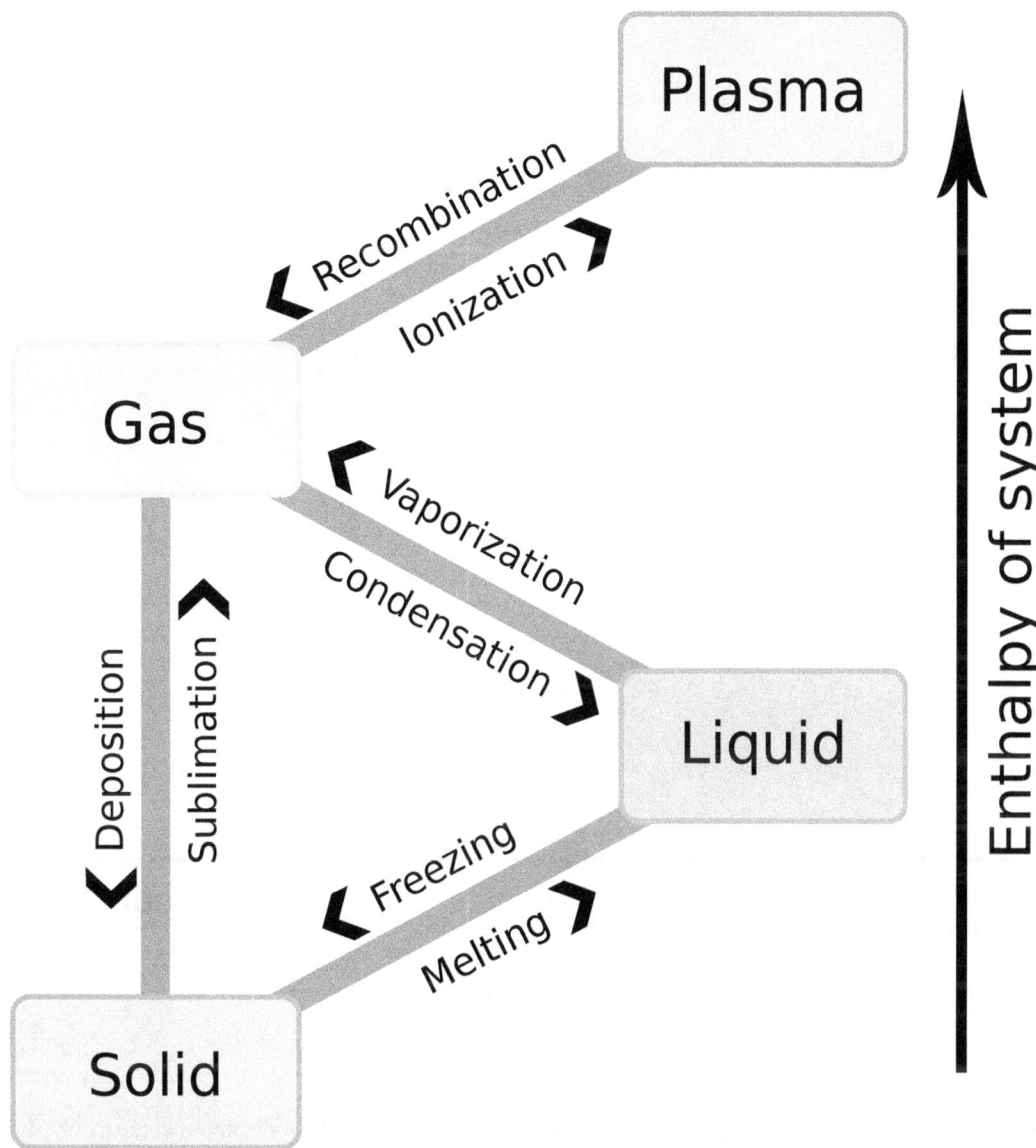

This diagram shows the nomenclature for the different phase transitions.

- Order-disorder transitions such as in alpha-titanium aluminides.

- The dependence of the adsorption geometry on coverage and temperature, such as for hydrogen on iron (110).

- The emergence of superconductivity in certain metals and ceramics when cooled below a critical temperature.

- The transition between different molecular structures (polymorphs, allotropes or polyamorphs), especially of solids, such as between an amorphous structure and a crystal structure, between two different crystal structures, or between two amorphous structures.

- Quantum condensation of bosonic fluids (Bose–Einstein condensation). The superfluid transition in liquid helium is an example of this.

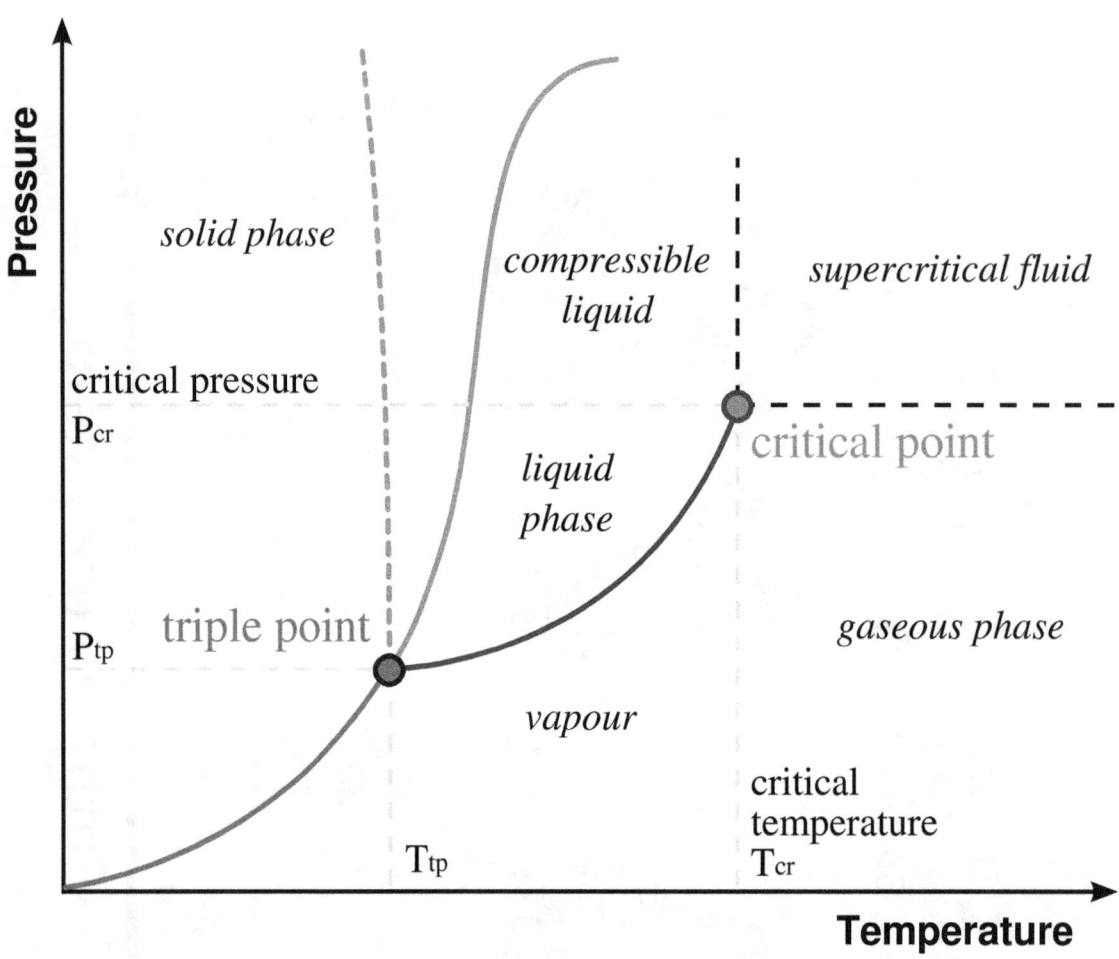

A typical phase diagram. The dotted line gives the anomalous behavior of water.

- The breaking of symmetries in the laws of physics during the early history of the universe as its temperature cooled.

- Isotope fractionation occurs during a phase transition, the ratio of light to heavy isotopes in the involved molecules changes. When water vapor condenses (an equilibrium fractionation), the heavier water isotopes (18O and 2H) become enriched in the liquid phase while the lighter isotopes (16O and 1H) tend toward the vapor phase.[1]

Phase transitions occur when the thermodynamic free energy of a system is non-analytic for some choice of thermodynamic variables (cf. phases). This condition generally stems from the interactions of a large number of particles in a system, and does not appear in systems that are too small. It is important to note that phase transitions can occur and are defined for non-thermodynamic systems, where temperature is not a parameter. Examples include: quantum phase transitions, dynamic phase transitions, and topological (structural) phase transitions. In these types of systems other parameters take the place of temperature. For instance, connection probability replaces temperature for percolating networks.

At the phase transition point (for instance, boiling point) the two phases of a substance, liquid and vapor, have identical free energies and therefore are equally likely to exist. Below the boiling point, the liquid is the more stable state of the two, whereas above the gaseous form is preferred.

It is sometimes possible to change the state of a system diabatically (as opposed to adiabatically) in such a way that it can be brought past a phase transition point without undergoing a phase transition. The resulting state is metastable, i.e., less stable than the phase to which the transition would have occurred, but not unstable either. This occurs in superheating, supercooling, and supersaturation, for example.

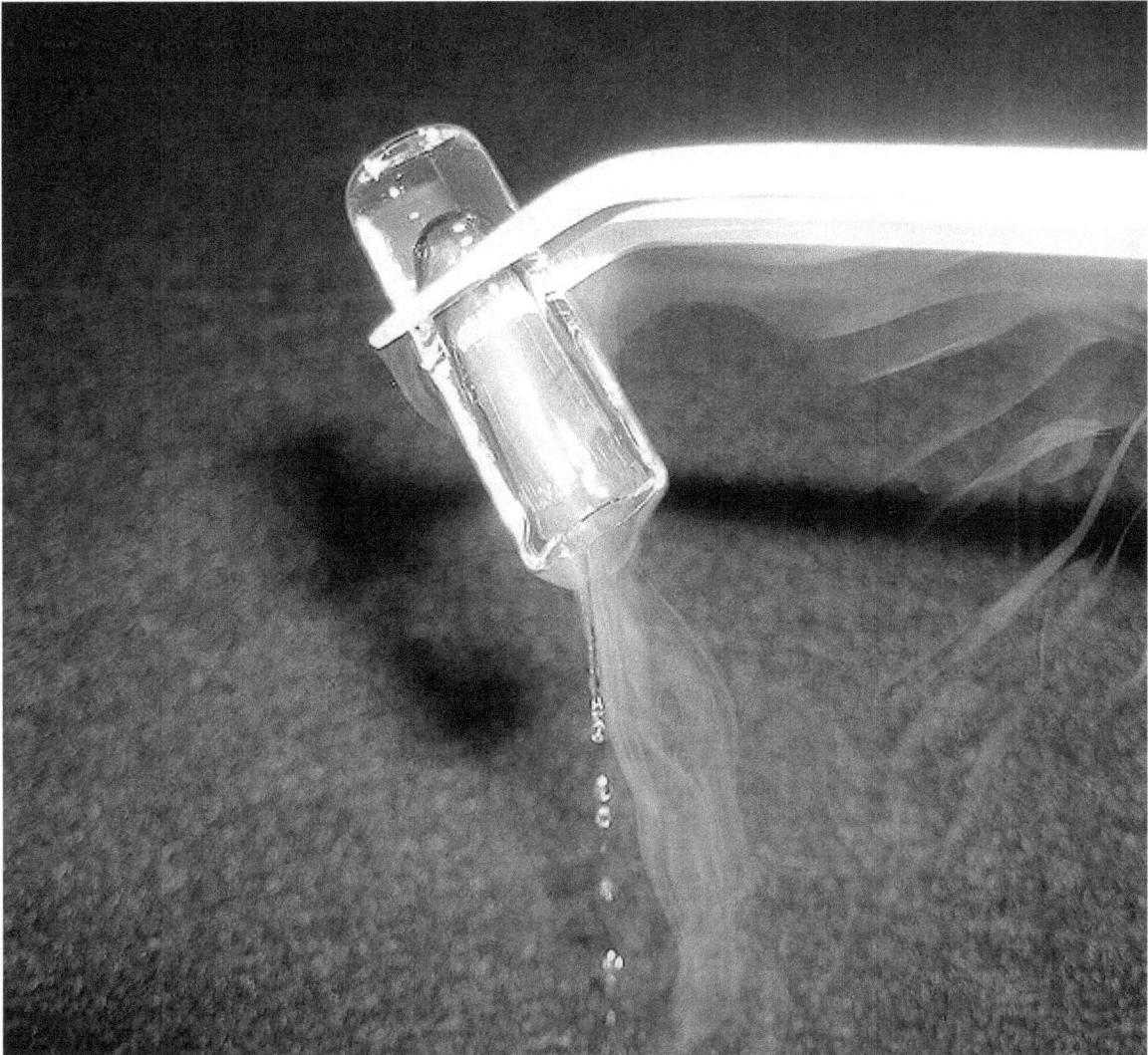

A small piece of rapidly melting solid argon simultaneously shows the transitions from solid to liquid and liquid to gas.

11.2 Classifications

11.2.1 Ehrenfest classification

Paul Ehrenfest classified phase transitions based on the behavior of the thermodynamic free energy as a function of other thermodynamic variables.[2] Under this scheme, phase transitions were labeled by the lowest derivative of the free energy that is discontinuous at the transition. *First-order phase transitions* exhibit a discontinuity in the first derivative of the free energy with respect to some thermodynamic variable.[3] The various solid/liquid/gas transitions are classified as first-order transitions because they involve a discontinuous change in density, which is the (inverse of the) first derivative of the free energy with respect to pressure. *Second-order phase transitions* are continuous in the first derivative (the order parameter, which is the first derivative of the free energy with respect to the external field, is continuous across the transition) but exhibit discontinuity in a second derivative of the free energy.[3] These include the ferromagnetic phase transition in materials such as iron, where the magnetization, which is the first derivative of the free energy with respect to the applied magnetic field strength, increases continuously from zero as the temperature is lowered below the Curie temperature. The magnetic susceptibility, the second derivative of the free energy with the field, changes discontinuously. Under the Ehrenfest classification scheme, there could in principle be third, fourth, and higher-order phase transitions.

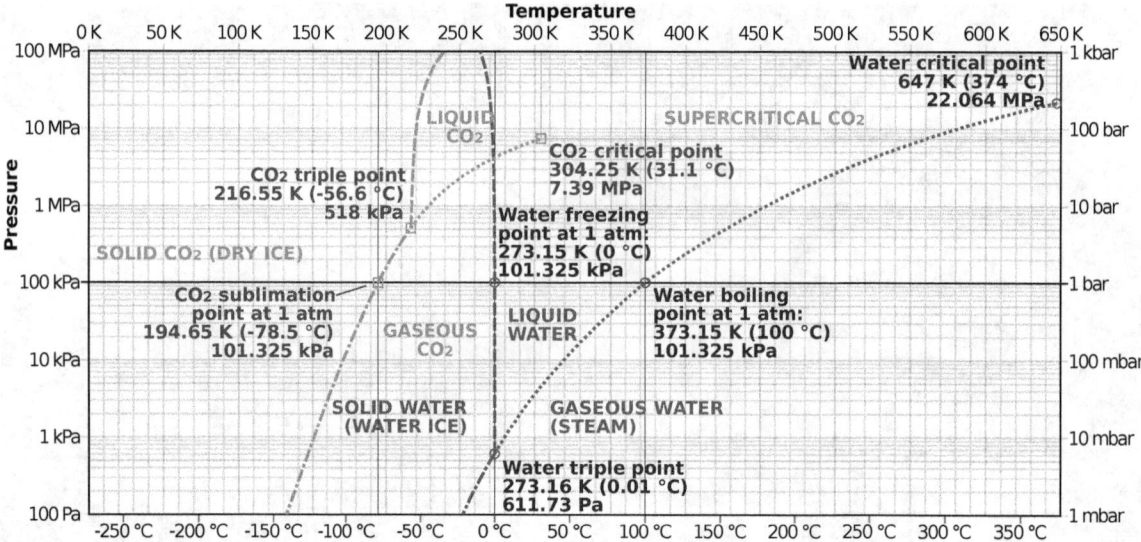

Comparison of phase diagrams of carbon dioxide (red) and water (blue) explaining their different phase transitions at 1 atmosphere

Though useful, Ehrenfest's classification has been found to be an incomplete method of classifying phase transitions, for it does not take into account the case where a derivative of free energy diverges (which is only possible in the thermodynamic limit). For instance, in the ferromagnetic transition, the heat capacity diverges to infinity.

11.2.2 Modern classifications

In the modern classification scheme, phase transitions are divided into two broad categories, named similarly to the Ehrenfest classes:[2]

First-order phase transitions are those that involve a latent heat. During such a transition, a system either absorbs or releases a fixed (and typically large) amount of energy per volume. During this process, the temperature of the system will stay constant as heat is added: the system is in a "mixed-phase regime" in which some parts of the system have completed the transition and others have not. Familiar examples are the melting of ice or the boiling of water (the water does not instantly turn into vapor, but forms a turbulent mixture of liquid water and vapor bubbles). Imry and Wortis showed that quenched disorder can broaden a first-order transition in that the transformation is completed over a finite range of temperatures, but phenomena like supercooling and superheating survive and hysteresis is observed on thermal cycling.[4][5][6]

Second-order phase transitions are also called *continuous phase transitions*. They are characterized by a divergent susceptibility, an infinite correlation length, and a power-law decay of correlations near criticality. Examples of second-order phase transitions are the ferromagnetic transition, superconducting transition (for a Type-I superconductor the phase transition is second-order at zero external field and for a Type-II superconductor the phase transition is second-order for both normal state-mixed state and mixed state-superconducting state transitions) and the superfluid transition. In contrast to viscosity, thermal expansion and heat capacity of amorphous materials show a relatively sudden change at the glass transition temperature[7] which enables accurate detection using differential scanning calorimetry measurements. Lev Landau gave a phenomenological theory of second-order phase transitions.

Apart from isolated, simple phase transitions, there exist transition lines as well as multicritical points, when varying external parameters like the magnetic field or composition.

Several transitions are known as the *infinite-order phase transitions*. They are continuous but break no symmetries. The most famous example is the Kosterlitz–Thouless transition in the two-dimensional XY model. Many quantum phase transitions, e.g., in two-dimensional electron gases, belong to this class.

The liquid–glass transition is observed in many polymers and other liquids that can be supercooled far below the melting point of the crystalline phase. This is atypical in several respects. It is not a transition between thermodynamic ground states: it is widely believed that the true ground state is always crystalline. Glass is a *quenched disorder* state, and its entropy, density, and so on, depend on the thermal history. Therefore, the glass transition is primarily a dynamic phenomenon: on cooling a liquid, internal degrees of freedom successively fall out of equilibrium. Some theoretical methods predict an underlying phase transition in the hypothetical limit of infinitely long relaxation times.[8][9] No direct experimental evidence supports the existence of these transitions.

11.3 Characteristic properties

11.3.1 Phase coexistence

A disorder-broadened first-order transition occurs over a finite range of temperatures where the fraction of the low-temperature equilibrium phase grows from zero to one (100%) as the temperature is lowered. This continuous variation of the coexisting fractions with temperature raised interesting possibilities. On cooling, some liquids vitrify into a glass rather than transform to the equilibrium crystal phase. This happens if the cooling rate is faster than a critical cooling rate, and is attributed to the molecular motions becoming so slow that the molecules cannot rearrange into the crystal positions.[10] This slowing down happens below a glass-formation temperature Tg, which may depend on the applied pressure.,[7][11] If the first-order freezing transition occurs over a range of temperatures, and Tg falls within this range, then there is an interesting possibility that the transition is arrested when it is partial and incomplete. Extending these ideas to first-order magnetic transitions being arrested at low temperatures, resulted in the observation of incomplete magnetic transitions, with two magnetic phases coexisting, down to the lowest temperature. First reported in the case of a ferromagnetic to anti-ferromagnetic transition,[12] such persistent phase coexistence has now been reported across a variety of first-order magnetic transitions. These include colossal-magnetoresistance manganite materials,[13][14] magnetocaloric materials,[15] magnetic shape memory materials,[16] and other materials.[17] The interesting feature of these observations of Tg falling within the temperature range over which the transition occurs is that the first-order magnetic transition is influenced by magnetic field, just like the structural transition is influenced by pressure. The relative ease with which magnetic field can be controlled, in contrast to pressure, raises the possibility that one can study the interplay between Tg and Tc in an exhaustive way. Phase coexistence across first-order magnetic transitions will then enable the resolution of outstanding issues in understanding glasses.

11.3.2 Critical points

In any system containing liquid and gaseous phases, there exists a special combination of pressure and temperature, known as the critical point, at which the transition between liquid and gas becomes a second-order transition. Near the critical point, the fluid is sufficiently hot and compressed that the distinction between the liquid and gaseous phases is almost non-existent. This is associated with the phenomenon of critical opalescence, a milky appearance of the liquid due to density fluctuations at all possible wavelengths (including those of visible light).

11.3.3 Symmetry

Phase transitions often involve a symmetry breaking process. For instance, the cooling of a fluid into a crystalline solid breaks continuous translation symmetry: each point in the fluid has the same properties, but each point in a crystal does not have the same properties (unless the points are chosen from the lattice points of the crystal lattice). Typically, the high-temperature phase contains more symmetries than the low-temperature phase due to spontaneous symmetry breaking, with the exception of certain accidental symmetries (e.g. the formation of heavy virtual particles, which only occurs at low temperatures).[18]

11.3.4   Order parameters

An order parameter is a measure of the degree of order across the boundaries in a phase transition system; it normally ranges between zero in one phase (usually above the critical point) and nonzero in the other.[19] At the critical point, the order parameter susceptibility will usually diverge.

An example of an order parameter is the net magnetization in a ferromagnetic system undergoing a phase transition. For liquid/gas transitions, the order parameter is the difference of the densities.

From a theoretical perspective, order parameters arise from symmetry breaking. When this happens, one needs to introduce one or more extra variables to describe the state of the system. For example, in the ferromagnetic phase, one must provide the net magnetization, whose direction was spontaneously chosen when the system cooled below the Curie point. However, note that order parameters can also be defined for non-symmetry-breaking transitions. Some phase transitions, such as superconducting and ferromagnetic, can have order parameters for more than one degree of freedom. In such phases, the order parameter may take the form of a complex number, a vector, or even a tensor, the magnitude of which goes to zero at the phase transition.

There also exist dual descriptions of phase transitions in terms of disorder parameters. These indicate the presence of line-like excitations such as vortex- or defect lines.

11.3.5   Relevance in cosmology

Symmetry-breaking phase transitions play an important role in cosmology. It has been speculated by Lee Smolin and Benjamin and Jeremy Bernstein that, in the hot early universe, the vacuum (i.e. the various quantum fields that fill space) possessed a large number of symmetries. As the universe expanded and cooled, the vacuum underwent a series of symmetry-breaking phase transitions. For example, the electroweak transition broke the SU(2)×U(1) symmetry of the electroweak field into the U(1) symmetry of the present-day electromagnetic field. This transition is important to understanding the asymmetry between the amount of matter and antimatter in the present-day universe (see electroweak baryogenesis.)

Progressive phase transitions in an expanding universe are implicated in the development of order in the universe, as is illustrated by the work of Eric Chaisson[20] and David Layzer.[21] See also Relational order theories.

See also: Order-disorder

11.3.6   Critical exponents and universality classes

Main article: critical exponent

Continuous phase transitions are easier to study than first-order transitions due to the absence of latent heat, and they have been discovered to have many interesting properties. The phenomena associated with continuous phase transitions are called critical phenomena, due to their association with critical points.

It turns out that continuous phase transitions can be characterized by parameters known as critical exponents. The most important one is perhaps the exponent describing the divergence of the thermal correlation length by approaching the transition. For instance, let us examine the behavior of the heat capacity near such a transition. We vary the temperature T of the system while keeping all the other thermodynamic variables fixed, and find that the transition occurs at some critical temperature Tc . When T is near Tc , the heat capacity C typically has a power law behavior,

$$C \propto |T_c - T|^{-\alpha}.$$

Such a behaviour has the heat capacity of amorphous materials near the glass transition temperature where the universal critical exponent $\alpha = 0.59$[22] A similar behavior, but with the exponent ν instead of α, applies for the correlation length.

The exponent ν is positive. This is different with α. Its actual value depends on the type of phase transition we are considering.

For $-1 < \alpha < 0$, the heat capacity has a "kink" at the transition temperature. This is the behavior of liquid helium at the lambda transition from a normal state to the superfluid state, for which experiments have found $\alpha = -0.013 \pm 0.003$. At least one experiment was performed in the zero-gravity conditions of an orbiting satellite to minimize pressure differences in the sample.[23] This experimental value of α agrees with theoretical predictions based on variational perturbation theory.[24]

For $0 < \alpha < 1$, the heat capacity diverges at the transition temperature (though, since $\alpha < 1$, the enthalpy stays finite). An example of such behavior is the 3D ferromagnetic phase transition. In the three-dimensional Ising model for uniaxial magnets, detailed theoretical studies have yielded the exponent $\alpha \sim +0.110$.

Some model systems do not obey a power-law behavior. For example, mean field theory predicts a finite discontinuity of the heat capacity at the transition temperature, and the two-dimensional Ising model has a logarithmic divergence. However, these systems are limiting cases and an exception to the rule. Real phase transitions exhibit power-law behavior.

Several other critical exponents, β, γ, δ, ν, and η, are defined, examining the power law behavior of a measurable physical quantity near the phase transition. Exponents are related by scaling relations, such as

$$\beta = \gamma/(\delta - 1), \qquad \nu = \gamma/(2 - \eta)$$

It can be shown that there are only two independent exponents, e.g. ν and η.

It is a remarkable fact that phase transitions arising in different systems often possess the same set of critical exponents. This phenomenon is known as *universality*. For example, the critical exponents at the liquid–gas critical point have been found to be independent of the chemical composition of the fluid.

More impressively, but understandably from above, they are an exact match for the critical exponents of the ferromagnetic phase transition in uniaxial magnets. Such systems are said to be in the same universality class. Universality is a prediction of the renormalization group theory of phase transitions, which states that the thermodynamic properties of a system near a phase transition depend only on a small number of features, such as dimensionality and symmetry, and are insensitive to the underlying microscopic properties of the system. Again, the divergency of the correlation length is the essential point.

11.3.7 Critical slowing down and other phenomena

There are also other critical phenomena; e.g., besides *static functions* there is also *critical dynamics*. As a consequence, at a phase transition one may observe critical slowing down or *speeding up*. The large *static universality classes* of a continuous phase transition split into smaller *dynamic universality* classes. In addition to the critical exponents, there are also universal relations for certain static or dynamic functions of the magnetic fields and temperature differences from the critical value.

11.3.8 Percolation theory

Another phenomenon which shows phase transitions and critical exponents is percolation. The simplest example is perhaps percolation in a two dimensional square lattice. Sites are randomly occupied with probability p. For small values of p the occupied sites form only small clusters. At a certain threshold p_c a giant cluster is formed and we have a second-order phase transition.[25] The behavior of $P\infty$ near p_c is, $P\infty \sim (p-p_c)^\beta$, where β is a critical exponent.

11.3.9 Phase transitions in biological systems

Phase transitions play many important roles in biological systems. Examples include the lipid bilayer formation, the coil-globule transition in the process of protein folding and DNA melting, liquid crystal-like transitions in the process of DNA condensation, and cooperative ligand binding to DNA and proteins with the character of phase transition.[26]

In *biological membranes*, gel to liquid crystalline phase transitions play a very critical role in physiological functioning of biomembranes. In gel phase, due to low fluidity of membrane lipid fatty-acyl chains, membrane proteins have restricted movement and thus are restrained in exercise of their physiological role. Plants depend critically on photosynthesis by chloroplast thylakoid membranes which are exposed cold environmental temperatures. Thylakoid membranes retain innate fluidity even at relatively low temperatures because of high degree of fatty-acyl disorder allowed by their high content of linolenic acid, 18-carbon chain with 3-double bonds.[27] Gel-to-liquid crystalline phase transition temperature of biological membranes can be determined by many techniques including calorimetry, flouorescence, spin label electron paramagnetic resonance and NMR by recording measurements of the concerned parameter by at series of sample temperatures. A simple method for its determination from 13-C NMR line intensities has also been proposed.[28]

The relevance of phase transitions in neural networks has been pointed out, because of the complex and emergent nature of neural interactions. A point of view can be found in the very recent paper by Tkačik et al.[29]

11.4   See also

- Allotropy

- Autocatalytic reactions and order creation

- Crystal growth

  - Abnormal grain growth

- Differential scanning calorimetry

- Diffusionless transformations

- Ehrenfest equations

- Jamming (physics)

- Kelvin probe force microscope

- Landau theory of second order phase transitions

- Laser-heated pedestal growth

- List of states of matter

- Micro-Pulling-Down

- Percolation theory

  - Continuum percolation theory

- Superfluid film

- Superradiant phase transition

11.5   References

[1] Carol Kendall (2004). "Fundamentals of Stable Isotope Geochemistry". USGS. Retrieved 10 April 2014.

[2] Jaeger, Gregg (1 May 1998). "The Ehrenfest Classification of Phase Transitions: Introduction and Evolution". *Archive for History of Exact Sciences* **53** (1): 51–81. doi:10.1007/s004070050021.

[3] Blundell, Stephen J.; Katherine M. Blundell (2008). *Concepts in Thermal Physics*. Oxford University Press. ISBN 978-0-19-856770-7.

[4] Imry, Y.; Wortis, M. (1979). "Influence of quenched impurities on first-order phase transitions". *Phys. Rev. B* **19** (7): 3580–3585. Bibcode:1979PhRvB..19.3580I. doi:10.1103/physrevb.19.3580.

[5] Kumar, K.; et al. (2006). "Relating supercooling and glass-like arrest of kinetics for phase separated systems: Doped-CeFe2and(La,Pr,Ca)MnO3". *Phys. Rev. B* **73** (18): 184435. arXiv:cond-mat/0602627. Bibcode:2006PhRvB..73r4435K. doi:10.1103/PhysRevB.73.184435.

[6] Pasquini, G.; et al. (2008). "Single-Qubit Lasing and Cooling at the Rabi Frequency". *Phys. Rev. Lett* **100** (3): 247003. arXiv:cond-mat/0701041. Bibcode:2008PhRvL.100c7003H. doi:10.1103/PhysRevLett.100.037003.

[7] Ojovan, M.I. (2013). "Ordering and structural changes at the glass-liquid transition". *J. Non-Cryst. Solids* **382**: 79–86. Bibcode:2013JNCS..382...79O. doi:10.1016/j.jnoncrysol.2013.10.016.

[8] Gotze, Wolfgang. "Complex Dynamics of Glass-Forming Liquids: A Mode-Coupling Theory."

[9] Lubchenko, V. Wolynes; Wolynes, Peter G. (2007). "Theory of Structural Glasses and Supercooled Liquids". *Annual Review of Physical Chemistry* **58**: 235–266. arXiv:cond-mat/0607349. Bibcode:2007ARPC...58..235L. doi:10.1146/annurev.physchem.58.032806.104653.PMID 17067282.

[10] Greer, A. L. (1995). "Metalli c Glasses". *Science* **267** (5206): 1947–1953. Bibcode:1995Sci...267.1947G. doi:10.1126/

[11] Tarjus, G. (2007). "Materials science: Metal turned to glass". *Nature* **448** (7155): 758–759. Bibcode:2007Natur.448..758T. doi:10.1038/448758a. PMID 17700684.

[12] Manekar, M. A.; et al. (2001). "Nonequilibrium relaxation study of Ising spin glass models". *Physical Review B* **64** (2): 104416. Bibcode:2001PhRvB..64b4416O. doi:10.1103/PhysRevB.64.024416.

[13] Banerjee, A; Pramanik, A K; Kumar, Kranti; Chaddah, P (2006). "Coexisting tunable fractions of glassy and equilibrium long-range-order phases in manganites". *Journal of Physics: Condensed Matter* **18** (49): L605. arXiv:cond-mat/0611152. Bibcode:2006JPCM...18L.605B. doi:10.1088/0953-8984/18/49/L02.

[14] Wu, W.; et al. (2006). "Magnetic imaging of a supercooling glass transition in a weakly disordered ferromagnet". *Nature Materials* **5** (11): 881–886. Bibcode:2006NatMa...5..881W. doi:10.1038/nmat1743.

[15] Roy, S. B.; et al. (2006). "Evidence of a magnetic glass state in the magnetocaloric materialGd5Ge4". *Physical Review B* **74**: 012403. Bibcode:2006PhRvB..74a2403R. doi:10.1103/PhysRevB.74.012403.

[16] Lakhani, A.; et al. (2012). "Magnetic glass in shape memory alloy: Ni45Co5Mn38Sn12". *J. Phys. Condens. Matter* **24** (38): 386004. arXiv:1206.2024. Bibcode:2012JPCM...24L6004L. doi:10.1088/0953-8984/24/38/386004.

[17] Kushwaha, P.; et al. (2009). "Non-Korringa nuclear relaxation in the ferromagneti c phase of the bilayered manganiteLa1.2Sr *Physical Review B* **80** (2): 174413. Bibcode:2009PhRvB..80b4413H. doi:10.1103/PhysRevB.80.024413.

[18] Ivancevic, Vladimir G.; Ivancevic, Tijiana, T. (2008). *Complex Nonlinearity*. Berlin: Springer. pp. 176–177. ISBN 978-3-540-79357-1. Retrieved 12 October 2014.

[19] A. D. McNaught and A. Wilkinson (ed.). "Compendium of Chemical Terminology". IUPAC. ISBN 0-86542-684-8. Retrieved 2007-10-23.

[20] Chaisson, *Cosmic Evolution*, Harvard, 2001

[21] David Layzer, *Cosmogenesis, The Development of Order in the Universe*, Oxford Univ. Press, 1991

[22] Ojovan, Michael I; Lee, William E (2006). "Topologically disordered systems at the glass transition" (PDF). *Journal of Physics: Condensed Matter* **18** (50): 11507–11520. Bibcode:2006JPCM...1811507O. doi:10.1088/0953-8984/18/50/007.

[23] Lipa, J.; Nissen, J.; Stricker, D.; Swanson, D.; Chui, T. (2003). "Specific heat of liquid helium in zero gravity very near the lambda point". *Physical Review B* **68** (17): 174518. arXiv:cond-mat/0310163. Bibcode:2003PhRvB..68q4518L. doi:10.

[24] Kleinert, Hagen (1999). "Critical exponents from seven-loop strong-coupling φ4 theory in three dimensions". *Physical Review D* **60** (8): 085001. arXiv:hep-th/9812197. Bibcode:1999PhRvD..60h5001K. doi:10.1103/PhysRevD.60.085001.

[25] Armin Bunde and Shlomo Havlin (1996). *Fractals and Disordered Systems*. Springer.

[26] D.Y. Lando and V.B. Teif (2000). "Long-range interactions between ligands bound to a DNA molecule give rise to adsorption with the character of phase transition of the first kind". *J. Biomol. Struct. Dynam.* **17** (5): 903–911. doi:10.1080/07

[27] YashRoy, R.C. (1987). "13-C NMR studies of lipid fatty acyl chains of chloroplast membranes". *Indian Journal of Biochemistry and Biophysics* **24** (6): 177–178.

[28] YashRoy, R C (1990). "Determination of membrane lipid phase transition temperature from 13-C NMR intensities". *Journal of Biochemical and Biophysical Methods* **20** (4): 353–356. doi:10.1016/0165-022X(90)90097-V. PMID 2365951.

[29] Tkacik, Gasper; Mora, Thierry; Marre, Olivier; Amodei, Dario; Berry II, Michael J.; Bialek, William (2014). "Thermodynamics for a network of neurons: Signatures of criticality". arXiv:1407.5946 [q-bio.NC].

11.5.1   Further reading

- Anderson, P.W., *Basic Notions of Condensed Matter Physics*, Perseus Publishing (1997).

- Fisher, M.E. (1974). "The renormalization group in the theory of critical behavior". *Rev. Mod. Phys.* **46**: 597–616. Bibcode:1974RvMP...46..597F. doi:10.1103/revmodphys.46.597.

- Goldenfeld, N., *Lectures on Phase Transitions and the Renormalization Group*, Perseus Publishing (1992).

- Ivancevic, Vladimir G; Ivancevic, Tijana T (2008), *Chaos, Phase Transitions, Topology Change and Path Integrals*, Berlin: Springer, ISBN 978-3-540-79356-4, retrieved 14 March 2013 e-ISBN 978-3-540-79357-1

- Kogut, J.; Wilson, K (1974). "The Renormalization Group and the epsilon-Expansion". *Phys. Rep.* **12**: 75. Bibcode:1974PhR....12...75W. doi:10.1016/0370-1573(74)90023-4.

- Krieger, Martin H., *Constitutions of matter : mathematically modelling the most everyday of physical phenomena*, University of Chicago Press, 1996. Contains a detailed pedagogical discussion of Onsager's solution of the 2-D Ising Model.

- Landau, L.D. and Lifshitz, E.M., *Statistical Physics Part 1*, vol. 5 of *Course of Theoretical Physics*, Pergamon Press, 3rd Ed. (1994).

- Kleinert, H., *Gauge Fields in Condensed Matter*, Vol. I, "Superfluid and Vortex lines; Disorder Fields, Phase Transitions,", pp. 1–742, World Scientific (Singapore, 1989); Paperback ISBN 9971-5-0210-0 *(readable online physik.fu-berlin.de)*

- Kleinert, H. and Verena Schulte-Frohlinde, *Critical Properties of φ^4-Theories*, World Scientific (Singapore, 2001); Paperback ISBN 981-02-4659-5 *(readable online here)*.

- Mussardo G., "Statistical Field Theory. An Introduction to Exactly Solved Models of Statistical Physics", Oxford University Press, 2010.

- Schroeder, Manfred R., *Fractals, chaos, power laws : minutes from an infinite paradise*, New York: W. H. Freeman, 1991. Very well-written book in "semi-popular" style—not a textbook—aimed at an audience with some training in mathematics and the physical sciences. Explains what scaling in phase transitions is all about, among other things.

- Yeomans J. M., *Statistical Mechanics of Phase Transitions*, Oxford University Press, 1992.

- H. E. Stanley, *Introduction to Phase Transitions and Critical Phenomena* (Oxford University Press, Oxford and New York 1971).

11.6   External links

- Interactive Phase Transitions on lattices with Java applets

Chapter 12

Higgs mechanism

In the Standard Model of particle physics, the **Higgs mechanism** is essential to explain the generation mechanism of the property "mass" for gauge bosons. Without the Higgs mechanism, or some other effect like it, all bosons (a type of fundamental particle) would be massless, but measurements show that the W^+, W^-, and Z bosons actually have relatively large masses of around 80 GeV/c^2. The Higgs field resolves this conundrum. The simplest description of the mechanism adds a quantum field (the Higgs field) that permeates all space, to the Standard Model. Below some extremely high temperature, the field causes spontaneous symmetry breaking during interactions. The breaking of symmetry triggers the Higgs mechanism, causing the bosons it interacts with to have mass. In the Standard Model, the phrase "Higgs mechanism" refers specifically to the generation of masses for the $W^\pm$, and Z weak gauge bosons through electroweak symmetry breaking.[1] The Large Hadron Collider at CERN announced results consistent with the Higgs particle on March 14, 2013, making it extremely likely that the field, or one like it, exists, and explaining how the Higgs mechanism takes place in nature.

The mechanism was proposed in 1962 by Philip Warren Anderson,[2] following work in the late 1950s on symmetry breaking in superconductivity and a 1960 paper by Yoichiro Nambu that discussed its application within particle physics. A theory able to finally explain mass generation without "breaking" gauge theory was published almost simultaneously by three independent groups in 1964: by Robert Brout and François Englert;[3] by Peter Higgs;[4] and by Gerald Guralnik, C. R. Hagen, and Tom Kibble.[5][6][7] The Higgs mechanism is therefore also called the **Brout–Englert–Higgs mechanism** or **Englert–Brout–Higgs–Guralnik–Hagen–Kibble mechanism**,[8] **Anderson–Higgs mechanism**,[9] **Anderson–Higgs-Kibble mechanism**,[10] **Higgs–Kibble mechanism** by Abdus Salam[11] and **ABEGHHK'tH mechanism** [for Anderson, Brout, Englert, Guralnik, Hagen, Higgs, Kibble and 't Hooft] by Peter Higgs.[11]

On October 8, 2013, following the discovery at CERN's Large Hadron Collider of a new particle that appeared to be the long-sought Higgs boson predicted by the theory, it was announced that Peter Higgs and François Englert had been awarded the 2013 Nobel Prize in Physics (Englert's co-author Robert Brout had died in 2011 and the Nobel Prize is not usually awarded posthumously).[12]

12.1   Standard model

The Higgs mechanism was incorporated into modern particle physics by Steven Weinberg and Abdus Salam, and is an essential part of the standard model.

In the standard model, at temperatures high enough that electroweak symmetry is unbroken, all elementary particles are massless. At a critical temperature the Higgs field becomes tachyonic, the symmetry is spontaneously broken by condensation, and the W and Z bosons acquire masses. (*EWSB*, electroweak symmetry breaking, is an abbreviation used for this.)

Fermions, such as the leptons and quarks in the Standard Model, can also acquire mass as a result of their interaction with the Higgs field, but not in the same way as the gauge bosons.

12.1.1   Structure of the Higgs field

In the standard model, the Higgs field is an **SU**(2) doublet, a complex scalar with four real components (or equivalently with two complex components). Its (weak hypercharge) **U**(1) charge is 1. That means that it transforms as a spinor under **SU**(2). Under **U**(1) rotations, it is multiplied by a phase, which thus mixes the real and imaginary parts of the complex spinor into each other—so this is *not the same* as two complex spinors mixing under **U**(1) (which would have eight real components between them), but instead is the spinor representation of the group **U**(2).

The Higgs field, through the interactions specified (summarized, represented, or even simulated) by its potential, induces spontaneous breaking of three out of the four generators ("directions") of the gauge group **SU**(2) × **U**(1): three out of its four components would ordinarily amount to Goldstone bosons, if they were not coupled to gauge fields.

However, after symmetry breaking, these three of the four degrees of freedom in the Higgs field mix with the three W and Z bosons (W+, W− and Z), and are only observable as spin components of these weak bosons, which are now massive; while the one remaining degree of freedom becomes the Higgs boson—a new scalar particle.

12.1.2   The photon as the part that remains massless

The gauge group of the electroweak part of the standard model is **SU**(2) × **U**(1). The group **SU**(2) is the group of all 2-by-2 unitary matrices with unit determinant; all the orthonormal changes of coordinates in a complex two dimensional vector space.

Rotating the coordinates so that the second basis vector points in the direction of the Higgs boson makes the vacuum expectation value of H the spinor $(0, v)$. The generators for rotations about the x, y, and z axes are by half the Pauli matrices σx, σy, and σz, so that a rotation of angle θ about the z-axis takes the vacuum to

$$(0, ve^{-i\theta/2}).$$

While the T_x and T_y generators mix up the top and bottom components of the spinor, the T_z rotations only multiply each by opposite phases. This phase can be undone by a **U**(1) rotation of angle $1/2\theta$. Consequently, under both an **SU**(2) T_z-rotation and a **U**(1) rotation by an amount $1/2\theta$, *the vacuum is invariant.*

This combination of generators

$$Q = T_z + \frac{Y}{2}$$

defines the unbroken part of the gauge group, where Q is the electric charge, T_z is the generator of rotations around the z-axis in the **SU**(2) and Y is the hypercharge generator of the **U**(1). This combination of generators (a z rotation in the **SU**(2) and a simultaneous **U**(1) rotation by half the angle) preserves the vacuum, and defines the unbroken gauge group in the standard model, namely *the electric charge* group. The part of the gauge field in this direction stays massless, and amounts to the physical photon.

12.1.3   Consequences for fermions

In spite of the introduction of spontaneous symmetry breaking, the mass terms oppose the chiral gauge invariance. For these fields the mass terms should always be replaced by a gauge-invariant "Higgs" mechanism. One possibility is some kind of "Yukawa coupling" (see below) between the fermion field ψ and the Higgs field Φ, with unknown couplings $G\psi$, which after symmetry breaking (more precisely: after expansion of the Lagrange density around a suitable ground state) again results in the original mass terms, which are now, however (i.e. by introduction of the Higgs field) written in a gauge-invariant way. The Lagrange density for the "Yukawa" interaction of a fermion field ψ and the Higgs field Φ is

$$\mathcal{L}_{\text{Fermion}}(\phi, A, \psi) = \overline{\psi}\gamma^{\mu}D_{\mu}\psi + G_{\psi}\overline{\psi}\phi\psi,$$

where again the gauge field A only enters $D\mu$ (i.e., it is only indirectly visible). The quantities γ^{μ} are the Dirac matrices, and $G\psi$ is the already-mentioned "Yukawa" coupling parameter. Already now the mass-generation follows the same principle as above, namely from the existence of a finite expectation value $|\langle\phi\rangle|$, as described above. Again, this is crucial for the existence of the property "mass".

12.2 History of research

12.2.1 Background

Spontaneous symmetry breaking offered a framework to introduce bosons into relativistic quantum field theories. However, according to Goldstone's theorem, these bosons should be massless.[13] The only observed particles which could be approximately interpreted as Goldstone bosons were the pions, which Yoichiro Nambu related to chiral symmetry breaking.

A similar problem arises with Yang–Mills theory (also known as non-abelian gauge theory), which predicts massless spin−1 gauge bosons. Massless weakly interacting gauge bosons lead to long-range forces, which are only observed for electromagnetism and the corresponding massless photon. Gauge theories of the weak force needed a way to describe massive gauge bosons in order to be consistent.

12.2.2 Discovery

The mechanism was proposed in 1962 by Philip Warren Anderson,[2] who discussed its consequences for particle physics but did not work out an explicit relativistic model. The relativistic model was developed in 1964 by three independent groups – Robert Brout and François Englert;[3] Peter Higgs;[4] and Gerald Guralnik, Carl Richard Hagen, and Tom Kibble.[5][6][7] Slightly later, in 1965, but independently from the other publications[14][15][16][17][18][19] the mechanism was also proposed by Alexander Migdal and Alexander Polyakov,[20] at that time Soviet undergraduate students. However, the paper was delayed by the Editorial Office of JETP, and was published only in 1966.

The mechanism is closely analogous to phenomena previously discovered by Yoichiro Nambu involving the "vacuum structure" of quantum fields in superconductivity.[21] A similar but distinct effect (involving an affine realization of what is now recognized as the Higgs field), known as the Stueckelberg mechanism, had previously been studied by Ernst Stueckelberg.

These physicists discovered that when a gauge theory is combined with an additional field that spontaneously breaks the symmetry group, the gauge bosons can consistently acquire a nonzero mass. In spite of the large values involved (see below) this permits a gauge theory description of the weak force, which was independently developed by Steven Weinberg and Abdus Salam in 1967. Higgs's original article presenting the model was rejected by Physics Letters. When revising the article before resubmitting it to Physical Review Letters, he added a sentence at the end,[22] mentioning that it implies the existence of one or more new, massive scalar bosons, which do not form complete representations of the symmetry group; these are the Higgs bosons.

The three papers by Brout and Englert; Higgs; and Guralnik, Hagen, and Kibble were each recognized as "milestone letters" by *Physical Review Letters* in 2008.[23] While each of these seminal papers took similar approaches, the contributions and differences among the 1964 PRL symmetry breaking papers are noteworthy. All six physicists were jointly awarded the 2010 J. J. Sakurai Prize for Theoretical Particle Physics for this work.[24]

Benjamin W. Lee is often credited with first naming the "Higgs-like" mechanism, although there is debate around when this first occurred.[25][26][27] One of the first times the *Higgs* name appeared in print was in 1972 when Gerardus 't Hooft and Martinus J. G. Veltman referred to it as the "Higgs–Kibble mechanism" in their Nobel winning paper.[28][29]

Philip W. Anderson, the first to propose the mechanism in 1962.

12.3 Examples

The Higgs mechanism occurs whenever a charged field has a vacuum expectation value. In the nonrelativistic context, this is the Landau model of a charged Bose–Einstein condensate, also known as a superconductor. In the relativistic condensate, the condensate is a scalar field, and is relativistically invariant.

Five of the six 2010 APS Sakurai Prize Winners – (L to R) Tom Kibble, Gerald Guralnik, Carl Richard Hagen, François Englert, and Robert Brout

12.3.1 Landau model

The Higgs mechanism is a type of superconductivity which occurs in the vacuum. It occurs when all of space is filled with a sea of particles which are charged, or, in field language, when a charged field has a nonzero vacuum expectation value. Interaction with the quantum fluid filling the space prevents certain forces from propagating over long distances (as it does in a superconducting medium; e.g., in the Ginzburg–Landau theory).

A superconductor expels all magnetic fields from its interior, a phenomenon known as the Meissner effect. This was mysterious for a long time, because it implies that electromagnetic forces somehow become short-range inside the superconductor. Contrast this with the behavior of an ordinary metal. In a metal, the conductivity shields electric fields by rearranging charges on the surface until the total field cancels in the interior. But magnetic fields can penetrate to any distance, and if a magnetic monopole (an isolated magnetic pole) is surrounded by a metal the field can escape without collimating into a string. In a superconductor, however, electric charges move with no dissipation, and this allows for permanent surface currents, not just surface charges. When magnetic fields are introduced at the boundary of a superconductor, they produce surface currents which exactly neutralize them. The Meissner effect is due to currents in a thin surface layer, whose thickness, the London penetration depth, can be calculated from a simple model (the Ginzburg–Landau theory).

This simple model treats superconductivity as a charged Bose–Einstein condensate. Suppose that a superconductor contains bosons with charge q. The wavefunction of the bosons can be described by introducing a quantum field, ψ, which obeys the Schrödinger equation as a field equation (in units where the reduced Planck constant, $\hbar$, is set to 1):

$$i\frac{\partial}{\partial t}\psi = \frac{(\nabla - iqA)^2}{2m}\psi.$$

The operator $\psi(x)$ annihilates a boson at the point x, while its adjoint $\psi^\dagger$ creates a new boson at the same point. The

Number six: Peter Higgs 2009

wavefunction of the Bose–Einstein condensate is then the expectation value ψ of $\psi(x)$, which is a classical function that obeys the same equation. The interpretation of the expectation value is that it is the phase that one should give to a newly created boson so that it will coherently superpose with all the other bosons already in the condensate.

When there is a charged condensate, the electromagnetic interactions are screened. To see this, consider the effect of a gauge transformation on the field. A gauge transformation rotates the phase of the condensate by an amount which changes from point to point, and shifts the vector potential by a gradient:

$$\psi \to e^{iq\phi(x)}\psi$$
$$A \to A + \nabla\phi.$$

When there is no condensate, this transformation only changes the definition of the phase of ψ at every point. But when there is a condensate, the phase of the condensate defines a preferred choice of phase.

The condensate wave function can be written as

$$\psi(x) = \rho(x)\,e^{i\theta(x)},$$

where ρ is real amplitude, which determines the local density of the condensate. If the condensate were neutral, the flow would be along the gradients of θ, the direction in which the phase of the Schrödinger field changes. If the phase θ changes slowly, the flow is slow and has very little energy. But now θ can be made equal to zero just by making a gauge transformation to rotate the phase of the field.

The energy of slow changes of phase can be calculated from the Schrödinger kinetic energy,

$$H = \frac{1}{2m}|(qA + \nabla)\psi|^2,$$

and taking the density of the condensate ρ to be constant,

$$H \approx \frac{\rho^2}{2m}(qA + \nabla\theta)^2.$$

Fixing the choice of gauge so that the condensate has the same phase everywhere, the electromagnetic field energy has an extra term,

$$\frac{q^2\rho^2}{2m}A^2.$$

When this term is present, electromagnetic interactions become short-ranged. Every field mode, no matter how long the wavelength, oscillates with a nonzero frequency. The lowest frequency can be read off from the energy of a long wavelength A mode,

$$E \approx \frac{\dot{A}^2}{2} + \frac{q^2\rho^2}{2m}A^2.$$

This is a harmonic oscillator with frequency

$$\sqrt{\frac{1}{m}q^2\rho^2}.$$

The quantity $|\psi|^2$ ($=\rho^2$) is the density of the condensate of superconducting particles.

In an actual superconductor, the charged particles are electrons, which are fermions not bosons. So in order to have superconductivity, the electrons need to somehow bind into Cooper pairs. The charge of the condensate q is therefore twice the electron charge e. The pairing in a normal superconductor is due to lattice vibrations, and is in fact very weak; this means that the pairs are very loosely bound. The description of a Bose–Einstein condensate of loosely bound pairs is actually more difficult than the description of a condensate of elementary particles, and was only worked out in 1957 by Bardeen, Cooper and Schrieffer in the famous BCS theory.

12.3.2 Abelian Higgs mechanism

Gauge invariance means that certain transformations of the gauge field do not change the energy at all. If an arbitrary gradient is added to A, the energy of the field is exactly the same. This makes it difficult to add a mass term, because a mass term tends to push the field toward the value zero. But the zero value of the vector potential is not a gauge invariant idea. What is zero in one gauge is nonzero in another.

So in order to give mass to a gauge theory, the gauge invariance must be broken by a condensate. The condensate will then define a preferred phase, and the phase of the condensate will define the zero value of the field in a gauge-invariant way. The gauge-invariant definition is that a gauge field is zero when the phase change along any path from parallel transport is equal to the phase difference in the condensate wavefunction.

The condensate value is described by a quantum field with an expectation value, just as in the Ginzburg-Landau model.

In order for the phase of the vacuum to define a gauge, the field must have a phase (also referred to as 'to be charged'). In order for a scalar field Φ to have a phase, it must be complex, or (equivalently) it should contain two fields with a symmetry which rotates them into each other. The vector potential changes the phase of the quanta produced by the field when they move from point to point. In terms of fields, it defines how much to rotate the real and imaginary parts of the fields into each other when comparing field values at nearby points.

The only renormalizable model where a complex scalar field Φ acquires a nonzero value is the Mexican-hat model, where the field energy has a minimum away from zero. The action for this model is

$$S(\phi) = \int \frac{1}{2}|\partial\phi|^2 - \lambda\left(|\phi|^2 - \Phi^2\right)^2,$$

which results in the Hamiltonian

$$H(\phi) = \frac{1}{2}|\dot{\phi}|^2 + |\nabla\phi|^2 + V(|\phi|).$$

The first term is the kinetic energy of the field. The second term is the extra potential energy when the field varies from point to point. The third term is the potential energy when the field has any given magnitude.

This potential energy, $V(z, \Phi) = \lambda(|z|^2 - \Phi^2)^2$,[30] has a graph which looks like a Mexican hat, which gives the model its name. In particular, the minimum energy value is not at $z = 0$, but on the circle of points where the magnitude of z is Φ.

When the field $\Phi(x)$ is not coupled to electromagnetism, the Mexican-hat potential has flat directions. Starting in any one of the circle of vacua and changing the phase of the field from point to point costs very little energy. Mathematically, if

$$\phi(x) = \Phi e^{i\theta(x)}$$

with a constant prefactor, then the action for the field $\theta(x)$, i.e., the "phase" of the Higgs field $\Phi(x)$, has only derivative terms. This is not a surprise. Adding a constant to $\theta(x)$ is a symmetry of the original theory, so different values of $\theta(x)$ cannot have different energies. This is an example of Goldstone's theorem: spontaneously broken continuous symmetries normally produce massless excitations.

The Abelian Higgs model is the Mexican-hat model coupled to electromagnetism:

$$S(\phi, A) = \int -\frac{1}{4}F^{\mu\nu}F_{\mu\nu} + |(\partial - iqA)\phi|^2 - \lambda(|\phi|^2 - \Phi^2)^2.$$

The classical vacuum is again at the minimum of the potential, where the magnitude of the complex field φ is equal to Φ. But now the phase of the field is arbitrary, because gauge transformations change it. This means that the field $\theta(x)$ can be set to zero by a gauge transformation, and does not represent any actual degrees of freedom at all.

Furthermore, choosing a gauge where the phase of the vacuum is fixed, the potential energy for fluctuations of the vector field is nonzero. So in the abelian Higgs model, the gauge field acquires a mass. To calculate the magnitude of the mass, consider a constant value of the vector potential A in the x direction in the gauge where the condensate has constant phase. This is the same as a sinusoidally varying condensate in the gauge where the vector potential is zero. In the gauge where A is zero, the potential energy density in the condensate is the scalar gradient energy:

$$E = \frac{1}{2}\left|\partial\left(\Phi e^{iqAx}\right)\right|^2 = \frac{1}{2}q^2\Phi^2 A^2.$$

This energy is the same as a mass term $1/2m^2A^2$ where $m = q\Phi$.

12.3.3   Nonabelian Higgs mechanism

The Nonabelian Higgs model has the following action:

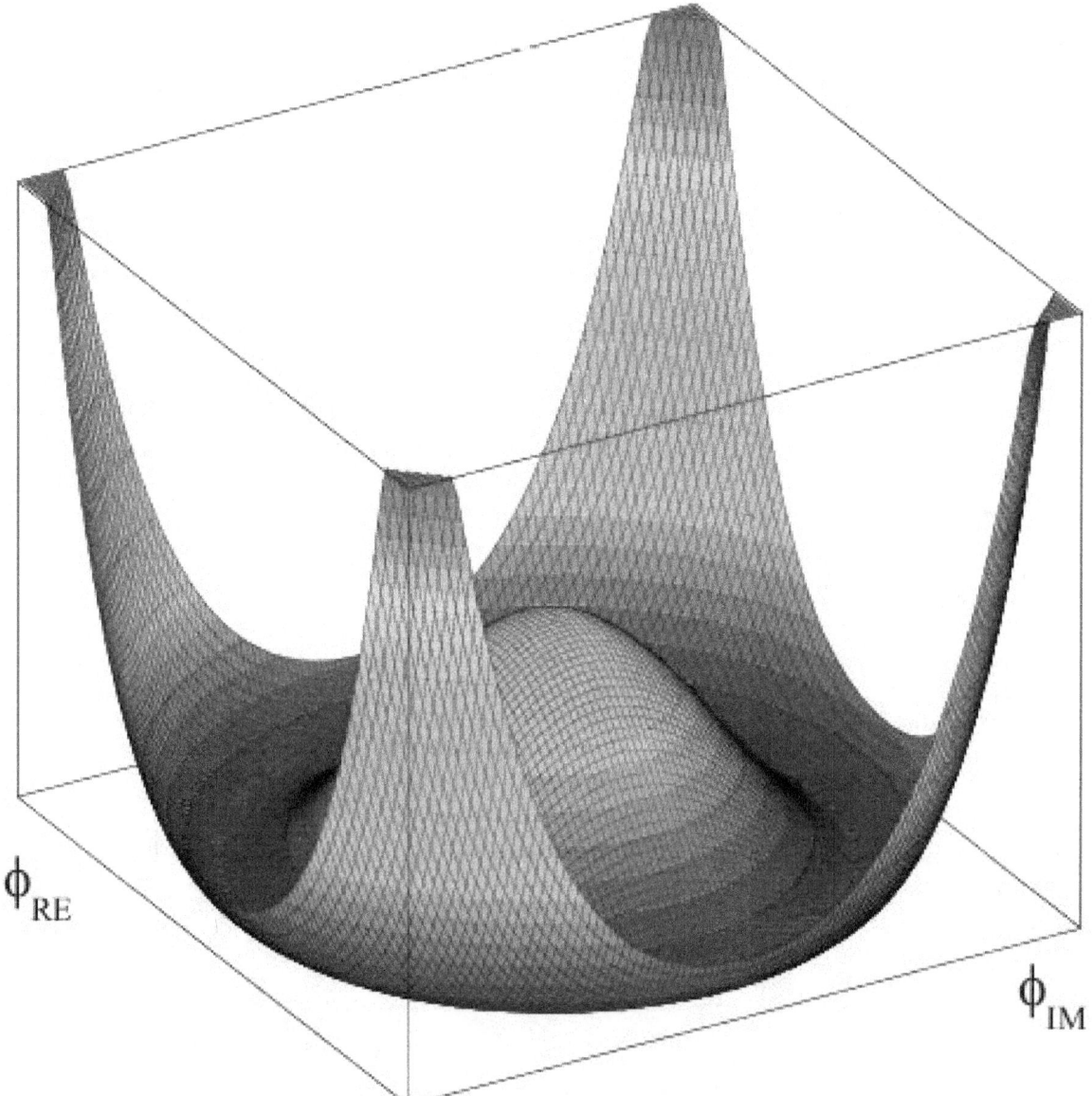

Higgs potential V. *For a fixed value of* λ *the potential is presented upwards against the real and imaginary parts of* Φ. *The* Mexican-hat *or* champagne-bottle *profile at the ground should be noted.*

$$S(\phi, \mathbf{A}) = \int \frac{1}{4g^2} \mathrm{tr}(F^{\mu\nu} F_{\mu\nu}) + |D\phi|^2 + V(|\phi|)$$

where now the nonabelian field $\mathbf{A}$ is contained in D and in the tensor components $F^{\mu\nu}$ and $F_{\mu\nu}$ (the relation between $\mathbf{A}$ and those components is well-known from the Yang–Mills theory).

It is exactly analogous to the Abelian Higgs model. Now the field φ is in a representation of the gauge group, and the gauge covariant derivative is defined by the rate of change of the field minus the rate of change from parallel transport using the gauge field A as a connection.

$$D\phi = \partial\phi - iA^k t_k \phi$$

Again, the expectation value of Φ defines a preferred gauge where the vacuum is constant, and fixing this gauge, fluctuations in the gauge field A come with a nonzero energy cost.

Depending on the representation of the scalar field, not every gauge field acquires a mass. A simple example is in the renormalizable version of an early electroweak model due to Julian Schwinger. In this model, the gauge group is **SO**(3) (or **SU**(2) – there are no spinor representations in the model), and the gauge invariance is broken down to **U**(1) or **SO**(2) at long distances. To make a consistent renormalizable version using the Higgs mechanism, introduce a scalar field φ^a which transforms as a vector (a triplet) of **SO**(3). If this field has a vacuum expectation value, it points in some direction in field space. Without loss of generality, one can choose the z-axis in field space to be the direction that φ is pointing, and then the vacuum expectation value of φ is $(0, 0, A)$, where A is a constant with dimensions of mass ($c = \hbar = 1$).

Rotations around the z-axis form a **U**(1) subgroup of **SO**(3) which preserves the vacuum expectation value of φ, and this is the unbroken gauge group. Rotations around the x and y-axis do not preserve the vacuum, and the components of the **SO**(3) gauge field which generate these rotations become massive vector mesons. There are two massive W mesons in the Schwinger model, with a mass set by the mass scale A, and one massless **U**(1) gauge boson, similar to the photon.

The Schwinger model predicts magnetic monopoles at the electroweak unification scale, and does not predict the Z meson. It doesn't break electroweak symmetry properly as in nature. But historically, a model similar to this (but not using the Higgs mechanism) was the first in which the weak force and the electromagnetic force were unified.

12.3.4   Affine Higgs mechanism

Ernst Stueckelberg discovered[31] a version of the Higgs mechanism by analyzing the theory of quantum electrodynamics with a massive photon. Effectively, Stueckelberg's model is a limit of the regular Mexican hat Abelian Higgs model, where the vacuum expectation value H goes to infinity and the charge of the Higgs field goes to zero in such a way that their product stays fixed. The mass of the Higgs boson is proportional to H, so the Higgs boson becomes infinitely massive and decouples, so is not present in the discussion. The vector meson mass, however, equals to the product eH, and stays finite.

The interpretation is that when a **U**(1) gauge field does not require quantized charges, it is possible to keep only the angular part of the Higgs oscillations, and discard the radial part. The angular part of the Higgs field θ has the following gauge transformation law:

$$\theta \to \theta + e\alpha$$

$$A \to A + \alpha.$$

The gauge covariant derivative for the angle (which is actually gauge invariant) is:

$$D\theta = \partial\theta - eA.$$

In order to keep θ fluctuations finite and nonzero in this limit, θ should be rescaled by H, so that its kinetic term in the action stays normalized. The action for the theta field is read off from the Mexican hat action by substituting $\phi = He^{\frac{1}{H}i\theta}$
.

$$S = \int \frac{1}{4}F^2 + \frac{1}{2}(D\theta)^2 = \int \frac{1}{4}F^2 + \frac{1}{2}(\partial\theta - HeA)^2 = \int \frac{1}{4}F^2 + \frac{1}{2}(\partial\theta - mA)^2$$

since eH is the gauge boson mass. By making a gauge transformation to set $\theta = 0$, the gauge freedom in the action is eliminated, and the action becomes that of a massive vector field:

$$S = \int \frac{1}{4}F^2 + \frac{1}{2}m^2A^2.$$

To have arbitrarily small charges requires that the **U**(1) is not the circle of unit complex numbers under multiplication, but the real numbers **R** under addition, which is only different in the global topology. Such a **U**(1) group is *non-compact*. The field θ transforms as an affine representation of the gauge group. Among the allowed gauge groups, only non-compact **U**(1) admits affine representations, and the **U**(1) of electromagnetism is experimentally known to be compact, since charge quantization holds to extremely high accuracy.

The Higgs condensate in this model has infinitesimal charge, so interactions with the Higgs boson do not violate charge conservation. The theory of quantum electrodynamics with a massive photon is still a renormalizable theory, one in which electric charge is still conserved, but magnetic monopoles are not allowed. For nonabelian gauge theory, there is no affine limit, and the Higgs oscillations cannot be too much more massive than the vectors.

12.4 See also

- Electromagnetic mass

- Higgs bundle

- Mass generation

- QCD vacuum

- Quantum triviality

- Top quark condensate

- Yang–Mills–Higgs equations

12.5 References

[1] G. Bernardi, M. Carena, and T. Junk: "Higgs bosons: theory and searches", Reviews of Particle Data Group: Hypothetical particles and Concepts, 2007, http://pdg.lbl.gov/2008/reviews/higgs_s055.pdf

[2] P. W. Anderson (1962). "Plasmons, Gauge Invariance, and Mass". *Physical Review* **130** (1): 439–442. Bibcode:1963PhRv doi:10.1103/PhysRev.130.439.

[3] F. Englert and R. Brout (1964). "Broken Symmetry and the Mass of Gauge Vector Mesons". *Physical Review Letters* **13** (9): 321–323. Bibcode:1964PhRvL..13..321E. doi:10.1103/PhysRevLett.13.321.

[4] Peter W. Higgs (1964). "Broken Symmetries and the Masses of Gauge Bosons". *Physical Review Letters* **13** (16): 508–509. Bibcode:1964PhRvL..13..508H. doi:10.1103/PhysRevLett.13.508.

[5] G. S. Guralnik, C. R. Hagen, and T. W. B. Kibble (1964). "Global Conservation Laws and Massless Particles". *Physical Review Letters* **13** (20): 585–587. Bibcode:1964PhRvL..13..585G. doi:10.1103/PhysRevLett.13.585.

[6] Gerald S. Guralnik (2009). "The History of the Guralnik, Hagen and Kibble development of the Theory of Spontaneous Symmetry Breaking and Gauge Particles". *International Journal of Modern Physics* **A24** (14): 2601–2627. arXiv:0907.3466. Bibcode:2009IJMPA..24.2601G. doi:10.1142/S0217751X09045431.

[7] History of Englert–Brout–Higgs–Guralnik–Hagen–Kibble Mechanism. Scholarpedia.

[8] "Englert–Brout–Higgs–Guralnik–Hagen–Kibble Mechanism". Scholarpedia. Retrieved 2012-06-16.

[9] Liu, G. Z.; Cheng, G. (2002). "Extension of the Anderson-Higgs mechanism". *Physical Review B* **65** (13): 132513. arXiv:cond-mat/0106070. Bibcode:2002PhRvB..65m2513L. doi:10.1103/PhysRevB.65.132513.

[10] Matsumoto, H.; Papastamatiou, N. J.; Umezawa, H.; Vitiello, G. (1975). "Dynamical rearrangement in the Anderson-Higgs-Kibble mechanism". *Nuclear Physics B* **97**: 61. doi:10.1016/0550-3213(75)90215-1.

[11] Close, Frank (2011). *The Infinity Puzzle: Quantum Field Theory and the Hunt for an Orderly Universe*. Oxford: Oxford University Press. ISBN 978-0-19-959350-7.

[12] "Press release from Royal Swedish Academy of Sciences" (PDF). 8 October 2013. Retrieved 8 October 2013.

[13] "Guralnik, G S; Hagen, C R and Kibble, T W B (1967). Broken Symmetries and the Goldstone Theorem. Advances in Physics, vol. 2" (PDF).

[14] A.M. Polyakov, A View From The Island, 1992

[15] Farhi, E., & Jackiw, R. W. (1982). *Dynamical Gauge Symmetry Breaking: A Collection Of Reprints*. Singapore: World Scientific Pub. Co.

[16] Frank Close. "The Infinity Puzzle." 2011, p.158

[17] Norman Dombey, "Higgs Boson: Credit Where It's Due". The Guardian, July 6, 2012

[18] Cern Courier, Mar 1, 2006

[19] Sean Carrol, "The Particle At The End Of The Universe: The Hunt For The Higgs And The Discovery Of A New World", 2012, p.228

[20] A. A. Migdal and A. M. Polyakov, "Spontaneous Breakdown of Strong Interaction Symmetry and Absence of Massless Particles", *JETP* **51**, 135, July 1966 (English translation: *Soviet Physics JETP*, **24**, 1, January 1967)

[21] Nambu, Y (1960). "Quasiparticles and Gauge Invariance in the Theory of Superconductivity". *Physical Review* **117** (3): 648–663. Bibcode:1960PhRv..117..648N. doi:10.1103/PhysRev.117.648.

[22] Higgs, Peter (2007). "Prehistory of the Higgs boson". *Comptes Rendus Physique* **8** (9): 970–972. Bibcode:2007CRPhy...8..970H. doi:10.1016/j.crhy.2006.12.006.

[23] "Physical Review Letters – 50th Anniversary Milestone Papers". Prl.aps.org. Retrieved 2012-06-16.

[24] "American Physical Society – J. J. Sakurai Prize Winners". Aps.org. Retrieved 2012-06-16.

[25] Department of Physics and Astronomy. "Rochester's Hagen Sakurai Prize Announcement". Pas.rochester.edu. Retrieved 2012-06-16.

[26] FermiFred (2010-02-15). "C.R. Hagen discusses naming of Higgs Boson in 2010 Sakurai Prize Talk". Youtube.com. Retrieved 2012-06-16.

[27] Sample, Ian (2009-05-29). "Anything but the God particle by Ian Sample". Guardian. Retrieved 2012-06-16.

[28] G. 't Hooft and M. Veltman (1972). "Regularization and Renormalization of Gauge Fields". *Nuclear Physics B* **44** (1): 189–219. Bibcode:1972NuPhB..44..189T. doi:10.1016/0550-3213(72)90279-9.

[29] "Regularization and Renormalization of Gauge Fields by t'Hooft and Veltman (PDF)" (PDF). Retrieved 2012-06-16.

[30] Goldstone, J. (1961). "Field theories with " Superconductor " solutions". *Il Nuovo Cimento* **19**: 154–164. doi:10.1007/BF 0281

[31] Stueckelberg, E. C. G. (1938), "Die Wechselwirkungskräfte in der Elektrodynamik und in der Feldtheorie der Kräfte", *Helv. Phys. Acta*. **11:** 225

12.6   Further reading

- Schumm, Bruce A. (2004) *Deep Down Things*. Johns Hopkins Univ. Press. Chpt. 9.

- Englert-Brout-Higgs-Guralnik-Hagen-Kibble mechanism Tom W B Kibble Scholarpedia, 4(1):6441. doi:10.4249/scholarpedia.6441

12.7 External links

- Guralnik, G.S.; Hagen, C.R.; Kibble, T.W.B. (1964). "Global Conservation Laws and Massless Particles". *Physical Review Letters* **13** (20): 585–87. Bibcode:1964PhRvL..13..585G. doi:10.1103/PhysRevLett.13.585.

- Mark D. Roberts (1999) "A Generalized Higgs Model"

- 2010 Sakurai Prize - All Events - YouTube

- From BCS to the LHC - CERN Courier Jan 21, 2008, Steven Weinberg, University of Texas at Austin.

- Higgs, dark matter and supersymmetry: What the Large Hadron Collider will tell us (Steven Weinberg) - YouTube on YouTube 06-11-2009

- Gerry Guralnik speaks at Brown University about the 1964 PRL papers

- Guralnik, Gerald (2013). "Heretical Ideas that Provided the Cornerstone for the Standard Model of Particle Physics". SPG MITTEILUNGEN March 2013, No. 39, (p. 14)

- Steven Weinberg Praises Teams for Higgs Boson Theory

- Physical Review Letters – 50th Anniversary Milestone Papers

- Imperial College London on PRL 50th Anniversary Milestone Papers

- Englert–Brout–Higgs–Guralnik–Hagen–Kibble Mechanism on Scholarpedia

- History of Englert–Brout–Higgs–Guralnik–Hagen–Kibble Mechanism on Scholarpedia

- The Hunt for the Higgs at Tevatron

- The Mystery of Empty Space on YouTube. A lecture with UCSD physicist Kim Griest (43 minutes)

Chapter 13

QCD vacuum

The **QCD vacuum** is the vacuum state of quantum chromodynamics (QCD). It is an example of a *non-perturbative* vacuum state, characterized by infinitely many non-vanishing condensates such as the gluon condensate or the quark condensate. These condensates characterize the **normal phase** or the **confined phase** of quark matter.

Another field-theoretic vacuum is the QED vacuum of quantum electrodynamics.

13.1 Symmetries and symmetry breaking

13.1.1 Symmetries of the QCD Lagrangian

Like any relativistic quantum field theory, QCD enjoys Poincaré symmetry including the discrete symmetries CPT (each of which is realized). Apart from these space-time symmetries, it also has internal symmetries. Since QCD is an SU(3) gauge theory, it has local SU(3) gauge symmetry.

Since it has many flavours of quarks, it has approximate flavour and chiral symmetry. This approximation is said to involve the **chiral limit** of QCD. Of these chiral symmetries, the baryon number symmetry is exact. Some of the broken symmetries include the axial U(1) symmetry of the flavour group. This is broken by the chiral anomaly. The presence of instantons implied by this anomaly also breaks CP symmetry.

In summary, the QCD Lagrangian has the following symmetries:

- Poincaré symmetry and CPT invariance

- SU(3) local gauge symmetry

- approximate global $SU(N_f) \times SU(N_f)$ flavour chiral symmetry and the U(1) baryon number symmetry

The following classical symmetries are broken in the QCD Lagrangian:

- scale, i.e., conformal symmetry (through the scale anomaly), giving rise to asymptotic freedom

- the axial part of the U(1) flavour chiral symmetry (through the chiral anomaly), giving rise to the strong CP problem.

13.1.2 Spontaneous symmetry breaking

Main article: Spontaneous symmetry breaking

When the Hamiltonian of a system (or the Lagrangian) has a certain symmetry, but the ground state (i.e., the vacuum) does not, then one says that **spontaneous symmetry breaking** (SSB) has taken place.

A familiar example of SSB is in ferromagnetic materials. Microscopically, the material consists of atoms with a non-vanishing spin, each of which acts like a tiny bar magnet, i.e., a magnetic dipole. The Hamiltonian of the material, describing the interaction of neighbouring dipoles, is invariant under rotations. At high temperature, there is no magnetization of a large sample of the material. Then one says that the symmetry of the Hamiltonian is realized by the system. However, at low temperature, there could be an overall magnetization. This magnetization has a *preferred direction*, since one can tell the north magnetic pole of the sample from the south magnetic pole. In this case, there is spontaneous symmetry breaking of the rotational symmetry of the Hamiltonian.

When a continuous symmetry is spontaneously broken, massless bosons appear, corresponding to the remaining symmetry. This is called the **Goldstone phenomenon** and the bosons are called Goldstone bosons.

13.1.3 Symmetries of the QCD vacuum

The $SU(N_f) \times SU(N_f)$ chiral flavour symmetry of the QCD Lagrangian is broken in the vacuum state of the theory. The symmetry of the vacuum state is the diagonal $SU(N_f)$ part of the chiral group. The diagnostic for this is the formation of a non-vanishing chiral condensate $\langle \bar{\psi}_i \psi_i \rangle$, where ψ_i is the quark field operator, and the flavour index **i** is summed. The Goldstone bosons of the symmetry breaking are the pseudoscalar mesons.

When $N_f=2$, i.e., only the **u** and **d** quarks are treated as massless, the three pions are the Goldstone bosons. When the **s** quark is also treated as massless, i.e., $N_f=3$, all eight pseudoscalar mesons of the quark model become Goldstone bosons. The actual masses of these mesons are obtained in chiral perturbation theory through an expansion in the (small) actual masses of the quarks.

In other phases of quark matter the full chiral flavour symmetry may be recovered, or broken in completely different ways.

13.1.4 Evidence: experimental consequences

The evidence for QCD condensates comes from two eras, the pre-QCD era 1950–1973 and the post-QCD era, after 1974. The pre-QCD results established that the strong interactions vacuum contains a quark chiral condensate, while the post-QCD results established that the vacuum also contains a gluon condensate.

Pre-QCD: gradient coupling

In the 1950s, there were many attempts to produce a field theory to describe the interactions of pions and nucleons. The obvious renormalizable interaction between the two objects is the Yukawa coupling to a pseudoscalar:

$$L_I = \bar{N} \gamma_5 \pi N$$

And this is clearly theoretically correct, since it is leading order and it takes all the symmetries into account. But it doesn't match experiment. The interaction that does couples the nucleons to the *gradient* of the pion field.

$$g \bar{N} \gamma^\mu \partial_\mu \pi N$$

This is the **gradient-coupling model**. This interaction has a very different dependence on the energy of the pion—it vanishes at zero momentum.

This type of coupling means that a coherent state of low momentum pions barely interacts at all. This is a manifestation of an approximate symmetry, a **shift symmetry** of the pion field. The replacement

$$\pi \to \pi + C$$

leaves the gradient coupling alone, but not the pseudoscalar coupling.

The modern explanation for the shift symmetry was first proposed by Yoichiro Nambu. The pion field is a Goldstone boson, and the shift symmetry is the lowest order approximation to moving along the flat directions.

Pre-QCD: Goldberger–Treiman relation

There is a mysterious relationship between the strong interaction coupling of the pions to the nucleons, the coefficient g in the gradient coupling model, and the axial vector current coefficient of the nucleon which determines the weak decay rate of the neutron. The relation is

$$g_{\pi NN} F_\pi = G_A M_N$$

and it is obeyed to 10% accuracy.

The constant G_A is the coefficient that determines the neutron decay rate. It gives the normalization of the weak interaction matrix elements for the nucleon. On the other hand, the pion-nucleon coupling is a phenomenological constant describing the scattering of bound states of quarks and gluons.

The weak interactions are current-current interactions ultimately because they come from a nonabelian gauge theory. The Goldberger Treiman relation suggests that the pions for some reason interact as if they are related to the same symmetry current.

PCAC   The phenomenon which gives rise to the Goldberger Treiman relation was called the "Partially Conserved Axial Current" hypothesis, or PCAC. Partially conserved is an archaic term for spontaneously broken, and the axial current is now called the chiral symmetry current.

The idea is that the symmetry current which performs axial rotations on the fundamental fields does not preserve the vacuum. This means that the current J applied to the vacuum produces particles. The particles must be scalars, otherwise the vacuum wouldn't be Lorentz invariant. By index matching, the matrix element is:

$$J_\mu |0\rangle = k_\mu |\pi\rangle \,,$$

where k_μ is the momentum carried by the created pion. Since the divergence of the axial current operator is zero, we must have

$$\partial_\mu J^\mu |0\rangle = k^\mu k_\mu |\pi\rangle = m_\pi^2 |\pi\rangle = 0 \,.$$

Hence the pions are massless, $m_\pi^2 = 0$, in accordance with Goldstone's theorem.

Now if the scattering matrix element is considered, we have

$$k_\mu \langle N(p)|\pi(k)N(p')\rangle = \langle N(p)|J_\mu|N(p')\rangle \,.$$

Up to a momentum factor, which is the gradient in the coupling, it takes the same form as the axial current turning a neutron into a proton in the current-current form of the weak interaction.

$$\langle N|J^\mu|N\rangle\langle e|J_\mu|\nu\rangle$$

Pre-QCD: soft pion emission

Extensions of the PCAC ideas allowed Steven Weinberg to calculate the amplitudes for collisions which emit low energy pions from the amplitude for the same process with no pions. The amplitudes are those given by acting with symmetry currents on the external particles of the collision.

These successes established the basic properties of the strong interaction vacuum well before QCD.

Pseudo-Goldstone bosons

Experimentally it is seen that the masses of the octet of pseudoscalar mesons is very much lighter than the next lightest states; i.e., the octet of vector mesons (such as the rho). The most convincing evidence for SSB of the chiral flavour symmetry of QCD is the appearance of these pseudo-Goldstone bosons. These would have been strictly massless in the chiral limit. There is convincing demonstration that the observed masses are compatible with chiral perturbation theory. The internal consistency of this argument is further checked by lattice QCD computations which allow one to vary the quark mass and check that the variation of the pseudoscalar masses with the quark mass is as required by chiral perturbation theory.

The η'

This pattern of SSB solves one of the earlier "mysteries" of the quark model, where all the pseudoscalar mesons should have been of nearly the same mass. Since $N_f = 3$, there should have been nine of these. However, one (the SU(3) singlet η') has quite a larger mass than the SU(3) octet. In the quark model, this has no natural explanation— a mystery named the **$\eta-\eta'$ mass splitting** (the η is one member of the octet, which should have been degenerate in mass with the η').

In QCD, one realizes that the η' is associated with the axial U(1) which is explicitly broken through the chiral anomaly, and thus its mass is not "protected" to be small, like that of the η. The η-η' mass splitting can be explained[1] [2] [3] through the 't Hooft instanton mechanism,[4] whose 1/N realization is also known as Witten-Veneziano mechanism.[5] [6]

Current algebra and QCD sum rules

PCAC and current algebra also provide evidence for this pattern of SSB. Direct estimates of the chiral condensate also come from such analysis.

Another method of analysis of correlation functions in QCD is through an operator product expansion (OPE). This writes the vacuum expectation value of a non-local operator as a sum over VEVs of local operators, i.e., condensates. The value of the correlation function then dictates the values of the condensates. Analysis of many separate correlation functions gives consistent results for several condensates, including the gluon condensate, the quark condensate, and many mixed and higher order condensates. In particular one obtains

$$\langle (gG)^2 \rangle \stackrel{\text{def}}{=} \langle g^2 G_{\mu\nu} G^{\mu\nu} \rangle \simeq 0.5 \text{ GeV}^4$$

$$\langle \overline{\psi}\psi \rangle \simeq (-0.23)^3 \text{ GeV}^3$$

$$\langle (gG)^4 \rangle \simeq 5 : 10 \langle (gG)^2 \rangle^2$$

Here **G** refers to the gluon field tensor, ψ to the quark field, and g to the QCD coupling.

These analyses are being refined further through improved sum rule estimates and direct estimates in lattice QCD. They provide the *raw data* which must be explained by models of the QCD vacuum.

13.2   Models of the QCD vacuum

A full solution of QCD would automatically give a full description of the vacuum, confinement and the hadron spectrum. Lattice QCD is making rapid progress towards providing the solution as a systematically improvable numerical computation. However, approximate models of the QCD vacuum remain useful in more restricted domains. The purpose of these models is to make quantitative sense of some set of condensates and hadron properties such as masses and form factors.

This section is devoted to models. Opposed to these are systematically improvable computational procedures such as large N QCD and lattice QCD, which are described in their own articles.

13.2.1   The Savvidy vacuum, instabilities and structure

The Savvidy vacuum is a model of the QCD vacuum which at a basic level is a statement that it cannot be the conventional Fock vacuum empty of particles and fields. In 1977, George Savvidy showed[7] that the QCD vacuum with zero field strength is unstable, and decays into a state with a calculable non vanishing value of the field. Since condensates are scalar, it seems like a good first approximation that the vacuum contains some non-zero but homogeneous field which gives rise to these condensates. This would then be a more complicated version of the Higgs mechanism. However, Stanley Mandelstam showed that a homogeneous vacuum field is also unstable. The instability of a homogeneous gluon field was argued by Niels Kjær Nielsen and Poul Olesen in their 1978 paper.[8] These arguments suggest that the scalar condensates are an effective long-distance description of the vacuum, and at short distances, below the QCD scale, the vacuum may have structure.

13.2.2   The dual superconducting model

In a type II superconductor, electric charges condense into Cooper pairs. As a result, magnetic flux is squeezed into tubes. In the dual superconductor picture of the QCD vacuum, chromomagnetic monopoles condense into dual Cooper pairs, causing chromoelectric flux to be squeezed into tubes. As a result, confinement and the *string picture* of hadrons follows. This **dual superconductor** picture is due to Gerard 't Hooft and Stanley Mandelstam. 't Hooft showed further that an Abelian projection of a non-Abelian gauge theory contains magnetic monopoles.

While the vortices in a type II superconductor are neatly arranged into a hexagonal or occasionally square lattice, as is reviewed in Olesen's 1980 seminar[9] one may expect a much more complicated and possibly dynamical structure in QCD. For example, nonabelian Abrikosov-Nielsen-Olesen vortices may vibrate wildly or be knotted.

13.2.3   String models

String models of confinement and hadrons have a long history. They were first invented to explain certain aspects of crossing symmetry in the scattering of two mesons. They were also found to be useful in the description of certain properties of the Regge trajectory of the hadrons. These early developments took on a life of their own called the dual resonance model (later renamed string theory). However, even after the development of QCD string models continued to play a role in the physics of strong interactions. These models are called *non-fundamental strings* or QCD strings, since they should be derived from QCD, as they are, in certain approximations such as the strong coupling limit of lattice QCD.

The model states that the colour electric flux between a quark and an antiquark collapses into a string, rather than spreading out into a Coulomb field as the normal electric flux does. This string also obeys a different force law. It behaves as if the string had constant tension, so that separating out the ends (quarks) would give a potential energy increasing linearly with the separation. When the energy is higher than that of a meson, the string breaks and the two new ends become a quark-antiquark pair, thus describing the creation of a meson. Thus confinement is incorporated naturally into the model.

In the form of the Lund model Monte Carlo program, this picture has had remarkable success in explaining experimental data collected in electron-electron and hadron-hadron collisions.

13.2.4 Bag models

Strictly, these models are not models of the QCD vacuum, but of physical single particle quantum states — the hadrons. The model proposed originally in 1974 by A. Chodos *et al.* [10] consists of inserting a quark model in a *perturbative vacuum* inside a volume of space called a **bag**. Outside this bag is the real QCD vacuum, whose effect is taken into account through the difference between energy density of the true QCD vacuum and the perturbative vacuum (bag constant B) and boundary conditions imposed on the quark wave functions and the gluon field. The hadron spectrum is obtained by solving the Dirac equation for quarks and the Yang–Mills equations for gluons. The wave functions of the quarks satisfy the boundary conditions of a fermion in an infinitely deep potential well of scalar type with respect to the Lorentz group. The boundary conditions for the gluon field are those of the dual color superconductor. The role of such a superconductor is attributed to the physical vacuum of QCD. Bag models strictly prohibit the existence of open color (free quarks, free gluons, etc.) and lead in particular to string models of hadrons.

The **chiral bag model** couples the axial vector current $\overline{\psi}\gamma_5\gamma_\mu\psi$ of the quarks at the bag boundary to a pionic field outside of the bag. In the most common formulation, the chiral bag model basically replaces the interior of the skyrmion with the bag of quarks. Very curiously, most physical properties of the nucleon become mostly insensitive to the bag radius. Prototypically, the baryon number of the chiral bag remains an integer, independent of bag radius: the exterior baryon number is identified with the topological winding number density of the Skyrme soliton, while the interior baryon number consists of the valence quarks (totaling to one) plus the spectral asymmetry of the quark eigenstates in the bag. The spectral asymmetry is just the vacuum expectation value $\langle\overline{\psi}\gamma_0\psi\rangle$ summed over all of the quark eigenstates in the bag. Other values, such as the total mass and the axial coupling constant g_A, are not precisely invariant like the baryon number, but are mostly insensitive to the bag radius, as long as the bag radius is kept below the nucleon diameter. Because the quarks are treated as free quarks inside the bag, the radius-independence in a sense validates the idea of asymptotic freedom.

13.2.5 Instanton ensemble

Main article: instanton fluid

Another view states that BPST-like instantons play an important role in the vacuum structure of QCD. These instantons were discovered in 1975 by Belavin, Polyakov, Schwartz and Tyupkin[11] as topologically stable solutions to the Yang-Mills field equations. They represent tunneling transitions from one vacuum state to another. These instantons are indeed found in lattice calculations. The first computations performed with instantons used the dilute gas approximation. The results obtained did not solve the infrared problem of QCD, making many physicists turn away from instanton physics. Later, though, an instanton liquid model was proposed, turning out to be more promising an approach.[12]

The **dilute instanton gas model** departs from the supposition that the QCD vacuum consists of a gas of BPST-like instantons. Although only the solutions with one or few instantons (or anti-instantons) are known exactly, a dilute gas of instantons and anti-instantons can be approximated by considering a superposition of one-instanton solutions at great distances from one another. 't Hooft calculated the effective action for such an ensemble,[13] and he found an infrared divergence for big instantons, meaning that an infinite amount of infinitely big instantons would populate the vacuum.

Later, an **instanton liquid model** was studied. This model starts from the assumption that an ensemble of instantons cannot be described by a mere sum of separate instantons. Various models have been proposed, introducing interactions between instantons or using variational methods (like the "valley approximation") endeavoring to approximate the exact multi-instanton solution as closely as possible. Many phenomenological successes have been reached.[12] Whether an instanton liquid can explain confinement in 3+1 dimensional QCD is not known, but many physicists think that it is unlikely.

13.2.6 Center vortex picture

A more recent picture of the QCD vacuum is one in which center vortices play an important role. These vortices are topological defects carrying a center element as charge. These vortices are usually studied using lattice simulations, and it has been found that the behavior of the vortices is closely linked with the confinement-deconfinement phase transition: in the confining phase vortices percolate and fill the space-time volume, in the deconfining phase they are much suppressed.[14] Also it has been shown that the string tension vanished upon removal of center vortices from the simulations,[15] hinting at an important role for center vortices.

13.3   See also

- Vacuum state and vacuum

- Spontaneous symmetry breaking

- Quantum chromodynamics and flavour

- Top quark condensate

- Goldstone boson

- Symmetry breaking

- Higgs mechanism

13.4   References and external links

[1] Del Debbio, Luigi; Giusti, Leonardo; Pica, Claudio. "Topological Susceptibility in SU(3) Gauge Theory" (PDF). *Phys. Rev. Lett.* **94** (032003). arXiv:hep-th/0407052. Bibcode:2005PhRvL..94c2003D. doi:10.1103/PhysRevLett.94.032003. Retrieved 4 March 2015.

[2] Lüscher, Martin; Palombi, Filippo (September 2010). "Universality of the topological susceptibility in the SU(3) gauge theory" (PDF). *Journal of High Energy Physics (JHEP)*. arXiv:1008.0732. Bibcode:2010JHEP...09..110L. doi:10.1007/JHEP09(2010)110. Retrieved 4 March 2015.

[3] Cè M, Consonni C, Engel G, Giusti L (30 October 2014). "Testing the Witten-Veneziano mechanism with the Yang-Mills gradient flow on the lattice" (PDF). v1. arXiv:1410.8358. Bibcode:2014arXiv1410.8358C. Retrieved 4 March 2015.

[4] 't Hooft, Gerard (5 July 1976). "Symmetry Breaking through Bell-Jackiw Anomalies". *Phys. Rev. Lett.* **37** (1): 8–11. Bibcode:1976PhRvL..37....8T. doi:10.1103/PhysRevLett.37.8.

[5] Witten, Edward (17 April 1979). "Current algebra theorems for the U(1) "Goldstone boson"". *Nuclear Physics B* **156** (2): 269–283. Bibcode:1979NuPhB.156..269W. doi:10.1016/0550-3213(79)90031-2.

[6] Veneziano, Gabriele (14 May 1979). "U(1) without instantons". *Nuclear Physics B* **159** (1-2): 213–224. Bibcode:1979NuPhB. doi:10.1016/0550-3213(79)90332-8.

[7] Savvidy, G. K. (1977). "Infrared instability of the vacuum state of gauge theories and asymptotic freedom". *Phys.Lett.* **B** (1): 133. Bibcode:1977PhLB...71..133S. doi:10.1016/0370-2693(77)90759-6.

[8] Nielsen, Niels Kjær; Olesen, Poul (1978). "An unstable Yang–Mills field mode". *Nucl.Phys.* **B** (144): 376. Bibcode:1978 doi:10.1016/0550-3213(78)90377-2.

[9] Olesen, P. (1981). "On the QCD vacuum". *Phys.Scripta* **23** (23): 1000. Bibcode:1981PhyS...23.1000O. doi:10.1088/0031-8949/23/5B/018.

[10] Chodos, A., Jaffe, R. L., Johnson, K., Thorn, C. B., Weisskopf, V. F. (1974). New extended model of hadrons, *Phys. Rev. D9, 3471*. doi:10.1103/PhysRevD.9.3471

[11] Belavin, A.A.; A.M. Polyakov; A.S. Schwartz & Yu.S. Tyupkin (1975). "Pseudoparticle solutions of the Yang-Mills equations". *Phys. Lett.* **59B** (1): 85–87. Bibcode:1975PhLB...59...85B. doi:10.1016/0370-2693(75)90163-X.

[12] Hutter, Marcus (1995). "Instantons in QCD: Theory and application of the instanton liquid model". arXiv:hep-ph/0107098.

[13] 't Hooft, Gerard (1976). "Computation of the quantum effects due to a four-dimensional pseudoparticle". *Phys. Rev.* **D14** (12): 3432–3450. Bibcode:1976PhRvD..14.3432T. doi:10.1103/PhysRevD.14.3432.

[14] Engelhardt, M.; Langfeld, K.; Reinhardt, H.; Tennert, O. (2000). "Deconfinement in SU(2) Yang–Mills theory as a center vortex percolation transition". *Physical Review D* **61** (5): 054504. arXiv:hep-lat/9904004. Bibcode:2000PhRvD..61e4504E. doi:10.1103/PhysRevD.61.054504.

[15] Del Debbio, L.; Faber, M.; Greensite, J.; Olejník, Š. (1997). "Center dominance and Z_2 vortices in SU(2) lattice gauge theory". *Physical Review D* **55** (4): 2298. arXiv:hep-lat/9610005. Bibcode:1997PhRvD..55.2298D. doi:10.1103/PhysRevD.55.2298.

13.5 Bibliography

- *The quantum quark*, by Andrew Watson ISBN 0-521-82907-0

- *Handbook of QCD*, by M.A. Shifman ISBN 981-238-028-0

- *The QCD vacuum, hadrons and superdense matter*, by E.V. Shuryak ISBN 981-238-574-6

Chapter 14

Goldstone boson

In particle and condensed matter physics, **Goldstone bosons** or **Nambu–Goldstone bosons** (**NGBs**) are bosons that appear necessarily in models exhibiting spontaneous breakdown of continuous symmetries. They were discovered by Yoichiro Nambu in the context of the BCS superconductivity mechanism,[1] and subsequently elucidated by Jeffrey Goldstone,[2] and systematically generalized in the context of quantum field theory.[3]

These spinless bosons correspond to the spontaneously broken internal symmetry generators, and are characterized by the quantum numbers of these. They transform **nonlinearly** (shift) under the action of these generators, and can thus be excited out of the asymmetric vacuum by these generators. Thus, they can be thought of as the excitations of the field in the broken symmetry directions in group space—and are massless if the spontaneously broken symmetry is *not also broken explicitly*.

If, instead, the symmetry is not exact, i.e. if it is *explicitly broken as well as spontaneously broken*, then the Nambu–Goldstone bosons are not massless, though they typically remain relatively light; they are then called **pseudo-Goldstone bosons** or **pseudo-Nambu–Goldstone bosons** (abbreviated *PNGBs*).

14.1 Goldstone's theorem

Goldstone's theorem examines a generic continuous symmetry which is spontaneously broken; i.e., its currents are conserved, but the ground state is not invariant under the action of the corresponding charges. Then, necessarily, new massless (or light, if the symmetry is not exact) scalar particles appear in the spectrum of possible excitations. There is one scalar particle—called a Nambu–Goldstone boson—for each generator of the symmetry that is broken, i.e., that does not preserve the ground state. The Nambu–Goldstone mode is a long-wavelength fluctuation of the corresponding order parameter.

By virtue of their special properties in coupling to the vacuum of the respective symmetry-broken theory, vanishing momentum ("soft") Goldstone bosons involved in field-theoretic amplitudes make such amplitudes vanish ("Adler zeros").

In theories with gauge symmetry, the Goldstone bosons are "eaten" by the gauge bosons. The latter become massive and their new, longitudinal polarization is provided by the Goldstone boson.

14.2 Examples

14.2.1 Natural

- In fluids, the phonon is longitudinal and it is the Goldstone boson of the spontaneously broken Galilean symmetry. In solids, the situation is more complicated; the Goldstone bosons are the longitudinal and transverse phonons and they happen to be the Goldstone bosons of spontaneously broken Galilean, translational, and rotational symmetry with no simple one-to-one correspondence between the Goldstone modes and the broken symmetries.

- In magnets, the original rotational symmetry (present in the absence of an external magnetic field) is spontaneously broken such that the magnetization points into a specific direction. The Goldstone bosons then are the *magnons*, i.e., spin waves in which the local magnetization direction oscillates.

- The *pions* are the pseudo-Goldstone bosons that result from the spontaneous breakdown of the chiral-flavor symmetries of QCD effected by quark condensation due to the strong interaction. These symmetries are further explicitly broken by the masses of the quarks, so that the pions are not massless, but their mass is *significantly smaller* than typical hadron masses.

- The longitudinal polarization components of the W and Z bosons correspond to the Goldstone bosons of the spontaneously broken part of the electroweak symmetry *SU(2)⊗U(1)*, which, however, are not observable. Because this symmetry is gauged, the three would-be Goldstone bosons are "eaten" by the three gauge bosons corresponding to the three broken generators; this gives these three gauge bosons a mass, and the associated necessary third polarization degree of freedom. This is described in the Standard Model through the Higgs mechanism. An analogous phenomenon occurs in superconductivity, which served as the original source of inspiration for Nambu, namely, the photon develops a dynamical mass (expressed as magnetic flux exclusion from a superconductor), cf. the Ginzburg–Landau theory.

14.2.2 Theory

Consider a complex scalar field φ, with the constraint that $\varphi^*\varphi = v^2$, a constant. One way to impose a constraint of this sort is by including a potential interaction term in its Lagrangian density,

$$\lambda(\phi^*\phi - v^2)^2 \,,$$

and taking the limit as $\lambda \to \infty$ (this is called the "Abelian nonlinear σ-model". It corresponds to the Goldstone sombrero potential where the tip and the sides shoot to infinity, preserving the location of the minimum at its base).

The constraint, and the action, below, are invariant under a $U(1)$ phase transformation, $\delta\varphi = i\varepsilon\varphi$. The field can be redefined to give a real scalar field (i.e., a spin-zero particle) θ without any constraint by

$$\phi = ve^{i\theta}$$

where θ is the Nambu–Goldstone boson (actually $v\theta$ is), and the $U(1)$ symmetry transformation effects a shift on θ, namely

$$\delta\theta = \epsilon \,,$$

but does not preserve the ground state $|0⟩$, (i.e. the above infinitesimal transformation *does not annihilate it*—the hallmark of invariance), as evident in the charge of the current below.

Thus, the vacuum is degenerate and noninvariant under the action of the spontaneously broken symmetry.

The corresponding Lagrangian density is given by

$$\mathcal{L} = -\frac{1}{2}(\partial^\mu\phi^*)\partial_\mu\phi + m^2\phi^*\phi = -\frac{1}{2}(-ive^{-i\theta}\partial^\mu\theta)(ive^{i\theta}\partial_\mu\theta) + m^2v^2,$$

and thus

$$= -\frac{v^2}{2}(\partial^\mu\theta)(\partial_\mu\theta) + m^2v^2 \,.$$

Note that the constant term m^2v^2 in the Lagrangian density has no physical significance, and the other term in it is simply the kinetic term for a massless scalar.

The symmetry-induced conserved $U(1)$ current is

$$J_\mu = -v^2 \partial_\mu \theta \ .$$

The charge, Q, resulting from this current shifts θ and the ground state to a new, degenerate, ground state. Thus, a vacuum with $\langle\theta\rangle = 0$ will shift to a *different vacuum* with $\langle\theta\rangle = -\varepsilon$. The current connects the original vacuum with the Nambu–Goldstone boson state, $\langle 0|J_0(0)|\theta\rangle \neq 0$.

In general, in a theory with several scalar fields, φ_j, the Nambu–Goldstone mode φ_g is massless, and parameterises the curve of possible (degenerate) vacuum states. Its hallmark under the broken symmetry transformation is ***nonvanishing vacuum expectation*** $\langle\delta\varphi g\rangle$, an order parameter, for vanishing $\langle\varphi g\rangle = 0$, at some ground state $|0\rangle$ chosen at the minimum of the potential, $\langle\partial V/\partial\varphi_j\rangle = 0$. Symmetry dictates that all variations of the potential with respect to the fields in all symmetry directions vanish. The vacuum value of the first order variation in any direction vanishes as just seen; while the vacuum value of the second order variation must also vanish, as follows. Vanishing vacuum values of field symmetry transformation increments add no new information.

By contrast, however, *nonvanishing vacuum expectations of transformation increments*, $\langle\delta\varphi_g\rangle$, specify the relevant (Goldstone) *null eigenvectors of the mass matrix,*

and hence the corresponding zero-mass eigenvalues.

14.3   Goldstone's argument

The principle behind Goldstone's argument is that the ground state is not unique. Normally, by current conservation, the charge operator for any symmetry current is time-independent,

$$\frac{d}{dt}Q = \frac{d}{dt}\int_x J^0(x) = 0 \ .$$

Acting with the charge operator on the vacuum either *annihilates the vacuum*, if that is symmetric; else, if *not*, as is the case in spontaneous symmetry breaking, it produces a zero-frequency state out of it, through its shift transformation feature illustrated above. Actually, here, the charge itself is ill-defined. But its better behaved commutators with fields, that is, the transformation shifts, are still time-invariant, $d\langle\delta\varphi_g\rangle/dt=0$, thus generating a $\delta(k^0)$ in its Fourier transform.[4]

Thus, if the vacuum is not invariant under the symmetry, action of the charge operator produces a state which is different from the vacuum chosen, but which has zero frequency. This is a long-wavelength oscillation of a field which is nearly stationary: there are physical states with zero frequency, k^0, so that the theory cannot have a mass gap.

This argument is further clarified by taking the limit carefully. If an approximate charge operator acting in a huge but finite region A is applied to the vacuum,

$$\frac{d}{dt}Q_A = \frac{d}{dt}\int_x e^{\frac{-x^2}{2A^2}} J^0(x) = -\int_x e^{\frac{-x^2}{2A^2}} \nabla \cdot J = \int_x \nabla(e^{\frac{-x^2}{2A^2}}) \cdot J \ ,$$

a state with approximately vanishing time derivative is produced,

$$\|\frac{d}{dt}Q_A|0\rangle\| \approx \frac{1}{A}\|Q_A|0\rangle\|.$$

Assuming a nonvanishing mass gap m_0, the frequency of any state like the above, which is orthogonal to the vacuum, is at least m_0,

$$\|\frac{d}{dt}|\theta\rangle\| = \||H|\theta\rangle\| \geq m_0\| |\theta\rangle\| \, .$$

Letting A become large leads to a contradiction. Consequently $m_0 = 0$.

Exception: This argument fails, however, when the symmetry is gauged, because then the symmetry generator is only performing a gauge transformation. A gauge transformed state is the same exact state, so that acting with a symmetry generator does not get one out of the vacuum. See Higgs mechanism.

14.4 Infraparticles

There is an arguable loophole in the theorem. If one reads the theorem carefully, it only states that there exist non-vacuum states with arbitrarily small energies. Take for example a chiral N = 1 super QCD model with a nonzero squark VEV which is conformal in the IR. The chiral symmetry is a global symmetry which is (partially) spontaneously broken. Some of the "Goldstone bosons" associated with this spontaneous symmetry breaking are charged under the unbroken gauge group and hence, these composite bosons have a continuous mass spectrum with arbitrarily small masses but yet there is no Goldstone boson with exactly zero mass. In other words, the Goldstone bosons are infraparticles.

14.5 Nonrelativistic theories

A version of Goldstone's theorem also applies to nonrelativistic theories (and also relativistic theories with spontaneously broken spacetime symmetries, such as Lorentz symmetry or conformal symmetry, rotational, or translational invariance).

It essentially states that, for each spontaneously broken symmetry, there corresponds some quasiparticle with no energy gap—the nonrelativistic version of the mass gap. (Note that the energy here is really $H - \mu N - \alpha \rightarrow \cdot P \rightarrow$ and not H.) However, two *different* spontaneously broken generators may now give rise to the *same* Nambu–Goldstone boson. For example, in a superfluid, both the *U(1)* particle number symmetry and Galilean symmetry are spontaneously broken. However, the phonon is the Goldstone boson for both.

In general, the phonon is effectively the Nambu–Goldstone boson for spontaneously broken Galilean/Lorentz symmetry. However, in contrast to the case of internal symmetry breaking, when spacetime symmetries are broken, the order parameter *need not* be a scalar field, but may be a tensor field, and the corresponding independent massless modes may now be *fewer* than the number of spontaneously broken generators, because the Goldstone modes may now be linearly dependent among themselves: e.g., the Goldstone modes for some generators might be expressed as gradients of Goldstone modes for other broken generators.

14.6 Nambu–Goldstone fermions

Spontaneously broken global fermionic symmetries, which occur in some supersymmetric models, lead to Nambu–Goldstone fermions, or *goldstinos*.[5][6] These have spin ½, instead of 0, and carry all quantum numbers of the respective supersymmetry generators broken spontaneously.

Spontaneous supersymmetry breaking smashes up ("reduces") supermultiplet structures into the characteristic nonlinear realizations of broken supersymmetry, so that goldstinos are superpartners of *all* particles in the theory, of *any spin*, and the only superpartners, at that. That is, to say, two non-goldstino particles are connected to only goldstinos through supersymmetry transformations, and not to each other, even if they were so connected before the breaking of supersymmetry. As a result, the masses and spin multiplicities of such particles are then arbitrary.

14.7   See also

- Pseudo-Goldstone boson

- Majoron

- Higgs mechanism

- Mermin–Wagner theorem

- Vacuum expectation value

- Noether's theorem

14.8   References

[1] Nambu, Y (1960). "Quasiparticles and Gauge Invariance in the Theory of Superconductivity". *Physical Review* **117**: 648–663. Bibcode:1960PhRv..117..648N. doi:10.1103/PhysRev.117.648.

[2] Goldstone, J (1961). "Field Theories with Superconductor Solutions". *Nuovo Cimento* **19**: 154–164. doi:10.1007/BF02812722.

[3] Goldstone, J; Salam, Abdus; Weinberg, Steven (1962). "Broken Symmetries". *Physical Review* **127**: 965–970. Bibcode:1962 doi:10.1103/PhysRev.127.965.

[4] Scholarpedia proof

[5] Volkov, D.V.; Akulov, V (1973). "Is the neutrino a goldstone particle?". *Physics Letters* **B46**: 109–110. Bibcode:1973PhLB... 46 doi:10.1016/0370-2693(73)90490-5.

[6] Salam, A; et al. (1974). "On Goldstone Fermion". *Physics Letters* **B49**: 465–467. Bibcode:1974PhLB...49..465S. doi:10-2693(74)90637-6.

Chapter 15

1964 PRL symmetry breaking papers

In 1964, three teams wrote scientific papers which proposed related but different approaches to explain how mass could arise in local gauge theories. These three now famous papers were written by

- Robert Brout and François Englert,[1][2]

- Peter Higgs,[3] and

- Gerald Guralnik, C. Richard Hagen, and Tom Kibble (GHK),[4][5]

and are credited with the theory of the Higgs mechanism and the prediction of the Higgs field and Higgs boson. Together, these provide a theoretical means by which Goldstone's theorem (a problematic limitation affecting early modern particle physics theories) can be avoided. They show how gauge bosons can acquire non-zero masses as a result of spontaneous symmetry breaking within gauge invariant models of the universe.[6]

As such, these form the key element of the electroweak theory that forms part of the Standard Model of particle physics, and of many models, such as the Grand Unified Theory, that go beyond it. The papers that introduce this mechanism were published in *Physical Review Letters* (*PRL*) and were each recognized as milestone papers by *PRL* 's 50th anniversary celebration.[7] All of the six physicists were awarded the 2010 J. J. Sakurai Prize for Theoretical Particle Physics for this work,[8] and in 2013 Englert and Higgs received the Nobel Prize in Physics.[9]

On 4 July 2012, the two main experiments at the LHC (ATLAS and CMS) both reported independently the confirmed existence of a previously unknown particle with a mass of about 125 GeV/c^2 (about 133 proton masses, on the order of 10^{-25} kg), which is "consistent with the Higgs boson" and widely believed to be the Higgs boson.[10]

15.1   Introduction

A gauge theory of elementary particles is a very attractive potential framework for constructing the ultimate theory. Such a theory has the very desirable property of being potentially renormalizable—shorthand for saying that all calculational infinities encountered can be consistently absorbed into a few parameters of the theory. However, as soon as one gives mass to the gauge fields, renormalizability is lost, and the theory rendered useless. Spontaneous symmetry breaking is a promising mechanism, which could be used to give mass to the vector gauge particles. A significant difficulty which one encounters, however, is Goldstone's theorem, which states that in any quantum field theory which has a spontaneously broken symmetry there must occur a zero-mass particle. So the problem arises—how can one break a symmetry and at the same time not introduce unwanted zero-mass particles. The resolution of this dilemma lies in the observation that in the case of gauge theories, the Goldstone theorem can be avoided by working in the so-called radiation gauge. This is because the proof of Goldstone's theorem requires manifest Lorentz covariance, a property not possessed by the radiation gauge.

15.2   History

Particle physicists study matter made from fundamental particles whose interactions are mediated by exchange particles known as force carriers. At the beginning of the 1960s a number of these particles had been discovered or proposed, along with theories suggesting how they relate to each other, some of which had already been reformulated as field theories in which the objects of study are not particles and forces, but quantum fields and their symmetries. However, attempts to unify known fundamental forces such as the electromagnetic force and the weak nuclear force were known to be incomplete. One known omission was that gauge invariant approaches, including non-abelian models such as Yang–Mills theory (1954), which held great promise for unified theories, also seemed to predict known massive particles as massless.[11] Goldstone's theorem, relating to continuous symmetries within some theories, also appeared to rule out many obvious solutions,[12] since it appeared to show that zero-mass particles would have to also exist that were "simply not seen".[13] According to Guralnik, physicists had "no understanding" how these problems could be overcome in 1964.[13] In 2014, Guralnik and Hagen wrote a paper that contended that even after 50 years there is still widespread misunderstanding, by physicists and the Nobel Committee, of the Goldstone boson role.[14] This paper, published in *Modern Physics Letters A*, turned out to be Guralnik's last published work.[15]

Particle physicist and mathematician Peter Woit summarised the state of research at the time:

> "Yang and Mills work on non-abelian gauge theory had one huge problem: in perturbation theory it has mass-less particles which don't correspond to anything we see. One way of getting rid of this problem is now fairly well-understood, the phenomenon of confinement realized in QCD, where the strong interactions get rid of the massless "gluon" states at long distances. By the very early sixties, people had begun to understand another source of massless particles: spontaneous symmetry breaking of a continuous symmetry. What Philip Anderson realized and worked out in the summer of 1962 was that, when you have *both* gauge symmetry *and* spontaneous symmetry breaking, the Nambu–Goldstone massless mode can combine with the massless gauge field modes to produce a physical massive vector field. This is what happens in superconductivity, a subject about which Anderson was (and is) one of the leading experts." *[text condensed]* [11]

The Higgs mechanism is a process by which vector bosons can get rest mass *without* explicitly breaking gauge invariance, as a byproduct of spontaneous symmetry breaking.[6][16] The mathematical theory behind spontaneous symmetry breaking was initially conceived and published within particle physics by Yoichiro Nambu in 1960,[17] the concept that such a mechanism could offer a possible solution for the "mass problem" was originally suggested in 1962 by Philip Anderson,[18]:4–5[19] and Abraham Klein and Benjamin Lee showed in March 1964 that Goldstone's theorem could be avoided this way in at least some non-relativistic cases and speculated it might be possible in truly relativistic cases.[20]

These approaches were quickly developed into a full relativistic model, independently and almost simultaneously, by three groups of physicists: by François Englert and Robert Brout in August 1964;[1] by Peter Higgs in October 1964;[3] and by Gerald Guralnik, Carl Hagen, and Tom Kibble (GHK) in November 1964.[4] Higgs also wrote a short but important[6] response published in September 1964 to an objection by Gilbert,[21] which showed that if calculating within the radiation gauge, Goldstone's theorem and Gilbert's objection would become inapplicable.[Note 1] (Higgs later described Gilbert's objection as prompting his own paper.[22]) Properties of the model were further considered by Guralnik in 1965,[23] by Higgs in 1966,[24] by Kibble in 1967,[25] and further by GHK in 1967.[26] The original three 1964 papers showed that when a gauge theory is combined with an additional field that spontaneously breaks the symmetry, the gauge bosons can consistently acquire a finite mass.[6][16][27] In 1967, Steven Weinberg[28] and Abdus Salam[29] independently showed how a Higgs mechanism could be used to break the electroweak symmetry of Sheldon Glashow's unified model for the weak and electromagnetic interactions[30] (itself an extension of work by Schwinger), forming what became the Standard Model of particle physics. Weinberg was the first to observe that this would also provide mass terms for the fermions.[31] [Note 2]

However, the seminal papers on spontaneous breaking of gauge symmetries were at first largely ignored, because it was widely believed that the (non-Abelian gauge) theories in question were a dead-end, and in particular that they could not be renormalised. In 1971–72, Martinus Veltman and Gerard 't Hooft proved renormalisation of Yang–Mills was possible in two papers covering massless, and then massive, fields.[31] Their contribution, and others' work on the renormalization group, was eventually "enormously profound and influential",[32] but even with all key elements of the eventual theory published there was still almost no wider interest. For example, Coleman found in a study that "essentially no-one paid any attention" to Weinberg's paper prior to 1971[33] – now the most cited in particle physics[34] – and even in 1970 according to Politzer, Glashow's teaching of the weak interaction contained no mention of Weinberg's, Salem's, or Glashow's own work.[32] In practice, Politzer states, almost everyone learned of the theory due to physicist Benjamin Lee, who combined the work of Veltman and 't Hooft with insights by others, and popularised the completed theory.[32] In this way, from 1971, interest and acceptance "exploded" [32] and the ideas were quickly absorbed in the mainstream.[31][32]

15.2.1   The significance of requiring manifest covariance

Most students who have taken a course in electromagnetism have encountered the Coulomb potential. It basically states that two charged particles attract or repel each other by a force which varies according to the inverse square of their separation. This is fairly unambiguous for particles at rest, but if one or the other is following an arbitrary trajectory the question arises whether one should compute the force using the instantaneous positions of the particles or the so-called retarded positions. The latter recognizes that information cannot propagate instantaneously, rather it propagates at the speed of light. However, the radiation gauge says that one uses the instantaneous positions of the particles, but doesn't violate causality because there are compensating terms in the force equation. In contrast, the Lorenz gauge imposes manifest covariance (and thus causality) at all stages of a calculation. Predictions of observable quantities are identical in the two gauges, but the radiation gauge formulation of quantum field theory avoids Goldstone's theorem.[35]

15.2.2   Summary and impact of the *PRL* papers

The three papers written in 1964 were each recognised as milestone papers during *Physical Review Letters* 's 50th anniversary celebration.[27] Their six authors were also awarded the 2010 J. J. Sakurai Prize for Theoretical Particle Physics for this work.[36] (A controversy also arose the same year, because in the event of a Nobel Prize only up to three scientists could be recognised, with six being credited for the papers.[37]) Two of the three *PRL* papers (by Higgs and by GHK) contained equations for the hypothetical field that eventually would become known as the Higgs field and its hypothetical quantum, the Higgs boson.[3][4] Higgs's subsequent 1966 paper showed the decay mechanism of the boson; only a massive boson can decay and the decays can prove the mechanism.

Each of these papers is unique and demonstrates different approaches to showing how mass arise in gauge particles. Over the years, the differences between these papers are no longer widely understood, due to the passage of time and acceptance of end-results by the particle physics community. A study of citation indices is interesting—more than 40 years after the 1964 publication in *Physical Review Letters* there is little noticeable pattern of preference among them, with the vast majority of researchers in the field mentioning all three milestone papers.

In the paper by Higgs the boson is massive, and in a closing sentence Higgs writes that "an essential feature" of the theory "is the prediction of incomplete multiplets of scalar and vector bosons".[3] (Frank Close comments that 1960s gauge theorists were focused on the problem of massless *vector* bosons, and the implied existence of a massive *scalar* boson was not seen as important; only Higgs directly addressed it.[38]:154, 166, 175) In the paper by GHK the boson is massless and decoupled from the massive states.[4] In reviews dated 2009 and 2011, Guralnik states that in the GHK model the boson is massless only in a lowest-order approximation, but it is not subject to any constraint and acquires mass at higher orders, and adds that the GHK paper was the only one to show that there are no massless Goldstone bosons in the model and to give a complete analysis of the general Higgs mechanism.[13]<ref name="[14][5] All three reached similar conclusions, despite their very different approaches: Higgs' paper essentially used classical techniques, Englert and Brout's involved calculating vacuum polarization in perturbation theory around an assumed symmetry-breaking vacuum state, and GHK used operator formalism and conservation laws to explore in depth the ways in which Goldstone's theorem explicitly fails.[6]

In addition to explaining how mass is acquired by vector bosons, the Higgs mechanism also predicts the ratio between the W boson and Z boson masses as well as their couplings with each other and with the Standard Model quarks and leptons. Subsequently, many of these predictions have been verified by precise measurements performed at the LEP and the SLC colliders, thus overwhelmingly confirming that some kind of Higgs mechanism does take place in nature,[39] but the exact manner by which it happens has not yet been discovered. The results of searching for the Higgs boson are expected to provide evidence about how this is realized in nature.

15.2.3   Consequences of the papers

The resulting electroweak theory and Standard Model have correctly predicted (among other discoveries) weak neutral currents, three bosons, the top and charm quarks, and with great precision, the mass and other properties of some of these.[Note 3] Many of those involved eventually won Nobel Prizes or other renowned awards. A 1974 paper in *Reviews of Modern Physics* commented that "while no one doubted the [mathematical] correctness of these arguments, no one quite believed that nature was diabolically clever enough to take advantage of them".[40] By 1986 and again in the 1990s it became possible to write that understanding and proving the Higgs sector of the Standard Model was "the central problem today in particle physics." [41][42]

15.3   See also

- Higgs mechanism

- Higgs boson

- Standard Model

- Symmetry breaking

- Large Hadron Collider

- Fermilab

- Tevatron

- J. J. Sakurai Prize for Theoretical Particle Physics

- *The God Particle*, a popular science book on the Higgs boson, written by Leon M. Lederman

15.4   Notes

[1] Goldstone's theorem only applies to gauges having manifest Lorentz covariance, a condition that took time to become questioned. But the process of quantisation requires a gauge to be fixed and at this point it becomes possible to choose a gauge such as the 'radiation' gauge which is not invariant over time, so that these problems can be avoided.

[2] A field with the "Mexican hat" potential $V(\phi) = \mu^2\phi^2 + \lambda\phi^4$ and $\mu^2 < 0$ has a minimum not at zero but at some non-zero value ϕ_0 . By expressing the action in terms of the field $\tilde{\phi} = \phi - \phi_0$ (where ϕ_0 is a constant independent of position), we find the Yukawa term has a component $g\phi_0\bar{\psi}\psi$. Since both g and ϕ_0 are constants, this looks exactly like the mass term for a fermion of mass $g\phi_0$. The field $\tilde{\phi}$ is then the Higgs field.

[3] The success of the Higgs based electroweak theory and Standard Model is illustrated by their predictions of the mass of two particles later detected: the W boson (predicted mass: 80.390 ± 0.018 GeV, experimental measurement: 80.387 ± 0.019 GeV), and the Z boson (predicted mass: 91.1874 ± 0.0021, experimental measurement: 91.1876 ± 0.0021 GeV). The existence of the Z boson was itself another prediction. Other correct predictions included the weak neutral current, the gluon, and the top and charm quarks, all later proven to exist as the theory said.

15.5   References

[1] Englert, François; Brout, Robert (1964). "Broken Symmetry and the Mass of Gauge Vector Mesons". *Physical Review Letters* **13** (9): 321–23. Bibcode:1964PhRvL..13..321E. doi:10.1103/PhysRevLett.13.321.

[2] Brout, R.; Englert, F. (1998). "Spontaneous Symmetry Breaking in Gauge Theories: A Historical Survey". arXiv:hep-th/9802142 [hep-th].

[3] Higgs, Peter (1964). "Broken Symmetries and the Masses of Gauge Bosons". *Physical Review Letters* **13** (16): 508–509. Bibcode:1964PhRvL..13..508H. doi:10.1103/PhysRevLett.13.508.

[4] Guralnik, Gerald; Hagen, C. R.; Kibble, T. W. B. (1964). "Global Conservation Laws and Massless Particles". *Physical Review Letters* **13** (20): 585–587. Bibcode:1964PhRvL..13..585G. doi:10.1103/PhysRevLett.13.585.

[5] G.S. Guralnik (2009). "The History of the Guralnik, Hagen and Kibble development of the Theory of Spontaneous Symmetry Breaking and Gauge Particles". *International Journal of Modern Physics A* **24** (14): 2601–2627. arXiv:0907.3466. Bibcode:2009IJMPA..24.2601G. doi:10.1142/S0217751X09045431.

[6] Kibble, T. (2009). "Englert-Brout-Higgs-Guralnik-Hagen-Kibble mechanism". *Scholarpedia* **4**: 6441–6410. Bibcode:2009 doi:10.4249/scholarpedia.6441.

[7] Blume, M.; Brown, S.; Millev, Y. (2008). "Letters from the past, a PRL retrospective (1964)". Physical Review Letters. Archived from the original on 10 January 2010. Retrieved 2010-01-30.

[8] "J. J. Sakurai Prize Winners". American Physical Society. 2010. Archived from the original on 12 February 2010. Retrieved 2010-01-30.

[9] http://www.nobelprize.org/nobel_prizes/physics/laureates/2013/

[10] "CERN experiments observe particle consistent with long-sought Higgs boson" (Press release). CERN. 4 July 2012. Retrieved 2015-06-02.

[11] Woit, P. (13 November 2010). "The Anderson–Higgs Mechanism". *Not Even Wrong*. Columbia University. Retrieved 2012-11-12.

[12] Goldstone, J.; Salam, A.; Weinberg, S. (1962). "Broken Symmetries". *Physical Review* **127** (3): 965–970. Bibcode:1962 doi:10.1103/PhysRev.127.965.

[13] Guralnik, G. S. (2011). "The Beginnings of Spontaneous Symmetry Breaking in Particle Physics — Derived From My on the Spot "Intellectual Battlefield Impressions"". arXiv:1110.2253v1 [physics.hist-ph].

[14] Guralnik, G.; Hagen, C. R. (2014). "Where have all the Goldstone bosons gone?". *Modern Physics Letters A* **29**: 1450046. arXiv:1401.6924. Bibcode:2014MPLA...2950046G. doi:10.1142/S0217732314500461.

[15] Hagen, C. R. (August 2014). "Obituaries - Gerald Stanford Guralnik". *Physics Today*. doi:10.1063/PT.3.2488.

[16] Kibble, T. W. B. (2009). "Englert–Brout–Higgs–Guralnik–Hagen–Kibble Mechanism (History)". *Scholarpedia* **4** (1): 8741. Bibcode:2009SchpJ...4.8741K. doi:10.4249/scholarpedia.8741.

[17] The Nobel Prize in Physics 2008 – official Nobel Prize website.

[18] Higgs, P. (24 November 2010). "My Life as a Boson" (PDF). Kings College London. Archived from the original (PDF) on 2014-05-01. – the original 2001 paper can be found at: Duff, M. J. and Liu; Liu, J. T., eds. (2003). *2001 A Spacetime Odyssey: Proceedings of the Inaugural Conference of the Michigan Center for Theoretical Physics*. World Scientific Publishing. pp. 86–88. ISBN 981-238-231-3.

[19] Anderson, P. (1963). "Plasmons, gauge invariance and mass". *Physical Review* **130**: 439. Bibcode:1963PhRv..130..439A. doi:10.1103/PhysRev.130.439.

[20] Klein, A.; Lee, B. (1964). "Does Spontaneous Breakdown of Symmetry Imply Zero-Mass Particles?". *Physical Review Letters* **12** (10): 266. Bibcode:1964PhRvL..12..266K. doi:10.1103/PhysRevLett.12.266.

[21] Higgs, Peter (1964). "Broken symmetries, massless particles and gauge fi elds". *Physics Letters* **12** (2): 132–133. Bibcode: doi:10.1016/0031-9163(64)91136-9.

[22] Higgs, Peter (2010-11-24). "My Life as a Boson" (PDF). Talk given by Peter Higgs at Kings College, London, Nov 24 2010. Retrieved 17 January 2013. Gilbert ... wrote a response to [Klein and Lee's paper] saying 'No, you cannot do that in a relativistic theory. You cannot have a preferred unit time-like vector like that.' This is where I came in, because the next month was when I responded to Gilbert's paper by saying 'Yes, you can have such a thing' but only in a gauge theory with a gauge field coupled to the current.

[23] G.S. Guralnik (2011). "Gauge Invariance and the Goldstone Theorem – 1965 Feldafing talk". *Modern Physics Letters A* **26** (19): 1381–1392. arXiv:1107.4592. Bibcode:2011MPLA...26.1381G. doi:10.1142/S0217732311036188.

[24] Higgs, Peter (1966). "Spontaneous Symmetry Breakdown without Massless Bosons". *Physical Review* **145** (4): 1156–1163. Bibcode:1966PhRv..145.1156H. doi:10.1103/PhysRev.145.1156.

[25] Kibble, Tom (1967). "Symmetry Breaking in Non-Abelian Gauge Theories". *Physical Review* **155** (5): 1554–1561. Bibcode doi:10.1103/PhysRev.155.1554.

[26] "Guralnik, G S; Hagen, C R and Kibble, T W B (1967). Broken Symmetries and the Goldstone Theorem. Advances in Physics, vol. 2" (PDF).

[27] "Physical Review Letters – 50th Anniversary Milestone Papers". Physical Review Letters.

[28] S. Weinberg (1967). "A Model of Leptons". *Physical Review Letters* **19** (21): 1264–1266. Bibcode:1967PhRvL..19.1264W. doi:10.1103/PhysRevLett.19.1264.

[29] A. Salam (1968). N. Svartholm, ed. *Elementary Particle Physics: Relativistic Groups and Analyticity.* Eighth Nobel Symposium. Stockholm: Almquvist and Wiksell. p. 367.

[30] S.L. Glasho w (1961). "Partial-symmetries of weak interactions". *Nuclear Physics* **22** (4): 579–588. Bibcode:1961NucPh..22 doi:10.1016/0029-5582(61)90469-2.

[31] Ellis, John; Gaillard, Mary K.; Nanopoulos, Dimitri V. (2012). "A Historical Profile of the Higgs Boson". arXiv:1201.6045 [hep-ph].

[32] Politzer, David. "The Dilemma of Attribution". *Nobel Prize lecture, 2004.* Nobel Prize. Retrieved 22 January 2013. Sidney Coleman published in *Science* magazine in 1979 a citation search he did documenting that essentially no one paid any attention to Weinberg's Nobel Prize winning paper until the work of 't Hooft (as explicated by Ben Lee). In 1971 interest in Weinberg's paper exploded. I had a parallel personal experience: I took a one-year course on weak interactions from Shelly Glashow in 1970, and he never even mentioned the Weinberg–Salam model or his own contributions.

[33] Coleman, Sidne y (1979-12-14). "The 1979 Nobe l Prize in Physics". *Science* **206** (4424): 1290–1292. Bibcode:1979Sci...206\ .1290C.doi:10.1126/science.206.4424.1290. Retrieved 22 January 2013. – discussed by Davi d Politzer in his 2004 Nobe l speech.[32]

[34] Letters from the Past – A PRL Retrospective (50 year celebration, 2008)

[35] G.S. Guralnik, C.R. Hagen, T.W.B. Kibble (1968). "Broken Symmetries and the Goldstone Theorem". In R. L. Cool, R. E. Marshak. *Advances in Particle Physics* **2**. Interscience Publishers. pp. 567–708. ISBN 0-470-17057-3.

[36] American Physical Society – "J. J. Sakurai Prize for Theoretical Particle Physics".

[37] Merali, Zeeya (4 August 2010). "Physicists get political over Higgs". *Nature Magazine.* Retrieved 28 December 2011.

[38] Close, Frank (2011). *The Infinity Puzzle: Quantum Field Theory and the Hunt for an Orderly Universe.* Oxford: Oxford University Press. ISBN 978-0-19-959350-7.

[39] "LEP Electroweak Working Group".

[40] Bernstein, Jeremy (January 1974). "Spontaneous symmetry breaking, gauge theories, the Higgs mechanism and all that" (PDF). *Reviews of Modern Physics* **46** (1): 7–48. Bibcode:1974RvMP...46....7B. doi:10.1103/revmodphys.46.7. Retrieved 2012-12-10.

[41] José Luis Lucio and Arnulfo Zepeda (1987). *Proceedings of the II Mexican School of Particles and Fields, Cuernavaca-Morelos, 1986.* World Scientific. p. 29. ISBN 9971504340.

[42] Gunion, Dawson, Kane, and Haber (199). *The Higgs Hunter's Guide (1st ed.).* pp. 11 (?). ISBN 9780786743186. – quoted as being in the first (1990) edition of the book by Peter Higgs in his talk "My Life as a Boson", 2001, ref#25.

15.6   Further reading

- Higgs, P. W. (1964). "Broken symmetries, massless particles and gauge fields". *Physics Letters* **12** (2): 132–201. Bibcode:1964PhL....12..132H. doi:10.1016/0031-9163(64)91136-9.

- Englert, F.; Brout, R. (1964). "Broken Symmetry and the Mass of Gauge Vector Mesons". *Physical Review Letters* **13** (9): 321. Bibcode:1964PhRvL..13..321E. doi:10.1103/PhysRevLett.13.321.

- Higgs, P. (1964). "Broken Symmetries and the Masses of Gauge Bosons". *Physical Review Letters* **13** (16): 508. Bibcode:1964PhRvL..13..508H. doi:10.1103/PhysRevLett.13.508.

- Guralnik, G.; Hagen, C.; Kibble, T. (1964). "Global Conservation Laws and Massless Particles". *Physical Review Letters* **13** (20): 585. Bibcode:1964PhRvL..13..585G. doi:10.1103/PhysRevLett.13.585.

- Higgs, P. (1966). "Spontaneous Symmetry Breakdown without Massless Bosons". *Physical Review* **145** (4): 1156. Bibcode:1966PhRv..145.1156H. doi:10.1103/PhysRev.145.1156.

- Nambu, Y.; Jona-Lasinio, G. (1961). "Dynamical Model of Elementary Particles Based on an Analogy with Superconductivity. I". *Physical Review* **122**: 345. Bibcode:1961PhRv..122..345N. doi:10.1103/PhysRev.122.345.

- Goldstone, J.; Salam, A.; Weinberg, S. (1962). "Broken Symmetries". *Physical Review* **127** (3): 965. Bibcode:196\\ doi:10.1103/PhysRev.127.965.

- Anderson, P. (1963). "Plasmons, Gauge Invariance, and Mass". *Physical Review* **130**: 439. Bibcode:1963PhRv. doi:10.1103/PhysRev.130.439.

- Klein, A.; Lee, B. (1964). "Does Spontaneous Breakdown of Symmetry Imply Zero-Mass Particles?". *Physical Review Letters* **12** (10): 266. Bibcode:1964PhRvL..12..266K. doi:10.1103/PhysRevLett.12.266.

- Gilbert, W. (1964). "Broken Symmetries and Massless Particles". *Physical Review Letters* **12** (25): 713. Bibcode doi:10.1103/PhysRevLett.12.713.

- Guralnik, Gerald (2009). "The History of the Guralnik, Hagen and Kibble development of the Theory of Spontaneous Symmetry Breaking and Gauge Particles". *International Journal of Modern Physics A* **24** (14): 2601–2627. arXiv:0907.3466. Bibcode:2009IJMPA..24.2601G. doi:10.1142/S0217751X09045431., Guralnik, Gerald (2011). "The Beginnings of Spontaneous Symmetry Breaking in Particle Physics. Proceedings of the DPF-2011 Conference, Providence, RI, 8–13 August 2011". arXiv:1110.2253v1 [physics.hist-ph]., and Guralnik, Gerald (2013). "Heretical Ideas that Provided the Cornerstone for the Standard Model of Particle Physics". SPG MITTEILUNGEN March 2013, No. 39, (p. 14)

- Kobayashi, M.; Maskawa, T. (1973). "*CP*-Violation in the Renormalizable Theory of Weak Interaction". *Progress of Theoretical Physics* **49** (2): 652–657. Bibcode:1973PThPh..49..652K. doi:10.1143/PTP.49.652.

- 't Hooft, G.; Veltman, M. (1972). "Regularization and renormalization of gauge fields". *Nuclear Physics B* **44**: 189. Bibcode:1972NuPhB..44..189T. doi:10.1016/0550-3213(72)90279-9.

- G.S. Guralnik, C.R. Hagen, T.W.B. Kibble (1968). "Broken Symmetries and the Goldstone Theorem". In R. L. Cool, R. E. Marshak. *Advances in Particle Physics* **2**. Interscience Publishers. pp. 567–708. ISBN 0-470-17057-3.

15.7 External links

- *Physical Review Letters* - 50th Anniversary Milestone Papers

- American Physical Society - J. J. Sakurai Prize Winners

- Gerry Guralnik speaks at Brown University about the 1964 *PRL* papers

- In CERN Courier, Steven Weinberg reflects on spontaneous symmetry breaking

- Steven Weinberg on LHC

- Englert-Brout-Higgs-Guralnik-Hagen-Kibble Mechanism on Scholarpedia

- History of Englert-Brout-Higgs-Guralnik-Hagen-Kibble Mechanism on Scholarpedia

- "The History of the Guralnik, Hagen and Kibble development of the Theory of Spontaneous Symmetry Breaking and Gauge Particles"

- *International Journal of Modern Physics A*: "The History of the Guralnik, Hagen and Kibble development of the Theory of Spontaneous Symmetry Breaking and Gauge Particles"

- G.S. Guralnik (2011) "Gauge Invariance and the Goldstone Theorem - 1965 Feldafing talk". *International Journal of Modern Physics A*

- Spontaneous Symmetry Breaking in Gauge Theories: a Historical Survey

- CERN Courier Letter from GHK – December 2008

- God Particle

- 2010 Sakurai Prize Videos

- Brown University Celebration of 2010 Sakurai Prize - Videos

- The Hunt for the Higgs at Tevatron

- Physicists get political over Higgs

- Ian Sample on Controversy and Nobel Reform

- Massive by Ian Sample

- Blog Not Even Wrong, Review of Massive by Ian Sample

- Blog Not Even Wrong, Anderson-Higgs Mechanism

Chapter 16

Symmetry (physics)

For other uses, see Symmetry (disambiguation).

In physics, a **symmetry** of a physical system is a physical or mathematical feature of the system (observed or intrinsic) that is preserved or remains unchanged under some transformation.

A family of particular transformations may be *continuous* (such as rotation of a circle) or *discrete* (e.g., reflection of a bilaterally symmetric figure, or rotation of a regular polygon). Continuous and discrete transformations give rise to corresponding types of symmetries. Continuous symmetries can be described by Lie groups while discrete symmetries are described by finite groups (see Symmetry group).

These two concepts, Lie and finite groups, are the foundation for the fundamental theories of modern physics. Symmetries are frequently amenable to mathematical formulations such as group representations and can, in addition, be exploited to simplify many problems.

Arguably the most important example of a symmetry in physics is that the speed of light has the same value in all frames of reference, which is known in mathematical terms as Poincare group, the symmetry group of special relativity. Another important example is the invariance of the form of physical laws under arbitrary differentiable coordinate transformations, which is an important idea in general relativity.

16.1 Symmetry as invariance

Invariance is specified mathematically by transformations that leave some quantity unchanged. This idea can apply to basic real-world observations. For example, temperature may be constant throughout a room. Since the temperature is independent of position within the room, the temperature is *invariant* under a shift in the measurer's position.

Similarly, a uniform sphere rotated about its center will appear exactly as it did before the rotation. The sphere is said to exhibit spherical symmetry. A rotation about any axis of the sphere will preserve how the sphere "looks".

16.1.1 Invariance in force

The above ideas lead to the useful idea of *invariance* when discussing observed physical symmetry; this can be applied to symmetries in forces as well.

For example, an electric field due to a wire is said to exhibit cylindrical symmetry, because the electric field strength at a given distance r from the electrically charged wire of infinite length will have the same magnitude at each point on the surface of a cylinder (whose axis is the wire) with radius r. Rotating the wire about its own axis does not change its position or charge density, hence it will preserve the field. The field strength at a rotated position is the same. Suppose some configuration of charges (may be non-stationary) produce an electric field in some direction, then rotating the configuration of the charges (without disturbing the internal dynamics that produces the particular field) will lead to a net

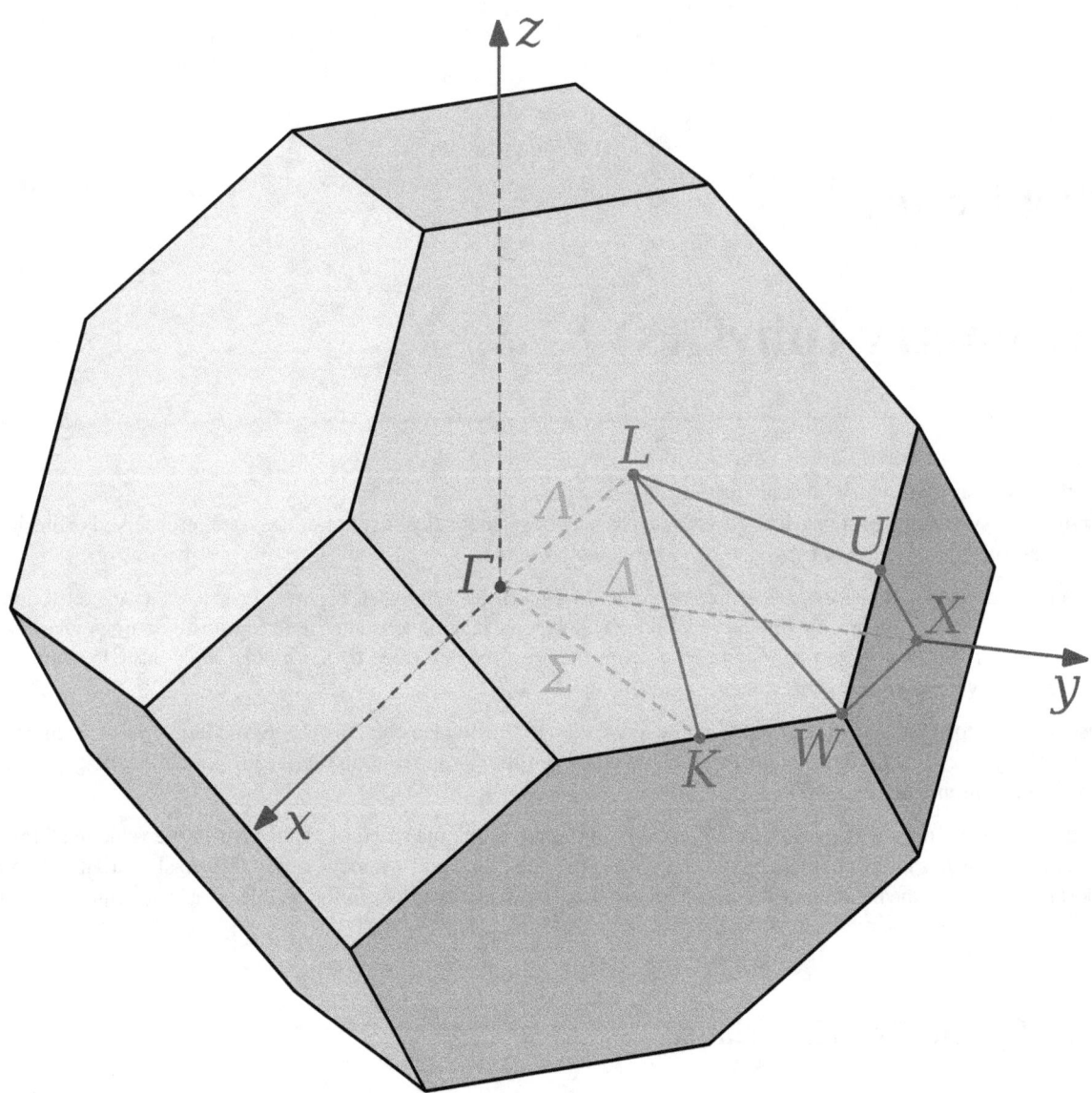

First Brillouin zone of FCC lattice showing symmetry labels

rotation of the direction of the electric field. These two properties are interconnected through the more general property that rotating *any* system of charges causes a corresponding rotation of the electric field.

In Newton's theory of mechanics, given two bodies, each with mass m, starting from rest at the origin and moving along the x-axis in opposite directions, one with speed v_1 and the other with speed v_2 the total kinetic energy of the system (as calculated from an observer at the origin) is $\frac{1}{2}m(v_1^2 + v_2^2)$ and remains the same if the velocities are interchanged. The total kinetic energy is preserved under a reflection in the y-axis.

The last example above illustrates another way of expressing symmetries, namely through the equations that describe some aspect of the physical system. The above example shows that the total kinetic energy will be the same if v_1 and v_2 are interchanged.

16.2   Local and global symmetries

Main articles: Global symmetry and Local symmetry

Symmetries may be broadly classified as *global* or *local*. A *global symmetry* is one that holds at all points of spacetime, whereas a *local symmetry* is one that has a different symmetry transformation at different points of spacetime; specifically a local symmetry transformation is parameterised by the spacetime co-ordinates. Local symmetries play an important role in physics as they form the basis for gauge theories.

16.3   Continuous symmetries

The two examples of rotational symmetry described above - spherical and cylindrical - are each instances of continuous symmetry. These are characterised by invariance following a continuous change in the geometry of the system. For example, the wire may be rotated through any angle about its axis and the field strength will be the same on a given cylinder. Mathematically, continuous symmetries are described by continuous or smooth functions. An important subclass of continuous symmetries in physics are spacetime symmetries.

16.3.1   Spacetime symmetries

Main article: Spacetime symmetries

Continuous *spacetime symmetries* are symmetries involving transformations of space and time. These may be further classified as *spatial symmetries*, involving only the spatial geometry associated with a physical system; *temporal symmetries*, involving only changes in time; or *spatio-temporal symmetries*, involving changes in both space and time.

- *Time translation*: A physical system may have the same features over a certain interval of time δt ; this is expressed mathematically as invariance under the transformation $t \rightarrow t + a$ for any real numbers t and a in the interval. For example, in classical mechanics, a particle solely acted upon by gravity will have gravitational potential energy mgh when suspended from a height h above the Earth's surface. Assuming no change in the height of the particle, this will be the total gravitational potential energy of the particle at all times. In other words, by considering the state of the particle at some time (in seconds) t_0 and also at $t_0 + 3$, say, the particle's total gravitational potential energy will be preserved.

- *Spatial translation*: These spatial symmetries are represented by transformations of the form $\vec{r} \rightarrow \vec{r} + \vec{a}$ and describe those situations where a property of the system does not change with a continuous change in location. For example, the temperature in a room may be independent of where the thermometer is located in the room.

- *Spatial rotation*: These spatial symmetries are classified as proper rotations and improper rotations. The former are just the 'ordinary' rotations; mathematically, they are represented by square matrices with unit determinant. The latter are represented by square matrices with determinant -1 and consist of a proper rotation combined with a spatial reflection (inversion). For example, a sphere has proper rotational symmetry. Other types of spatial rotations are described in the article *Rotation symmetry*.

- *Poincaré transformations*: These are spatio-temporal symmetries which preserve distances in Minkowski space-time, i.e. they are isometries of Minkowski space. They are studied primarily in special relativity. Those isometries that leave the origin fixed are called Lorentz transformations and give rise to the symmetry known as Lorentz co-variance.

- *Projective symmetries*: These are spatio-temporal symmetries which preserve the geodesic structure of spacetime. They may be defined on any smooth manifold, but find many applications in the study of exact solutions in general relativity.

- *Inversion transformations*: These are spatio-temporal symmetries which generalise Poincaré transformations to include other conformal one-to-one transformations on the space-time coordinates. Lengths are not invariant under inversion transformations but there is a cross-ratio on four points that is invariant.

Mathematically, spacetime symmetries are usually described by smooth vector fields on a smooth manifold. The underlying local diffeomorphisms associated with the vector fields correspond more directly to the physical symmetries, but the vector fields themselves are more often used when classifying the symmetries of the physical system.

Some of the most important vector fields are Killing vector fields which are those spacetime symmetries that preserve the underlying metric structure of a manifold. In rough terms, Killing vector fields preserve the distance between any two points of the manifold and often go by the name of isometries.

16.4   Discrete symmetries

Main article: Discrete symmetry

A **discrete symmetry** is a symmetry that describes non-continuous changes in a system. For example, a square possesses discrete rotational symmetry, as only rotations by multiples of right angles will preserve the square's original appearance. Discrete symmetries sometimes involve some type of 'swapping', these swaps usually being called *reflections* or *interchanges*.

- *Time reversal*: Many laws of physics describe real phenomena when the direction of time is reversed. Mathematically, this is represented by the transformation, $t \rightarrow -t$. For example, Newton's second law of motion still holds if, in the equation $F = m\ddot{r}$, t is replaced by $-t$. This may be illustrated by recording the motion of an object thrown up vertically (neglecting air resistance) and then playing it back. The object will follow the same parabolic trajectory through the air, whether the recording is played normally or in reverse. Thus, position is symmetric with respect to the instant that the object is at its maximum height.

- *Spatial inversion*: These are represented by transformations of the form $\vec{r} \rightarrow -\vec{r}$ and indicate an invariance property of a system when the coordinates are 'inverted'. Said another way, these are symmetries between a certain object and its mirror image.

- *Glide reflection*: These are represented by a composition of a translation and a reflection. These symmetries occur in some crystals and in some planar symmetries, known as wallpaper symmetries.

16.4.1   C, P, and T symmetries

The Standard model of particle physics has three related natural near-symmetries. These state that the actual universe about us is indistinguishable from one where:

- Every particle is replaced with its antiparticle. This is C-symmetry (charge symmetry);

- Everything appears as if reflected in a mirror. This is P-symmetry (parity symmetry);

- The direction of time is reversed. This is T-symmetry (time symmetry).

T-symmetry is counterintuitive (surely the future and the past are not symmetrical) but explained by the fact that the Standard model describes local properties, not global ones like entropy. To properly reverse the direction of time, one would have to put the big bang and the resulting low-entropy state in the "future." Since we perceive the "past" ("future") as having lower (higher) entropy than the present (see perception of time), the inhabitants of this hypothetical time-reversed universe would perceive the future in the same way as we perceive the past.

These symmetries are near-symmetries because each is broken in the present-day universe. However, the Standard Model predicts that the combination of the three (that is, the simultaneous application of all three transformations) must be a symmetry, called CPT symmetry. In <ref name=*qm*>G. Kalmbach H.E.: *Quantum Mathematics: WIGRIS.* RGN Publications, Delhi, 2014.</ref> the 4 dimensional matrix description of P,T is through a diagonal matrix, the negative identity, as well as C. Hence CPT is the identity operator. CP violation, the violation of the combination of C- and P-symmetry, is necessary for the presence of significant amounts of baryonic matter in the universe. CP violation is a fruitful area of current research in particle physics.

16.4.2 Supersymmetry

Main article: Supersymmetry

A type of symmetry known as supersymmetry has been used to try to make theoretical advances in the standard model. Supersymmetry is based on the idea that there is another physical symmetry beyond those already developed in the standard model, specifically a symmetry between bosons and fermions. Supersymmetry asserts that each type of boson has, as a supersymmetric partner, a fermion, called a superpartner, and vice versa. Supersymmetry has not yet been experimentally verified: no known particle has the correct properties to be a superpartner of any other known particle. If superpartners exist they must have masses greater than current particle accelerators can generate.

16.5 Mathematics of physical symmetry

Main article: Symmetry group
See also: Symmetry in quantum mechanics and Symmetries in general relativity

The transformations describing physical symmetries typically form a mathematical group. Group theory is an important area of mathematics for physicists.

Continuous symmetries are specified mathematically by *continuous groups* (called Lie groups). Many physical symmetries are isometries and are specified by symmetry groups. Sometimes this term is used for more general types of symmetries. The set of all proper rotations (about any angle) through any axis of a sphere form a Lie group called the special orthogonal group $SO(3)$. (The *3* refers to the three-dimensional space of an ordinary sphere.) Thus, the symmetry group of the sphere with proper rotations is $SO(3)$. Any rotation preserves distances on the surface of the ball. The set of all Lorentz transformations form a group called the Lorentz group (this may be generalised to the Poincaré group).

Discrete symmetries are described by discrete groups. For example, the symmetries of an equilateral triangle are described by the symmetric group S_3.

An important type of physical theory based on *local* symmetries is called a *gauge* theory and the symmetries natural to such a theory are called gauge symmetries. Gauge symmetries in the Standard model, used to describe three of the fundamental interactions, are based on the SU(3) × SU(2) × U(1) group. (Roughly speaking, the symmetries of the SU(3) group describe the strong force, the SU(2) group describes the weak interaction and the U(1) group describes the electromagnetic force.)

Also, the reduction by symmetry of the energy functional under the action by a group and spontaneous symmetry breaking of transformations of symmetric groups appear to elucidate topics in particle physics (for example, the unification of electromagnetism and the weak force in physical cosmology).

16.5.1 Conservation laws and symmetry

Main article: Noether's theorem

The symmetry properties of a physical system are intimately related to the conservation laws characterizing that system.

Noether's theorem gives a precise description of this relation. The theorem states that each continuous symmetry of a physical system implies that some physical property of that system is conserved. Conversely, each conserved quantity has a corresponding symmetry. For example, the isometry of space gives rise to conservation of (linear) momentum, and isometry of time gives rise to conservation of energy.

The following table summarizes some fundamental symmetries and the associated conserved quantity.

16.6 Mathematics

Continuous symmetries in physics preserve transformations. One can specify a symmetry by showing how a very small transformation affects various particle fields. The commutator of two of these infinitessimal transformations are equivalent to a third infinitessimal transformation of the same kind hence they form a Lie algebra.

A general coordinate transformation (also known as a diffeomorphism) has the infinitessimal effect on a scalar, spinor and vector field for example:

$$\delta\phi(x) = h^\mu(x)\partial_\mu\phi(x)$$
$$\delta\psi^\alpha(x) = h^\mu(x)\partial_\mu\psi^\alpha(x) + \partial_\mu h_\nu(x)\sigma^{\alpha\beta}_{\mu\nu}\psi^\beta(x)$$
$$\delta A_\mu(x) = h^\nu(x)\partial_\nu A_\mu(x) + A_\nu(x)\partial_\mu h^\nu(x)$$

for a general field, $h(x)$. Without gravity only the Poincaré symmetries are preserved which restricts $h(x)$ to be of the form:

$$h^\mu(x) = M^{\mu\nu}x_\nu + P^\mu$$

where **M** is an antisymmetric matrix (giving the Lorentz and rotational symmetries) and **P** is a general vector (giving the translational symmetries). Other symmetries affect multiple fields simultaneously. For example local gauge transformations apply to both a vector and spinor field:

$$\delta\psi^\alpha(x) = \lambda(x).\tau^{\alpha\beta}\psi^\beta(x)$$
$$\delta A_\mu(x) = \partial_\mu\lambda(x)$$

where τ are generators of a particular Lie group. So far the transformations on the right have only included fields of the same type. Supersymmetries are defined according to how the mix fields of *different* types.

Another symmetry which is part of some theories of physics and not in others is scale invariance which involve Weyl transformations of the following kind:

$$\delta\phi(x) = \Omega(x)\phi(x)$$

If the fields have this symmetry then it can be shown that the field theory is almost certainly conformally invariant also. This means that in the absence of gravity h(x) would restricted to the form:

$$h^\mu(x) = M^{\mu\nu}x_\nu + P^\mu + Dx_\mu + K^\mu|x|^2 - 2K^\nu x_\nu x_\mu$$

with **D** generating scale transformations and **K** generating special conformal transformations. For example N=4 super-Yang-Mills theory has this symmetry while General Relativity doesn't although other theories of gravity such as conformal gravity do. The 'action' of a field theory is an invariant under all the symmetries of the theory. Much of modern theoretical physics is to do with speculating on the various symmetries the Universe may have and finding the invariants to construct field theories as models.

In string theories, since a string can be decomposed into an infinite number of particle fields, the symmetries on the string world sheet is equivalent to special transformations which mix an infinite number of fields.

16.7 See also

- Conservation law
- Conserved current

- Coordinate-free

- Covariance and contravariance

- Diffeomorphism

- Fictitious force

- Galilean invariance

- Gauge theory

- General covariance

- Harmonic coordinate condition

- Inertial frame of reference

- Lie group

- List of mathematical topics in relativity

- Lorentz covariance

- Noether's theorem

- Poincaré group

- Special relativity

- Spontaneous symmetry breaking

- Standard model

- Standard model (mathematical formulation)

- Symmetry breaking

- Wheeler–Feynman Time-Symmetric Theory

16.8   References

16.8.1   General readers

- Leon Lederman and Christopher T. Hill (2005) *Symmetry and the Beautiful Universe.* Amherst NY: Prometheus Books.

- Schumm, Bruce (2004) *Deep Down Things.* Johns Hopkins Univ. Press.

- Victor J. Stenger (2000) *Timeless Reality: Symmetry, Simplicity, and Multiple Universes.* Buffalo NY: Prometheus Books. Chpt. 12 is a gentle introduction to symmetry, invariance, and conservation laws.

- Anthony Zee (2007) *Fearful Symmetry: The search for beauty in modern physics,* 2nd ed. Princeton University Press. ISBN 978-0-691-00946-9. 1986 1st ed. published by Macmillan.

16.8.2   Technical readers

- Brading, K., and Castellani, E., eds. (2003) *Symmetries in Physics: Philosophical Reflections.* Cambridge Univ. Press.

- -------- (2007) "Symmetries and Invariances in Classical Physics" in Butterfield, J., and John Earman, eds., *Philosophy of Physic Part B*. North Holland: 1331-68.

- Debs, T. and Redhead, M. (2007) *Objectivity, Invariance, and Convention: Symmetry in Physical Science.* Harvard Univ. Press.

- John Earman (2002) "Laws, Symmetry, and Symmetry Breaking: Invariance, Conservations Principles, and Objectivity." Address to the 2002 meeting of the Philosophy of Science Association.

- G. Kalmbach H.E.: *Quantum Mathematics: WIGRIS.* RGN Publications, Delhi, 2014

- Mainzer, K. (1996) *Symmetries of nature.* Berlin: De Gruyter.

- Mouchet, A. "Reflections on the four facets of symmetry: how physics exemplifies rational thinking". European Physical Journal H 38 (2013) 661 hal.archives-ouvertes.fr:hal-00637572

- Thompson, William J. (1994) *Angular Momentum: An Illustrated Guide to Rotational Symmetries for Physical Systems.* Wiley. ISBN 0-471-55264-X.

- Bas Van Fraassen (1989) *Laws and symmetry.* Oxford Univ. Press.

- Eugene Wigner (1967) *Symmetries and Reflections.* Indiana Univ. Press.

16.9   External links

- Stanford Encyclopedia of Philosophy: "Symmetry"—by K. Brading and E. Castellani.

- Pedagogic Aids to Quantum Field Theory Click on link to Chapter 6: Symmetry, Invariance, and Conservation for a simplified, step-by-step introduction to symmetry in physics.

Chapter 17

Lie group

In mathematics, a **Lie group** /ˈliː/ is a group that is also a differentiable manifold, with the property that the group operations are compatible with the smooth structure. Lie groups are named after Sophus Lie, who laid the foundations of the theory of continuous transformation groups. The term *groupes de Lie* first appeared in French in 1893 in the thesis of Lie's student Arthur Tresse, page 3.[1]

Lie groups represent the best-developed theory of continuous symmetry of mathematical objects and structures, which makes them indispensable tools for many parts of contemporary mathematics, as well as for modern theoretical physics. They provide a natural framework for analysing the continuous symmetries of differential equations (differential Galois theory), in much the same way as permutation groups are used in Galois theory for analysing the discrete symmetries of algebraic equations. An extension of Galois theory to the case of continuous symmetry groups was one of Lie's principal motivations.

17.1 Overview

Lie groups are smooth differentiable manifolds and as such can be studied using differential calculus, in contrast with the case of more general topological groups. One of the key ideas in the theory of Lie groups is to replace the *global* object, the group, with its *local* or linearized version, which Lie himself called its "infinitesimal group" and which has since become known as its Lie algebra.

Lie groups play an enormous role in modern geometry, on several different levels. Felix Klein argued in his Erlangen program that one can consider various "geometries" by specifying an appropriate transformation group that leaves certain geometric properties invariant. Thus Euclidean geometry corresponds to the choice of the group E(3) of distance-preserving transformations of the Euclidean space $\mathbf{R}^3$, conformal geometry corresponds to enlarging the group to the conformal group, whereas in projective geometry one is interested in the properties invariant under the projective group. This idea later led to the notion of a G-structure, where G is a Lie group of "local" symmetries of a manifold.

On a "global" level, whenever a Lie group acts on a geometric object, such as a Riemannian or a symplectic manifold, this action provides a measure of rigidity and yields a rich algebraic structure. The presence of continuous symmetries expressed via a Lie group action on a manifold places strong constraints on its geometry and facilitates analysis on the manifold. Linear actions of Lie groups are especially important, and are studied in representation theory.

In the 1940s–1950s, Ellis Kolchin, Armand Borel, and Claude Chevalley realised that many foundational results concerning Lie groups can be developed completely algebraically, giving rise to the theory of algebraic groups defined over an arbitrary field. This insight opened new possibilities in pure algebra, by providing a uniform construction for most finite simple groups, as well as in algebraic geometry. The theory of automorphic forms, an important branch of modern number theory, deals extensively with analogues of Lie groups over adele rings; p-adic Lie groups play an important role, via their connections with Galois representations in number theory.

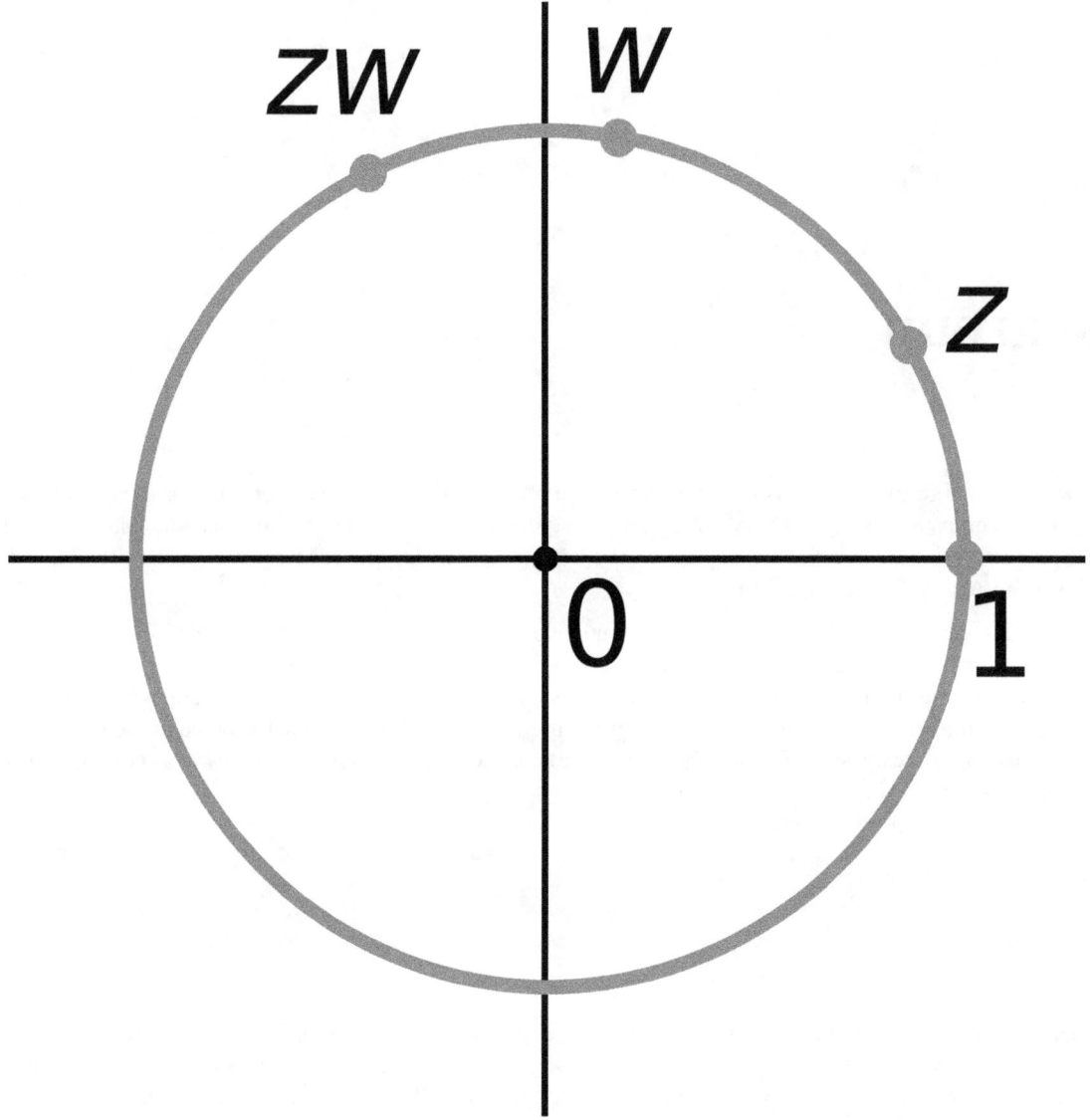

The circle of center 0 and radius 1 in the complex plane is a Lie group with complex multiplication.

17.2   Definitions and examples

A **real Lie group** is a group that is also a finite-dimensional real smooth manifold, in which the group operations of multiplication and inversion are smooth maps. Smoothness of the group multiplication

$$\mu : G \times G \to G \quad \mu(x, y) = xy$$

means that μ is a smooth mapping of the product manifold $G{\times}G$ into G. These two requirements can be combined to the single requirement that the mapping

$$(x, y) \mapsto x^{-1}y$$

be a smooth mapping of the product manifold into *G*.

17.2.1   First examples

- The 2×2 real invertible matrices form a group under multiplication, denoted by GL(2, **R**) or by GL2(**R**):

$$GL(2, \mathbf{R}) = \left\{ A = \begin{pmatrix} a & b \\ c & d \end{pmatrix} : \det A = ad - bc \neq 0 \right\}.$$

  This is a four-dimensional noncompact real Lie group. This group is disconnected; it has two connected components corresponding to the positive and negative values of the determinant.

- The rotation matrices form a subgroup of GL(2, **R**), denoted by SO(2, **R**). It is a Lie group in its own right: specifically, a one-dimensional compact connected Lie group which is diffeomorphic to the circle. Using the rotation angle φ as a parameter, this group can be parametrized as follows:

$$SO(2, \mathbf{R}) = \left\{ \begin{pmatrix} \cos \varphi & -\sin \varphi \\ \sin \varphi & \cos \varphi \end{pmatrix} : \varphi \in \mathbf{R}/2\pi\mathbf{Z} \right\}.$$

  Addition of the angles corresponds to multiplication of the elements of SO(2, **R**), and taking the opposite angle corresponds to inversion. Thus both multiplication and inversion are differentiable maps.

- The orthogonal group also forms an interesting example of a Lie group.

All of the previous examples of Lie groups fall within the class of classical groups.

17.2.2   Related concepts

A **complex Lie group** is defined in the same way using complex manifolds rather than real ones (example: SL(2, **C**)), and similarly, using an alternate metric completion of **Q**, one can define a **p-adic Lie group** over the p-adic numbers, a topological group in which each point has a p-adic neighborhood. Hilbert's fifth problem asked whether replacing differentiable manifolds with topological or analytic ones can yield new examples. The answer to this question turned out to be negative: in 1952, Gleason, Montgomery and Zippin showed that if *G* is a topological manifold with continuous group operations, then there exists exactly one analytic structure on *G* which turns it into a Lie group (see also Hilbert–Smith conjecture). If the underlying manifold is allowed to be infinite-dimensional (for example, a Hilbert manifold), then one arrives at the notion of an infinite-dimensional Lie group. It is possible to define analogues of many Lie groups over finite fields, and these give most of the examples of finite simple groups.

The language of category theory provides a concise definition for Lie groups: a Lie group is a group object in the category of smooth manifolds. This is important, because it allows generalization of the notion of a Lie group to Lie supergroups.

17.3   More examples of Lie groups

See also: Table of Lie groups and List of simple Lie groups

Lie groups occur in abundance throughout mathematics and physics. Matrix groups or algebraic groups are (roughly) groups of matrices (for example, orthogonal and symplectic groups), and these give most of the more common examples of Lie groups.

17.3.1   Examples with a specific number of dimensions

- The circle group $\mathbf{S}^1$ consisting of angles mod 2π under addition or, alternatively, the complex numbers with absolute value 1 under multiplication. This is a one-dimensional compact connected abelian Lie group.

- The 3-sphere $\mathbf{S}^3$ forms a Lie group by identification with the set of quaternions of unit norm, called versors. The only other spheres that admit the structure of a Lie group are the 0-sphere $\mathbf{S}^0$ (real numbers with absolute value 1) and the circle $\mathbf{S}^1$ (complex numbers with absolute value 1). For example, for even $n > 1$, $\mathbf{S}^n$ is not a Lie group because it does not admit a nonvanishing vector field and so *a fortiori* cannot be parallelizable as a differentiable manifold. Of the spheres only $\mathbf{S}^0, \mathbf{S}^1, \mathbf{S}^3$, and $\mathbf{S}^7$ are parallelizable. The last carries the structure of a Lie quasigroup (a nonassociative group), which can be identified with the set of unit octonions.

- The (3-dimensional) metaplectic group is a double cover of SL(2, $\mathbf{R}$) playing an important role in the theory of modular forms. It is a connected Lie group that cannot be faithfully represented by matrices of finite size, i.e., a nonlinear group.

- The Heisenberg group is a connected nilpotent Lie group of dimension 3, playing a key role in quantum mechanics.

- The Lorentz group is a 6-dimensional Lie group of linear isometries of the Minkowski space.

- The Poincaré group is a 10-dimensional Lie group of affine isometries of the Minkowski space.

- The group U(1)×SU(2)×SU(3) is a Lie group of dimension 1+3+8=12 that is the gauge group of the Standard Model in particle physics. The dimensions of the factors correspond to the 1 photon + 3 vector bosons + 8 gluons of the standard model

- The exceptional Lie groups of types G_2, F_4, E_6, E_7, E_8 have dimensions 14, 52, 78, 133, and 248. Along with the A-B-C-D series of simple Lie groups, the exceptional groups complete the list of simple Lie groups. There is also a Lie group named $E_7\tfrac{1}{2}$ of dimension 190, but it is not a *simple* Lie group.

17.3.2   Examples with n dimensions

- Euclidean space $\mathbf{R}^n$ with ordinary vector addition as the group operation becomes an n-dimensional noncompact abelian Lie group.

- The Euclidean group E(n, $\mathbf{R}$) is the Lie group of all Euclidean motions, i.e., isometric affine maps, of n-dimensional Euclidean space $\mathbf{R}^n$.

- The orthogonal group O(n, $\mathbf{R}$), consisting of all $n \times n$ orthogonal matrices with real entries is an $n(n-1)/2$-dimensional Lie group. This group is disconnected, but it has a connected subgroup SO(n, $\mathbf{R}$) of the same dimension consisting of orthogonal matrices of determinant 1, called the special orthogonal group (for $n = 3$, the rotation group SO(3)).

- The unitary group U(n) consisting of $n \times n$ unitary matrices (with complex entries) is a compact connected Lie group of dimension n^2. Unitary matrices of determinant 1 form a closed connected subgroup of dimension $n^2 - 1$ denoted SU(n), the special unitary group.

- Spin groups are double covers of the special orthogonal groups, used for studying fermions in quantum field theory (among other things).

- The group GL(n, $\mathbf{R}$) of invertible matrices (under matrix multiplication) is a Lie group of dimension n^2, called the general linear group. It has a closed connected subgroup SL(n, $\mathbf{R}$), the special linear group, consisting of matrices of determinant 1 which is also a Lie group.

- The symplectic group Sp($2n$, $\mathbf{R}$) consists of all $2n \times 2n$ matrices preserving a *symplectic form* on $\mathbf{R}^{2n}$. It is a connected Lie group of dimension $2n^2 + n$.

- The group of invertible upper triangular n by n matrices is a solvable Lie group of dimension $n(n + 1)/2$. (cf. Borel subgroup)

- The A-series, B-series, C-series and D-series, whose elements are denoted by An, Bn, Cn, and Dn, are infinite families of simple Lie groups.

17.3.3 Constructions

There are several standard ways to form new Lie groups from old ones:

- The product of two Lie groups is a Lie group.

- Any topologically closed subgroup of a Lie group is a Lie group. This is known as the Closed subgroup theorem or **Cartan's theorem**.

- The quotient of a Lie group by a closed normal subgroup is a Lie group.

- The universal cover of a connected Lie group is a Lie group. For example, the group **R** is the universal cover of the circle group **S**^{1}. In fact any covering of a differentiable manifold is also a differentiable manifold, but by specifying *universal* cover, one guarantees a group structure (compatible with its other structures).

17.3.4 Related notions

Some examples of groups that are *not* Lie groups (except in the trivial sense that any group can be viewed as a 0-dimensional Lie group, with the discrete topology), are:

- Infinite-dimensional groups, such as the additive group of an infinite-dimensional real vector space. These are not Lie groups as they are not *finite-dimensional* manifolds.

- Some totally disconnected groups, such as the Galois group of an infinite extension of fields, or the additive group of the p-adic numbers. These are not Lie groups because their underlying spaces are not real manifolds. (Some of these groups are "p-adic Lie groups".) In general, only topological groups having similar local properties to **R**^{n} for some positive integer n can be Lie groups (of course they must also have a differentiable structure).

17.4 Basic concepts

17.4.1 The Lie algebra associated with a Lie group

Main article: Lie group–Lie algebra correspondence

To every Lie group we can associate a Lie algebra whose underlying vector space is the tangent space of the Lie group at the identity element and which completely captures the local structure of the group. Informally we can think of elements of the Lie algebra as elements of the group that are "infinitesimally close" to the identity, and the Lie bracket of the Lie algebra is related to the commutator of two such infinitesimal elements. Before giving the abstract definition we give a few examples:

- The Lie algebra of the vector space **R**^{n} is just **R**^{n} with the Lie bracket given by
  $[A, B] = 0$.
  (In general the Lie bracket of a connected Lie group is always 0 if and only if the Lie group is abelian.)

- The Lie algebra of the general linear group GL(n, **R**) of invertible matrices is the vector space M(n, **R**) of square matrices with the Lie bracket given by
  $[A, B] = AB - BA$.
  If G is a closed subgroup of GL(n, **R**) then the Lie algebra of G can be thought of informally as the matrices m

of M(n, **R**) such that $1 + \varepsilon m$ is in G, where ε is an infinitesimal positive number with $\varepsilon^2 = 0$ (of course, no such real number ε exists). For example, the orthogonal group O(n, **R**) consists of matrices A with $AA^\mathrm{T} = 1$, so the Lie algebra consists of the matrices m with $(1 + \varepsilon m)(1 + \varepsilon m)^\mathrm{T} = 1$, which is equivalent to $m + m^\mathrm{T} = 0$ because $\varepsilon^2 = 0$.

- Formally, when working over the reals, as here, this is accomplished by considering the limit as $\varepsilon \to 0$; but the "infinitesimal" language generalizes directly to Lie groups over general rings.

The concrete definition given above is easy to work with, but has some minor problems: to use it we first need to represent a Lie group as a group of matrices, but not all Lie groups can be represented in this way, and it is not obvious that the Lie algebra is independent of the representation we use. To get around these problems we give the general definition of the Lie algebra of a Lie group (in 4 steps):

1. Vector fields on any smooth manifold M can be thought of as derivations X of the ring of smooth functions on the manifold, and therefore form a Lie algebra under the Lie bracket $[X, Y] = XY - YX$, because the Lie bracket of any two derivations is a derivation.

2. If G is any group acting smoothly on the manifold M, then it acts on the vector fields, and the vector space of vector fields fixed by the group is closed under the Lie bracket and therefore also forms a Lie algebra.

3. We apply this construction to the case when the manifold M is the underlying space of a Lie group G, with G acting on $G = M$ by left translations $Lg(h) = gh$. This shows that the space of left invariant vector fields (vector fields satisfying $Lg*Xh = Xgh$ for every h in G, where $Lg*$ denotes the differential of Lg) on a Lie group is a Lie algebra under the Lie bracket of vector fields.

4. Any tangent vector at the identity of a Lie group can be extended to a left invariant vector field by left translating the tangent vector to other points of the manifold. Specifically, the left invariant extension of an element v of the tangent space at the identity is the vector field defined by $v^\wedge g = Lg*v$. This identifies the tangent space TeG at the identity with the space of left invariant vector fields, and therefore makes the tangent space at the identity into a Lie algebra, called the Lie algebra of G, usually denoted by a Fraktur $\mathfrak{g}$. Thus the Lie bracket on $\mathfrak{g}$ is given explicitly by $[v, w] = [v^\wedge, w^\wedge]e$.

This Lie algebra $\mathfrak{g}$ is finite-dimensional and it has the same dimension as the manifold G. The Lie algebra of G determines G up to "local isomorphism", where two Lie groups are called **locally isomorphic** if they look the same near the identity element. Problems about Lie groups are often solved by first solving the corresponding problem for the Lie algebras, and the result for groups then usually follows easily. For example, simple Lie groups are usually classified by first classifying the corresponding Lie algebras.

We could also define a Lie algebra structure on Te using right invariant vector fields instead of left invariant vector fields. This leads to the same Lie algebra, because the inverse map on G can be used to identify left invariant vector fields with right invariant vector fields, and acts as -1 on the tangent space Te.

The Lie algebra structure on Te can also be described as follows: the commutator operation

$$(x, y) \to xyx^{-1}y^{-1}$$

on $G \times G$ sends (e, e) to e, so its derivative yields a bilinear operation on TeG. This bilinear operation is actually the zero map, but the second derivative, under the proper identification of tangent spaces, yields an operation that satisfies the axioms of a Lie bracket, and it is equal to twice the one defined through left-invariant vector fields.

17.4.2   Homomorphisms and isomorphisms

If G and H are Lie groups, then a Lie group homomorphism $f : G \to H$ is a smooth group homomorphism. In the case of complex Lie groups, such a homomorphism is required to be a holomorphic map. However, these requirements are a bit stringent; over real or complex numbers, every continuous homomorphism between Lie groups turns out to be (real or complex) analytic.

The composition of two Lie homomorphisms is again a homomorphism, and the class of all Lie groups, together with these morphisms, forms a category. Moreover, every Lie group homomorphism induces a homomorphism between the corresponding Lie algebras. Let $\phi: G \to H$ be a Lie group homomorphism and let ϕ_* be its derivative at the identity. If we identify the Lie algebras of G and H with their tangent spaces at the identity elements then ϕ_* is a map between the corresponding Lie algebras:

$$\phi_* : \mathfrak{g} \to \mathfrak{h}$$

One can show that ϕ_* is actually a Lie algebra homomorphism (meaning that it is a linear map which preserves the Lie bracket). In the language of category theory, we then have a covariant functor from the category of Lie groups to the category of Lie algebras which sends a Lie group to its Lie algebra and a Lie group homomorphism to its derivative at the identity.

Two Lie groups are called *isomorphic* if there exists a bijective homomorphism between them whose inverse is also a Lie group homomorphism. Equivalently, it is a diffeomorphism which is also a group homomorphism.

Ado's theorem says every finite-dimensional Lie algebra is isomorphic to a matrix Lie algebra. For every finite-dimensional matrix Lie algebra, there is a linear group (matrix Lie group) with this algebra as its Lie algebra. So every abstract Lie algebra is the Lie algebra of some (linear) Lie group.

The *global structure* of a Lie group is not determined by its Lie algebra; for example, if Z is any discrete subgroup of the center of G then G and G/Z have the same Lie algebra (see the table of Lie groups for examples). A *connected* Lie group is simple, semisimple, solvable, nilpotent, or abelian if and only if its Lie algebra has the corresponding property.

If we require that the Lie group be simply connected, then the global structure is determined by its Lie algebra: for every finite-dimensional Lie algebra $\mathfrak{g}$ over $\mathbf{F}$ there is a simply connected Lie group G with $\mathfrak{g}$ as Lie algebra, unique up to isomorphism. Moreover every homomorphism between Lie algebras lifts to a unique homomorphism between the corresponding simply connected Lie groups.

17.4.3   The exponential map

Main article: Exponential map (Lie theory)

The exponential map from the Lie algebra $M(n, \mathbf{R})$ of the general linear group $GL(n, \mathbf{R})$ to $GL(n, \mathbf{R})$ is defined by the usual power series:

$$\exp(A) = 1 + A + \frac{A^2}{2!} + \frac{A^3}{3!} + \cdots$$

for matrices A. If G is any subgroup of $GL(n, \mathbf{R})$, then the exponential map takes the Lie algebra of G into G, so we have an exponential map for all matrix groups.

The definition above is easy to use, but it is not defined for Lie groups that are not matrix groups, and it is not clear that the exponential map of a Lie group does not depend on its representation as a matrix group. We can solve both problems using a more abstract definition of the exponential map that works for all Lie groups, as follows.

Every vector v in $\mathfrak{g}$ determines a linear map from $\mathbf{R}$ to $\mathfrak{g}$ taking 1 to v, which can be thought of as a Lie algebra homomorphism. Because $\mathbf{R}$ is the Lie algebra of the simply connected Lie group $\mathbf{R}$, this induces a Lie group homomorphism $c : \mathbf{R} \to G$ so that

$$c(s + t) = c(s)c(t)$$

for all s and t. The operation on the right hand side is the group multiplication in G. The formal similarity of this formula with the one valid for the exponential function justifies the definition

$\exp(v) = c(1).$

This is called the **exponential map**, and it maps the Lie algebra $\mathfrak{g}$ into the Lie group G. It provides a diffeomorphism between a neighborhood of 0 in $\mathfrak{g}$ and a neighborhood of e in G. This exponential map is a generalization of the exponential function for real numbers (because $\mathbf{R}$ is the Lie algebra of the Lie group of positive real numbers with multiplication), for complex numbers (because $\mathbf{C}$ is the Lie algebra of the Lie group of non-zero complex numbers with multiplication) and for matrices (because $M(n, \mathbf{R})$ with the regular commutator is the Lie algebra of the Lie group $GL(n, \mathbf{R})$ of all invertible matrices).

Because the exponential map is surjective on some neighbourhood N of e, it is common to call elements of the Lie algebra **infinitesimal generators** of the group G. The subgroup of G generated by N is the identity component of G.

The exponential map and the Lie algebra determine the *local group structure* of every connected Lie group, because of the Baker–Campbell–Hausdorff formula: there exists a neighborhood U of the zero element of $\mathfrak{g}$, such that for u, v in U we have

$$\exp(u)\,\exp(v) = \exp\left(u + v + \tfrac{1}{2}[u,v] + \tfrac{1}{12}[\,[u,v],v] - \tfrac{1}{12}[\,[u,v],u] - \cdots\right),$$

where the omitted terms are known and involve Lie brackets of four or more elements. In case u and v commute, this formula reduces to the familiar exponential law $\exp(u)\,\exp(v) = \exp(u + v)$.

The exponential map relates Lie group homomorphisms. That is, if $\phi : G \to H$ is a Lie group homomorphism and $\phi_* : \mathfrak{g} \to \mathfrak{h}$ the induced map on the corresponding Lie algebras, then for all $x \in \mathfrak{g}$ we have

$\phi(\exp(x)) = \exp(\phi_*(x)).$

In other words the following diagram commutes,[Note 1]

(In short, exp is a natural transformation from the functor Lie to the identity functor on the category of Lie groups.)

The exponential map from the Lie algebra to the Lie group is not always onto, even if the group is connected (though it does map onto the Lie group for connected groups that are either compact or nilpotent). For example, the exponential map of $SL(2, \mathbf{R})$ is not surjective. Also, exponential map is not surjective nor injective for infinite-dimensional (see below) Lie groups modelled on C^∞ Fréchet space, even from arbitrary small neighborhood of 0 to corresponding neighborhood of 1.

See also: derivative of the exponential map and normal coordinates.

17.4.4   Lie subgroup

A **Lie subgroup** H of a Lie group G is a Lie group that is a subset of G and such that the inclusion map from H to G is an injective immersion and group homomorphism. According to Cartan's theorem, a closed subgroup of G admits a unique smooth structure which makes it an embedded Lie subgroup of G—i.e. a Lie subgroup such that the inclusion map is a smooth embedding.

Examples of non-closed subgroups are plentiful; for example take G to be a torus of dimension ≥ 2, and let H be a one-parameter subgroup of *irrational slope*, i.e. one that winds around in G. Then there is a Lie group homomorphism $\varphi : \mathbf{R} \to G$ with H as its image. The closure of H will be a sub-torus in G.

The exponential map gives a one-to-one correspondence between the connected Lie subgroups of a connected Lie group G and the subalgebras of the Lie algebra of G.[2] Typically, the subgroup corresponding to a subalgebra is not a closed subgroup. There is no criterion solely based on the structure of g which determines which subalgebras correspond to closed subgroups.

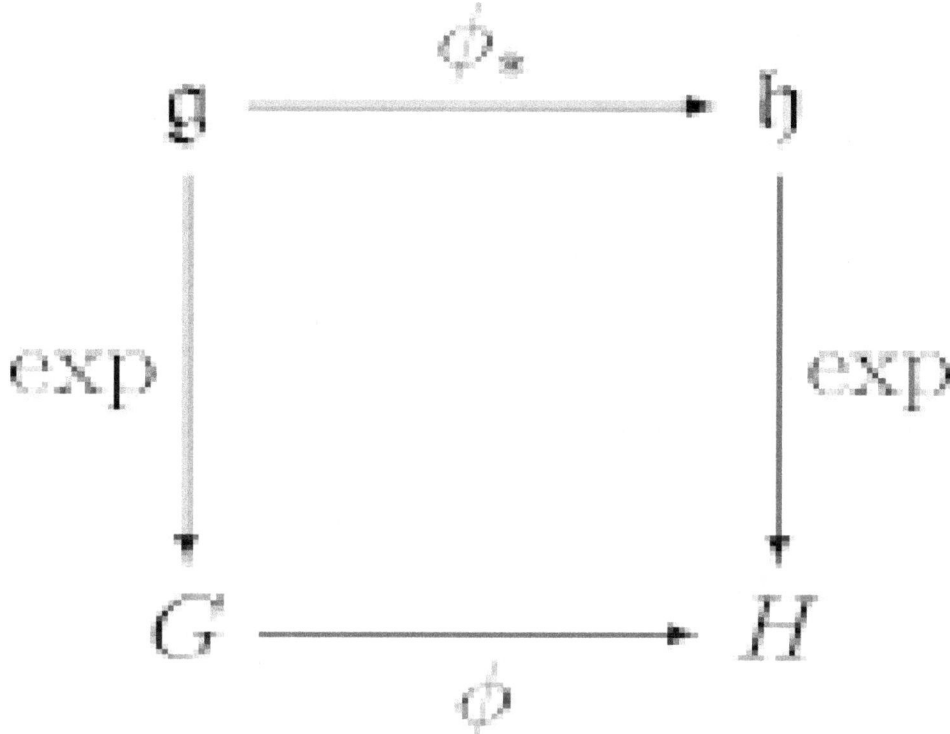

17.5   Early history

According to the most authoritative source on the early history of Lie groups (Hawkins, p. 1), Sophus Lie himself considered the winter of 1873–1874 as the birth date of his theory of continuous groups. Hawkins, however, suggests that it was "Lie's prodigious research activity during the four-year period from the fall of 1869 to the fall of 1873" that led to the theory's creation (*ibid*). Some of Lie's early ideas were developed in close collaboration with Felix Klein. Lie met with Klein every day from October 1869 through 1872: in Berlin from the end of October 1869 to the end of February 1870, and in Paris, Göttingen and Erlangen in the subsequent two years (*ibid*, p. 2). Lie stated that all of the principal results were obtained by 1884. But during the 1870s all his papers (except the very first note) were published in Norwegian journals, which impeded recognition of the work throughout the rest of Europe (*ibid*, p. 76). In 1884 a young German mathematician, Friedrich Engel, came to work with Lie on a systematic treatise to expose his theory of continuous groups. From this effort resulted the three-volume *Theorie der Transformationsgruppen*, published in 1888, 1890, and 1893.

Lie's ideas did not stand in isolation from the rest of mathematics. In fact, his interest in the geometry of differential equations was first motivated by the work of Carl Gustav Jacobi, on the theory of partial differential equations of first order and on the equations of classical mechanics. Much of Jacobi's work was published posthumously in the 1860s, generating enormous interest in France and Germany (Hawkins, p. 43). Lie's *idée fixe* was to develop a theory of symmetries of differential equations that would accomplish for them what Évariste Galois had done for algebraic equations: namely, to classify them in terms of group theory. Lie and other mathematicians showed that the most important equations for special functions and orthogonal polynomials tend to arise from group theoretical symmetries. In Lie's early work, the idea was to construct a theory of *continuous groups*, to complement the theory of discrete groups that had developed in the theory of modular forms, in the hands of Felix Klein and Henri Poincaré. The initial application that Lie had in mind was to the theory of differential equations. On the model of Galois theory and polynomial equations, the driving conception was of a theory capable of unifying, by the study of symmetry, the whole area of ordinary differential equations. However, the hope that Lie Theory would unify the entire field of ordinary differential equations was not fulfilled. Symmetry methods for ODEs continue to be studied, but do not dominate the subject. There is a differential Galois theory, but it was developed by others, such as Picard and Vessiot, and it provides a theory of quadratures, the indefinite integrals required to express solutions.

Additional impetus to consider continuous groups came from ideas of Bernhard Riemann, on the foundations of geometry, and their further development in the hands of Klein. Thus three major themes in 19th century mathematics were combined by Lie in creating his new theory: the idea of symmetry, as exemplified by Galois through the algebraic notion of a group; geometric theory and the explicit solutions of differential equations of mechanics, worked out by Poisson and Jacobi; and the new understanding of geometry that emerged in the works of Plücker, Möbius, Grassmann and others, and culminated in Riemann's revolutionary vision of the subject.

Although today Sophus Lie is rightfully recognized as the creator of the theory of continuous groups, a major stride in the development of their structure theory, which was to have a profound influence on subsequent development of mathematics, was made by Wilhelm Killing, who in 1888 published the first paper in a series entitled *Die Zusammensetzung der stetigen endlichen Transformationsgruppen* (*The composition of continuous finite transformation groups*) (Hawkins, p. 100). The work of Killing, later refined and generalized by Élie Cartan, led to classification of semisimple Lie algebras, Cartan's theory of symmetric spaces, and Hermann Weyl's description of representations of compact and semisimple Lie groups using highest weights.

In 1900 David Hilbert challenged Lie theorists with his Fifth Problem presented at the International Congress of Mathematicians in Paris.

Weyl brought the early period of the development of the theory of Lie groups to fruition, for not only did he classify irreducible representations of semisimple Lie groups and connect the theory of groups with quantum mechanics, but he also put Lie's theory itself on firmer footing by clearly enunciating the distinction between Lie's *infinitesimal groups* (i.e., Lie algebras) and the Lie groups proper, and began investigations of topology of Lie groups.[3] The theory of Lie groups was systematically reworked in modern mathematical language in a monograph by Claude Chevalley.

17.6   The concept of a Lie group, and possibilities of classification

Lie groups may be thought of as smoothly varying families of symmetries. Examples of symmetries include rotation about an axis. What must be understood is the nature of 'small' transformations, e.g., rotations through tiny angles, that link nearby transformations. The mathematical object capturing this structure is called a Lie algebra (Lie himself called them "infinitesimal groups"). It can be defined because Lie groups are manifolds, so have tangent spaces at each point.

The Lie algebra of any compact Lie group (very roughly: one for which the symmetries form a bounded set) can be decomposed as a direct sum of an abelian Lie algebra and some number of simple ones. The structure of an abelian Lie algebra is mathematically uninteresting (since the Lie bracket is identically zero); the interest is in the simple summands. Hence the question arises: what are the simple Lie algebras of compact groups? It turns out that they mostly fall into four infinite families, the "classical Lie algebras" An, Bn, Cn and Dn, which have simple descriptions in terms of symmetries of Euclidean space. But there are also just five "exceptional Lie algebras" that do not fall into any of these families. E$_8$ is the largest of these.

Lie groups are classified according to their algebraic properties (simple, semisimple, solvable, nilpotent, abelian), their connectedness (connected or simply connected) and their compactness.

- Compact Lie groups are all known: they are finite central quotients of a product of copies of the circle group $\mathbf{S}^1$ and simple compact Lie groups (which correspond to connected Dynkin diagrams).

- Any simply connected solvable Lie group is isomorphic to a closed subgroup of the group of invertible upper triangular matrices of some rank, and any finite-dimensional irreducible representation of such a group is 1-dimensional. Solvable groups are too messy to classify except in a few small dimensions.

- Any simply connected nilpotent Lie group is isomorphic to a closed subgroup of the group of invertible upper triangular matrices with 1's on the diagonal of some rank, and any finite-dimensional irreducible representation of such a group is 1-dimensional. Like solvable groups, nilpotent groups are too messy to classify except in a few small dimensions.

- Simple Lie groups are sometimes defined to be those that are simple as abstract groups, and sometimes defined to be connected Lie groups with a simple Lie algebra. For example, SL(2, **R**) is simple according to the second definition but not according to the first. They have all been classified (for either definition).

- Semisimple Lie groups are Lie groups whose Lie algebra is a product of simple Lie algebras.[4] They are central extensions of products of simple Lie groups.

The identity component of any Lie group is an open normal subgroup, and the quotient group is a discrete group. The universal cover of any connected Lie group is a simply connected Lie group, and conversely any connected Lie group is a quotient of a simply connected Lie group by a discrete normal subgroup of the center. Any Lie group G can be decomposed into discrete, simple, and abelian groups in a canonical way as follows. Write

G_{con} for the connected component of the identity

G_{sol} for the largest connected normal solvable subgroup

G_{nil} for the largest connected normal nilpotent subgroup

so that we have a sequence of normal subgroups

$$1 \subseteq G_{nil} \subseteq G_{sol} \subseteq G_{con} \subseteq G.$$

Then

G/G_{con} is discrete

G_{con}/G_{sol} is a central extension of a product of simple connected Lie groups.

G_{sol}/G_{nil} is abelian. A connected abelian Lie group is isomorphic to a product of copies of **R** and the circle group S^1.

$G_{nil}/1$ is nilpotent, and therefore its ascending central series has all quotients abelian.

This can be used to reduce some problems about Lie groups (such as finding their unitary representations) to the same problems for connected simple groups and nilpotent and solvable subgroups of smaller dimension.

- The diffeomorphism group of a Lie group acts transitively on the Lie group

- Every Lie group is parallelizable, and hence an orientable manifold (there is a bundle isomorphism between its tangent bundle and the product of itself with the tangent space at the identity)

17.7 Infinite-dimensional Lie groups

Lie groups are often defined to be finite-dimensional, but there are many groups that resemble Lie groups, except for being infinite-dimensional. The simplest way to define infinite-dimensional Lie groups is to model them on Banach spaces, and in this case much of the basic theory is similar to that of finite-dimensional Lie groups. However this is inadequate for many applications, because many natural examples of infinite-dimensional Lie groups are not Banach manifolds. Instead one needs to define Lie groups modeled on more general locally convex topological vector spaces. In this case the relation between the Lie algebra and the Lie group becomes rather subtle, and several results about finite-dimensional Lie groups no longer hold.

The literature is not entirely uniform in its terminology as to exactly which properties of infinite-dimensional groups qualify the group for the prefix *Lie* in *Lie group*. On the Lie algebra side of affairs, things are simpler since the qualifying criteria for the prefix *Lie* in *Lie algebra* are purely algebraic. For example, an infinite-dimensional Lie algebra may or may not have a corresponding Lie group. That is, there may be a group corresponding to the Lie algebra, but it might not be nice enough to be called a Lie group, or the connection between the group and the Lie algebra might not be nice enough (e.g failure of the exponential map to be onto a neighborhood of the identity). It is the "nice enough" that is not universally defined.

Some of the examples that have been studied include:

- The group of diffeomorphisms of a manifold. Quite a lot is known about the group of diffeomorphisms of the circle. Its Lie algebra is (more or less) the Witt algebra, which has a central extension called the Virasoro algebra, used in string theory and conformal field theory. Diffeomorphism groups of compact manifolds of larger dimension are regular Fréchet Lie groups; very little about their structure is known.

The diffeomorphism group of spacetime sometimes appears in attempts to quantize gravity.

- The group of smooth maps from a manifold to a finite-dimensional Lie group is an example of a gauge group (with operation of pointwise multiplication), and is used in quantum field theory and Donaldson theory. If the manifold is a circle these are called loop groups, and have central extensions whose Lie algebras are (more or less) Kac–Moody algebras.

- There are infinite-dimensional analogues of general linear groups, orthogonal groups, and so on. One important aspect is that these may have *simpler* topological properties: see for example Kuiper's theorem. In M-Theory theory, for example, a 10 dimensional SU(N) gauge theory becomes an 11 dimensional theory when N becomes infinite.

- A specific example is that $SU(\infty)$ is equal to the group of area preserving diffeomorphisms of a torus.

17.8   See also

- Lie subgroup

- E_8

- Adjoint representation of a Lie group

- Adjoint endomorphism

- Haar measure

- Homogeneous space

- List of Lie group topics

- List of simple Lie groups

- Moufang polygon

- Riemannian manifold

- Representations of Lie groups

- Table of Lie groups

- Lie algebra

- Symmetry in quantum mechanics

- Lie group action

17.9 Notes

17.9.1 Explanatory notes

[1] http://www.math.sunysb.edu/~{}vkiritch/MAT552/ProblemSet1.pdf

17.9.2 Citations

[1] Arthur Tresse (1893). "Sur les invariants différentiels des groupes continus de transformations". *Acta Mathematica* **18**: 1–88. doi:10.1007/bf02418270.

[2] Hall 2015 Theorem 5.20

[3] Borel (2001).

[4] Helgason, Sigurdur (1978). *Differential Geometry, Lie Groups, and Symmetric Spaces*. New York: Academic Press. p. 131. ISBN 0-12-338460-5.

17.10 References

- Adams, John Frank (1969), *Lectures on Lie Groups*, Chicago Lectures in Mathematics, Chicago: Univ. of Chicago Press, ISBN 0-226-00527-5, MR 0252560.

- Borel, Armand (2001), *Essays in the history of Lie groups and algebraic groups*, History of Mathematics **21**, Providence, R.I.: American Mathematical Society, ISBN 978-0-8218-0288-5, MR 1847105

- Bourbaki, Nicolas, *Elements of mathematics: Lie groups and Lie algebras*. Chapters 1–3 ISBN 3-540-64242-0, Chapters 4–6 ISBN 3-540-42650-7, Chapters 7–9 ISBN 3-540-43405-4

- Chevalley, Claude (1946), *Theory of Lie groups*, Princeton: Princeton University Press, ISBN 0-691-04990-4.

- P. M. Cohn (1957) *Lie Groups*, Cambridge Tracts in Mathematical Physics.

- J. L. Coolidge (1940) *A History of Geometrical Methods*, pp 304–17, Oxford University Press (Dover Publications 2003).

- Fulton, William; Harris, Joe (1991), *Representation theory. A first course*, Graduate Texts in Mathematics, Readings in Mathematics **129**, New York: Springer-Verlag, ISBN 978-0-387-97495-8, MR 1153249, ISBN 978-0-387-97527-6

- Robert Gilmore (2008) *Lie groups, physics, and geometry: an introduction for physicists, engineers and chemists*, Cambridge University Press ISBN 9780521884006 .

- Hall, Brian C. (2015), *Lie Groups, Lie Algebras, and Representations: An Elementary Introduction*, Graduate Texts in Mathematics **222** (2nd ed.), Springer, ISBN 0-387-40122-9.

- F. Reese Harvey (1990) *Spinors and calibrations*, Academic Press, ISBN 0-12-329650-1 .

- Hawkins, Thomas (2000), *Emergence of the theory of Lie groups*, Sources and Studies in the History of Mathematics and Physical Sciences, Berlin, New York: Springer-Verlag, ISBN 978-0-387-98963-1, MR 1771134 Borel's review

- Helgason, Sigurdur (2001), *Differential geometry, Lie groups, and symmetric spaces*, Graduate Studies in Mathematics **34**, Providence, R.I.: American Mathematical Society, ISBN 978-0-8218-2848-9, MR 1834454

- Knapp, Anthony W. (2002), *Lie Groups Beyond an Introduction*, Progress in Mathematics **140** (2nd ed.), Boston: Birkhäuser, ISBN 0-8176-4259-5.

- Nijenhuis, Albert (1959). "Review: *Lie groups*, by P. M. Cohn". *Bulletin of the American Mathematical Society* **65** (6): 338–341. doi:10.1090/s0002-9904-1959-10358-x.

- Rossmann, Wulf (2001), *Lie Groups: An Introduction Through Linear Groups*, Oxford Graduate Texts in Mathematics, Oxford University Press, ISBN 978-0-19-859683-7. The 2003 reprint corrects several typographical mistakes.

- Sattinger, David H.; Weaver, O. L. (1986). *Lie groups and algebras with applications to physics, geometry, and mechanics.* Springer-Verlag. ISBN 3-540-96240-9. MR 0835009.

- Serre, Jean-Pierre (1965), *Lie Algebras and Lie Groups: 1964 Lectures given at Harvard University*, Lecture notes in mathematics **1500**, Springer, ISBN 3-540-55008-9.

- Stillwell, John (2008). *Naive Lie Theory*. Springer. ISBN 0-387-98289-2.

- Heldermann Verlag Journal of Lie Theory

- Warner, Frank W. (1983), *Foundations of differentiable manifolds and Lie groups*, Graduate Texts in Mathematics **94**, New York Berlin Heidelberg: Springer-Verlag, ISBN 978-0-387-90894-6, MR 0722297

- Steeb, Willi-Hans (2007), *Continuous Symmetries, Lie algebras, Differential Equations and Computer Algebra: second edition*, World Scientific Publishing, ISBN 981-270-809-X, MR 2382250.

- Lie Groups. Representation Theory and Symmetric Spaces Wolfgang Ziller, Vorlesung 2010

Chapter 18

Symmetry group

Not to be confused with Symmetric group.

This article is about the abstract algebraic structures. For other meanings, see Symmetry group (disambiguation).

In abstract algebra, the **symmetry group** of an object (image, signal, etc.) is the group of all transformations under which the object is invariant with composition as the group operation. For a space with a metric, it is a subgroup of the isometry group of the space concerned. If not stated otherwise, this article considers symmetry groups in Euclidean geometry, but the concept may also be studied in more general contexts as expanded below.

18.1   Introduction

The "objects" may be geometric figures, images, and patterns, such as a wallpaper pattern. The definition can be made more precise by specifying what is meant by image or pattern, e.g., a function of position with values in a set of colors. For symmetry of physical objects, one may also want to take their physical composition into account. The group of isometries of space induces a group action on objects in it.

The symmetry group is sometimes also called **full symmetry group** in order to emphasize that it includes the orientation-reversing isometries (like reflections, glide reflections and improper rotations) under which the figure is invariant. The subgroup of orientation-preserving isometries (i.e. translations, rotations, and compositions of these) that leave the figure invariant is called its **proper symmetry group**. The proper symmetry group of an object is equal to its full symmetry group if and only if the object is chiral (and thus there are no orientation-reversing isometries under which it is invariant).

Any symmetry group whose elements have a common fixed point, which is true for all finite symmetry groups and also for the symmetry groups of bounded figures, can be represented as a subgroup of the orthogonal group $O(n)$ by choosing the origin to be a fixed point. The proper symmetry group is then a subgroup of the special orthogonal group $SO(n)$, and is therefore also called **rotation group** of the figure.

A **discrete symmetry group** is a symmetry group such that for every point of the space the set of images of the point under the isometries in the symmetry group is a discrete set.

Discrete symmetry groups come in three types: (1) finite **point groups**, which include only rotations, reflections, inversion and rotoinversion – they are just the finite subgroups of $O(n)$, (2) infinite **lattice groups**, which include only translations, and (3) infinite **space groups** which combines elements of both previous types, and may also include extra transformations like screw axis and glide reflection. There are also *continuous* symmetry groups, which contain rotations of arbitrarily small angles or translations of arbitrarily small distances. The group of all symmetries of a sphere $O(3)$ is an example of this, and in general such continuous symmetry groups are studied as Lie groups. With a categorization of subgroups of the Euclidean group corresponds a categorization of symmetry groups.

Two geometric figures are considered to be of the same symmetry type if their symmetry groups are conjugate subgroups of the Euclidean group $E(n)$ (the isometry group of $\mathbf{R}^n$), where two subgroups H_1, H_2 of a group G are *conjugate*, if there exists $g \in G$ such that $H_1 = g^{-1}H_2g$. For example:

- two 3D figures have mirror symmetry, but with respect to different mirror planes.

- two 3D figures have 3-fold rotational symmetry, but with respect to different axes.

- two 2D patterns have translational symmetry, each in one direction; the two translation vectors have the same length but a different direction.

When considering isometry groups, one may restrict oneself to those where for all points the set of images under the isometries is topologically closed. This includes all discrete isometry groups and also those involved in continuous symmetries, but excludes for example in 1D the group of translations by a rational number. A "figure" with this symmetry group is non-drawable and up to arbitrarily fine detail homogeneous, without being really homogeneous.

18.2   One dimension

The isometry groups in one dimension where for all points the set of images under the isometries is topologically closed are:

- the trivial group C_1

- the groups of two elements generated by a reflection in a point; they are isomorphic with C_2

- the infinite discrete groups generated by a translation; they are isomorphic with Z, the additive group of the integers

- the infinite discrete groups generated by a translation and a reflection in a point; they are isomorphic with the generalized dihedral group of Z, Dih(Z), also denoted by D∞ (which is a semidirect product of Z and C_2).

- the group generated by all translations (isomorphic with the additive group of the real numbers **R**); this group cannot be the symmetry group of a "pattern": it would be homogeneous, hence could also be reflected. However, a uniform one-dimensional vector field has this symmetry group.

- the group generated by all translations and reflections in points; they are isomorphic with the generalized dihedral group of **R**, Dih(**R**).

See also symmetry groups in one dimension.

18.3   Two dimensions

Up to conjugacy the discrete point groups in two-dimensional space are the following classes:

- cyclic groups C_1, C_2, C_3, C_4, ... where Cn consists of all rotations about a fixed point by multiples of the angle $360°/n$

- dihedral groups D_1, D_2, D_3, D_4, ..., where Dn (of order $2n$) consists of the rotations in Cn together with reflections in n axes that pass through the fixed point.

C_1 is the trivial group containing only the identity operation, which occurs when the figure has no symmetry at all, for example the letter **F**. C_2 is the symmetry group of the letter **Z**, C_3 that of a triskelion, C_4 of a swastika, and C_5, C_6, etc. are the symmetry groups of similar swastika-like figures with five, six, etc. arms instead of four.

D_1 is the 2-element group containing the identity operation and a single reflection, which occurs when the figure has only a single axis of bilateral symmetry, for example the letter **A**.

D_2, which is isomorphic to the Klein four-group, is the symmetry group of a non-equilateral rectangle. This figure has four symmetry operations: the identity operation, one twofold axis of rotation, and two nonequivalent mirror planes.

D_3, D_4 etc. are the symmetry groups of the regular polygons.

The actual symmetry groups in each of these cases have two degrees of freedom for the center of rotation, and in the case of the dihedral groups, one more for the positions of the mirrors.

The remaining isometry groups in two dimensions with a fixed point, where for all points the set of images under the isometries is topologically closed are:

- the special orthogonal group SO(2) consisting of all rotations about a fixed point; it is also called the circle group S^1, the multiplicative group of complex numbers of absolute value 1. It is the *proper* symmetry group of a circle and the continuous equivalent of Cn. There is no geometric figure that has as *full* symmetry group the circle group, but for a vector field it may apply (see the three-dimensional case below).

- the orthogonal group O(2) consisting of all rotations about a fixed point and reflections in any axis through that fixed point. This is the symmetry group of a circle. It is also called Dih(S^1) as it is the generalized dihedral group of S^1.

For non-bounded figures, the additional isometry groups can include translations; the closed ones are:

- the 7 frieze groups

- the 17 wallpaper groups

- for each of the symmetry groups in one dimension, the combination of all symmetries in that group in one direction, and the group of all translations in the perpendicular direction

- ditto with also reflections in a line in the first direction

18.4   Three dimensions

See also: Point groups in three dimensions

Up to conjugacy the set of three-dimensional point groups consists of 7 infinite series, and 7 separate ones. In crystallography they are restricted to be compatible with the discrete translation symmetries of a crystal lattice. This crystallographic restriction of the infinite families of general point groups results in 32 crystallographic point groups (27 from the 7 infinite series, and 5 of the 7 others).

The continuous symmetry groups with a fixed point include those of:

- cylindrical symmetry without a symmetry plane perpendicular to the axis, this applies for example often for a bottle

- cylindrical symmetry with a symmetry plane perpendicular to the axis

- spherical symmetry

For objects and scalar fields the cylindrical symmetry implies vertical planes of reflection. However, for vector fields it does not: in cylindrical coordinates with respect to some axis, $\mathbf{A} = A_\rho \hat{\boldsymbol{\rho}} + A_\phi \hat{\boldsymbol{\phi}} + A_z \hat{\boldsymbol{z}}$ has cylindrical symmetry with respect to the axis if and only if A_ρ, A_ϕ, and A_z have this symmetry, i.e., they do not depend on φ. Additionally there is reflectional symmetry if and only if $A_\phi = 0$.

For spherical symmetry there is no such distinction, it implies planes of reflection.

The continuous symmetry groups without a fixed point include those with a screw axis, such as an infinite helix. See also subgroups of the Euclidean group.

18.5   Symmetry groups in general

See also: Automorphism

In wider contexts, a **symmetry group** may be any kind of **transformation group**, or automorphism group. Once we know what kind of mathematical structure we are concerned with, we should be able to pinpoint what mappings preserve the structure. Conversely, specifying the symmetry can define the structure, or at least clarify what we mean by an invariant, geometric language in which to discuss it; this is one way of looking at the Erlangen programme.

For example, automorphism groups of certain models of finite geometries are not "symmetry groups" in the usual sense, although they preserve symmetry. They do this by preserving *families* of point-sets rather than point-sets (or "objects") themselves.

Like above, the group of automorphisms of space induces a group action on objects in it.

For a given geometric figure in a given geometric space, consider the following equivalence relation: two automorphisms of space are equivalent if and only if the two images of the figure are the same (here "the same" does not mean something like e.g. "the same up to translation and rotation", but it means "exactly the same"). Then the equivalence class of the identity is the symmetry group of the figure, and every equivalence class corresponds to one isomorphic version of the figure.

There is a bijection between every pair of equivalence classes: the inverse of a representative of the first equivalence class, composed with a representative of the second.

In the case of a finite automorphism group of the whole space, its order is the order of the symmetry group of the figure multiplied by the number of isomorphic versions of the figure.

Examples:

- Isometries of the Euclidean plane, the figure is a rectangle: there are infinitely many equivalence classes; each contains 4 isometries.

- The space is a cube with Euclidean metric; the figures include cubes of the same size as the space, with colors or patterns on the faces; the automorphisms of the space are the 48 isometries; the figure is a cube of which one face has a different color; the figure has a symmetry group of 8 isometries, there are 6 equivalence classes of 8 isometries, for 6 isomorphic versions of the figure.

Compare Lagrange's theorem (group theory) and its proof.

18.6   See also

- Crystallography

- Crystallographic point group

- Crystal system

- Euclidean plane isometry

- Fixed points of isometry groups in Euclidean space

- Group action

- Molecular symmetry

- Permutation group

- Point group

- Space group

- Symmetric group

- Symmetry

- Symmetry in quantum mechanics

18.7   Further reading

- Burns, G.; Glazer, A. M. (1990). *Space Groups for Scientists and Engineers* (2nd ed.). Boston: Academic Press, Inc. ISBN 0-12-145761-3.

- Clegg, W (1998). *Crystal Structure Determination (Oxford Chemistry Primer)*. Oxford: Oxford University Press. ISBN 0-19-855901-1.

- O'Keeffe, M.; Hyde, B. G. (1996). *Crystal Structures; I. Patterns and Symmetry*. Washington, DC: Mineralogical Society of America, *Monograph Series*. ISBN 0-939950-40-5.

- Miller, Willard Jr. (1972). *Symmetry Groups and Their Applications*. New York: Academic Press. OCLC 589081. Retrieved 2009-09-28.

18.8   External links

- Weisstein, Eric W., "Symmetry Group", *MathWorld*.

- Weisstein, Eric W., "Tetrahedral Group", *MathWorld*.

- Overview of the 32 crystallographic point groups - form the first parts (apart from skipping $n=5$) of the 7 infinite series and 5 of the 7 separate 3D point groups

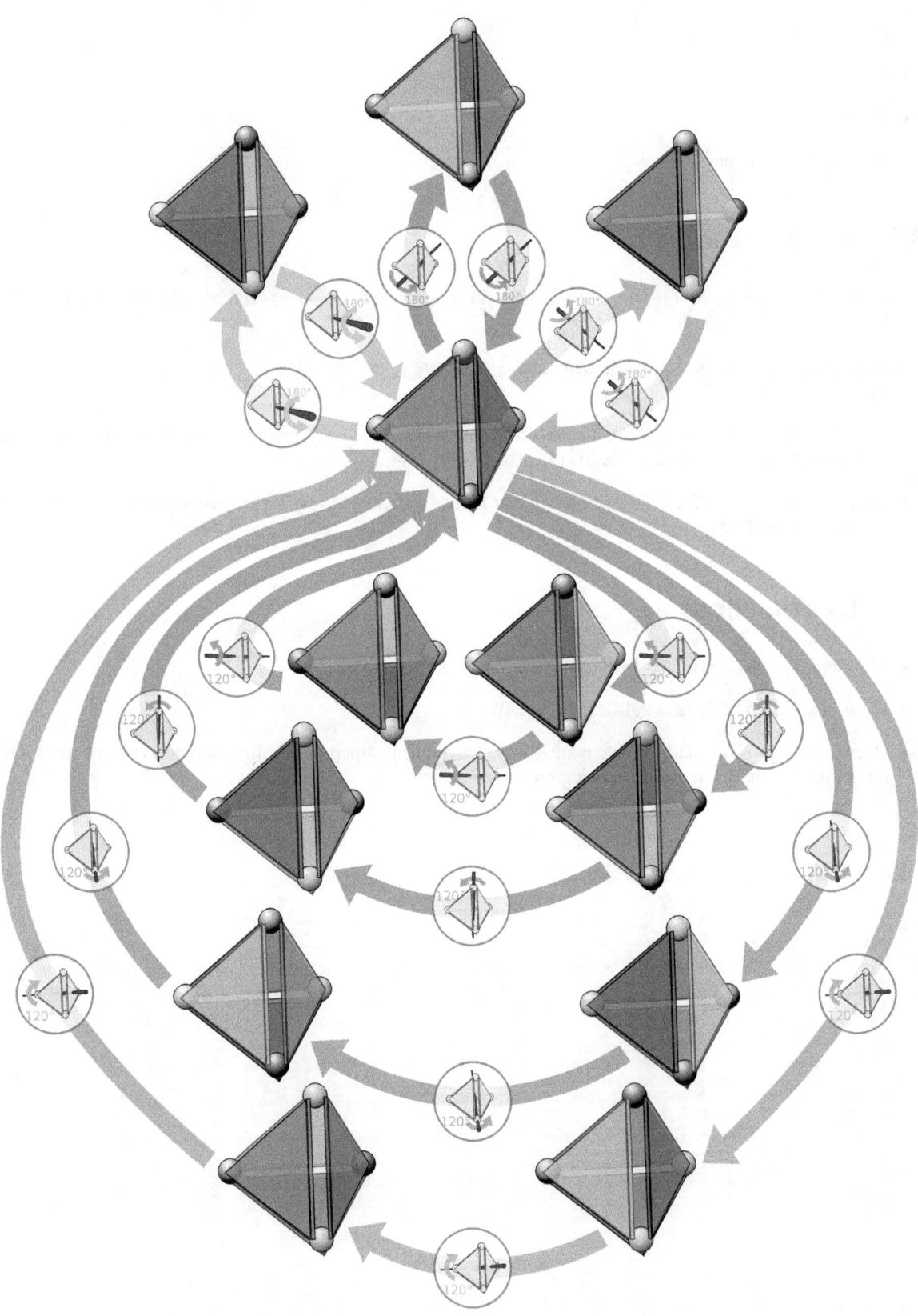

*A tetrahedron is invariant under 12 distinct rotations, reflections excluded. These are illustrated here in the cycle graph format, along with the 180° edge (blue arrows) and 120° vertex (reddish arrows) rotations that permute the tetrahedron through the positions. The 12 rotations form the **rotation (symmetry) group** of the figure.*

Chapter 19

Representation of a Lie group

In mathematics and theoretical physics, the idea of a **representation of a Lie group** plays an important role in the study of continuous symmetry. A great deal is known about such representations, a basic tool in their study being the use of the corresponding 'infinitesimal' representations of Lie algebras. The physics literature sometimes passes over the distinction between Lie groups and Lie algebras.

19.1   Representations on a complex finite-dimensional vector space

Let us first discuss representations acting on finite-dimensional complex vector spaces. A representation of a Lie group G on a finite-dimensional complex vector space V is a smooth group homomorphism $\Psi:G\rightarrow\mathrm{Aut}(V)$ from G to the automorphism group of V.

For n-dimensional V, the automorphism group of V is identified with a subset of the complex square matrices of order n. The automorphism group of V is given the structure of a smooth manifold using this identification. The condition that Ψ is smooth, in the definition above, means that Ψ is a smooth map from the smooth manifold G to the smooth manifold $\mathrm{Aut}(V)$.

If a basis for the complex vector space V is chosen, the representation can be expressed as a homomorphism into general linear group $\mathrm{GL}(n,\mathbf{C})$. This is known as a *matrix representation*.

19.2   Representations on a finite-dimensional vector space over an arbitrary field

A representation of a Lie group G on a vector space V (over a field K) is a smooth (i.e. respecting the differential structure) group homomorphism $G\rightarrow\mathrm{Aut}(V)$ from G to the automorphism group of V. If a basis for the vector space V is chosen, the representation can be expressed as a homomorphism into general linear group $\mathrm{GL}(n,K)$. This is known as a *matrix representation*. Two representations of G on vector spaces V, W are *equivalent* if they have the same matrix representations with respect to some choices of bases for V and W.

On the Lie algebra level, there is a corresponding linear mapping from the Lie algebra of G to $\mathrm{End}(V)$ preserving the Lie bracket [,]. See representation of Lie algebras for the Lie algebra theory.

If the homomorphism is in fact a monomorphism, the representation is said to be *faithful*.

A unitary representation is defined in the same way, except that G maps to unitary matrices; the Lie algebra will then map to skew-hermitian matrices.

If G is a compact Lie group, every finite-dimensional representation is equivalent to a unitary one.[1]

19.3   Representations on Hilbert spaces

A representation of a Lie group G on a complex Hilbert space V is a group homomorphism $\Psi:G \to B(V)$ from G to $B(V)$, the group of bounded linear operators of V which have a bounded inverse, such that the map $G{\times}V \to V$ given by $(g,v) \to \Psi(g)v$ is continuous.

This definition can handle representations on **infinite-dimensional** Hilbert spaces. Such representations can be found in e.g. quantum mechanics, but also in Fourier analysis as shown in the following example.

Let $G{=}\mathbf{R}$, and let the complex Hilbert space V be $L^2(\mathbf{R})$. We define the representation $\Psi:\mathbf{R} \to B(L^2(\mathbf{R}))$ by $\Psi(r)\{f(x)\} \to f(r^{-1}x)$.

See also Wigner's classification for representations of the Poincaré group.

19.4   Classification

If G is a semisimple group, its finite-dimensional representations can be decomposed as direct sums of irreducible representations.[2] The irreducibles are indexed by highest weight; the allowable (*dominant*) highest weights satisfy a suitable positivity condition.[3] In particular, there exists a set of *fundamental weights*, indexed by the vertices of the Dynkin diagram of G, such that dominant weights are simply non-negative integer linear combinations of the fundamental weights. The characters of the irreducible representations are given by the Weyl character formula.

If G is a commutative Lie group, then its irreducible representations are simply the continuous characters of G: see Pontryagin duality for this case.

A quotient representation is a quotient module of the group ring.

19.5   Formulaic examples

Let $\mathbf{F}q$ be a finite field of order q and characteristic p. Let G be a finite group of Lie type, that is, G is the $\mathbf{F}q$-rational points of a connected reductive group G defined over $\mathbf{F}q$. For example, if n is a positive integer $GL(n, \mathbf{F}q)$ and $SL(n, \mathbf{F}q)$ are finite groups of Lie type. Let $J = \left[\begin{smallmatrix} 0 & I_n \\ -I_n & 0 \end{smallmatrix} \right]$, where I_n is the $n{\times}n$ identity matrix. Let

$$Sp_2(\mathbb{F}_q) = \left\{ g \in GL_{2n}(\mathbb{F}_q) \mid {}^t gJg = J \right\}.$$

Then $Sp(2,\mathbf{F}q)$ is a symplectic group of rank n and is a finite group of Lie type. For $G = GL(n, \mathbf{F}q)$ or $SL(n, \mathbf{F}q)$ (and some other examples), the *standard Borel subgroup* B of G is the subgroup of G consisting of the upper triangular elements in G. A *standard parabolic subgroup* of G is a subgroup of G which contains the standard Borel subgroup B. If P is a standard parabolic subgroup of $GL(n, \mathbf{F}q)$, then there exists a partition $(n_1, \ldots, n_r)$ of n (a set of positive integers n_j such that $n_1 + \ldots + n_r = n$) such that $P = P_{(n_1,\ldots,n_r)} = M \times N$, where $M \simeq GL_{n_1}(\mathbb{F}_q) \times \ldots \times GL_{n_r}(\mathbb{F}_q)$ has the form

$$M = \left\{ \begin{pmatrix} A_1 & 0 & \cdots & 0 \\ 0 & A_2 & \cdots & 0 \\ \vdots & \ddots & \ddots & \vdots \\ 0 & \cdots & 0 & A_r \end{pmatrix} \middle| A_j \in GL_{n_j}(\mathbb{F}_q), 1 \leq j \leq r \right\},$$

and

$$N = \left\{ \begin{pmatrix} I_{n_1} & * & \cdots & * \\ 0 & I_{n_2} & \cdots & * \\ \vdots & \ddots & \ddots & \vdots \\ 0 & \cdots & 0 & I_{n_r} \end{pmatrix} \right\},$$

where $*$ denotes arbitrary entries in $\mathbb{F}_q$.

19.6   See also

- Representation theory of the Lorentz group

- Representation theory of Hopf algebras

- Adjoint representation of a Lie group

- List of Lie group topics

- Symmetry in quantum mechanics

19.7   References

[1] Hall 2015 Theorem 4.28

[2] Hall 2015 Theorem 10.9

[3] Hall 2015 Theorems 9.4 and 9.5

- Fulton, William; Harris, Joe (1991), *Representation theory. A first course*, Graduate Texts in Mathematics, Readings in Mathematics **129**, New York: Springer-Verlag, ISBN 978-0-387-97495-8, MR 1153249, ISBN 978-0-387-97527-6

- Hall, Brian C. (2015), *Lie Groups, Lie Algebras, and Representations: An Elementary Introduction*, Graduate Texts in Mathematics **222** (2nd ed.), Springer, ISBN 0-387-40122-9.

- Knapp, Anthony W. (2002), *Lie Groups Beyond an Introduction*, Progress in Mathematics **140** (2nd ed.), Boston: Birkhäuser.

- Rossmann, Wulf (2001), *Lie Groups: An Introduction Through Linear Groups*, Oxford Graduate Texts in Mathematics, Oxford University Press, ISBN 978-0-19-859683-7. The 2003 reprint corrects several typographical mistakes.

Chapter 20

Poincaré group

For the Poincaré group (fundamental group) of a topological space, see Fundamental group.

The **Poincaré group**, named after Henri Poincaré (1906),[1] was first defined by Minkowski (1908) being the group of Minkowski spacetime isometries.[2][3] It is a ten-generator non-abelian Lie group of fundamental importance in physics.

20.1 Overview

A Minkowski spacetime isometry has the property that the interval between events is left invariant. For example, if everything was postponed by two hours, including the two events and the path you took to go from one to the other, then the time interval between the events recorded by a stop-watch you carried with you would be the same. Or if everything was shifted five miles to the west, or turned 60 degrees to the right, you would also see no change in the interval. It turns out that the proper length of an object is also unaffected by such a shift. A time or space reversal (a reflection) is also an isometry of this group.

In Minkowski space (i.e. ignoring the effects of gravity), there are ten degrees of freedom of the isometries, which may be thought of as translation through time or space (four degrees, one per dimension); reflection through a plane (three degrees, the freedom in orientation of this plane); or a "boost" in any of the three spatial directions (three degrees). Composition of transformations is the operator of the Poincaré group, with proper rotations being produced as the composition of an even number of reflections.

In classical physics, the Galilean group is a comparable ten-parameter group that acts on absolute time and space. Instead of boosts, it features shear mappings to relate co-moving frames of reference.

20.2 Details

The Poincaré group is the group of Minkowski spacetime isometries. It is a ten-dimensional noncompact Lie group. The abelian group of translations is a normal subgroup, while the Lorentz group is also a subgroup, the stabilizer of the origin. The Poincaré group itself is the minimal subgroup of the affine group which includes all translations and Lorentz transformations. More precisely, it is a semidirect product of the translations and the Lorentz group,

$$\mathbf{R}^{1,3} \rtimes \mathrm{SO}(1,3).$$

Another way of putting this is that the Poincaré group is a group extension of the Lorentz group by a vector representation of it; it is sometimes dubbed, informally, as the "*inhomogeneous Lorentz group*". In turn, it can also be obtained as a group contraction of the de Sitter group SO(4,1) ~ Sp(2,2), as the de Sitter radius goes to infinity.

Its positive energy unitary irreducible representations are indexed by mass (nonnegative number) and spin (integer or half integer) and are associated with particles in quantum mechanics (see Wigner's classification).

In accordance with the Erlangen program, the geometry of Minkowski space is defined by the Poincaré group: Minkowski space is considered as a homogeneous space for the group.

The **Poincaré algebra** is the Lie algebra of the Poincaré group. It is a Lie algebra extension of the Lie algebra of the Lorentz group. More specifically, the proper (detΛ=1), orthochronous ($\Lambda^0{}_0 \geq 1$) part of the Lorentz subgroup (its identity component), SO$^+$(1, 3), is connected to the identity and is thus provided by the exponentiation $\exp(ia_\mu P^\mu) \exp(i\omega_{\mu\nu} M^{\mu\nu}/2)$ of this Lie algebra. In component form, the Poincaré algebra is given by the commutation relations:[4][5]

where P is the generator of translations, M is the generator of Lorentz transformations, and η is the (+,−,−,−) Minkowski metric (see Sign convention).

The bottom commutation relation is the ("homogeneous") Lorentz group, consisting of rotations, $Ji = -\epsilon_{imn} M^{mn}/2$, and boosts, $Ki = Mi0$. In this notation, the entire Poincaré algebra is expressible in noncovariant (but more practical) language as

$$[J_m, P_n] = i\epsilon_{mnk} P_k ,$$

$$[J_i, P_0] = 0 ,$$

$$[K_i, P_k] = i\eta_{ik} P_0 ,$$

$$[K_i, P_0] = -iP_i ,$$

$$[J_m, J_n] = i\epsilon_{mnk} J_k ,$$

$$[J_m, K_n] = i\epsilon_{mnk} K_k ,$$

$$[K_m, K_n] = -i\epsilon_{mnk} J_k ,$$

where the bottom line commutator of two boosts is often referred to as a "Wigner rotation". Note the important simplification $[Jm+i\ Km,\ Jn-i\ Kn] = 0$, which permits reduction of the Lorentz subalgebra to **su(2)**⊕**su(2)** and efficient treatment of its associated representations.

The Casimir invariants of this algebra are $P_\mu P^\mu$ and $W_\mu W^\mu$ where W_μ is the Pauli–Lubanski pseudovector; they serve as labels for the representations of the group.

The Poincaré group is the full symmetry group of any relativistic field theory. As a result, all elementary particles fall in representations of this group. These are usually specified by the *four-momentum* squared of each particle (i.e. its mass squared) and the intrinsic quantum numbers J^{PC}, where J is the spin quantum number, P is the parity and C is the charge-conjugation quantum number. In practice, charge conjugation and parity are violated by many quantum field theories; where this occurs, P and C are forfeited. Since CPT symmetry is invariant in quantum field theory, a time-reversal quantum number may be constructed from those given.

As a topological space, the group has four connected components: the component of the identity; the time reversed component; the spatial inversion component; and the component which is both time-reversed and spatially inverted.

20.3   Poincaré symmetry

Poincaré symmetry is the full symmetry of special relativity. It includes:

- *translations* (displacements) in time and space (**P**), forming the abelian Lie group of translations on space-time;

- *rotations* in space, forming the non-Abelian Lie group of three-dimensional rotations (**J**);

- *boosts*, transformations connecting two uniformly moving bodies (*K*).

The last two symmetries, *J* and *K*, together make the Lorentz group (see also Lorentz invariance); the semi-direct product of the translations group and the Lorentz group then produce the Poincaré group. Objects which are invariant under this group are then said to possess **Poincaré invariance** or **relativistic invariance**.

20.4   See also

- Euclidean group

- Representation theory of the Poincaré group

- Wigner's classification

- Symmetry in quantum mechanics

- Center of mass (relativistic)

- Pauli–Lubanski pseudovector

- Particle physics and representation theory

20.5   Notes

[1] Poincaré, Henri, "Sur la dynamique de l'électron", *Rendiconti del Circolo matematico di Palermo* **21**: 129–176, doi:10.1007/bf0 3013466(Wikisource translation: On the Dynamic s of the Electron). The group de fined in this paper would no w be described as thehomogeneous Lorentz group with scalar multipliers.

[2] Minkowski, Hermann, "Die Grundgleichungen für die elektromagnetischen Vorgänge in bewegten Körpern", *Nachrichten von der Gesellschaft der Wissenschaften zu Göttingen, Mathematisch-Physikalische Klasse*: 53–111 (Wikisource translation: The Fundamental Equations for Electromagnetic Processes in Moving Bodies).

[3] Minkowski, Hermann, "Raum und Zeit", *Physikalische Zeitschrift* **10**: 75–88

[4] N.N. Bogolubov (1989). *General Principles of Quantum Field Theory* (2nd ed.). Springer. p. 272. ISBN 0-7923-0540-X.

[5] T. Ohlsson (2011). *Relativistic Quantum Physics: From Advanced Quantum Mechanics to Introductory Quantum Field Theory*. Cambridge University Press. p. 10. ISBN 1-13950-4320.

20.6   References

- Wu-Ki Tung (1985). *Group Theory in Physics*. World Scientific Publishing. ISBN 9971-966-57-3.

- Weinberg, Steven (1995). *The Quantum Theory of Fields* **1**. Cambridge: Cambridge University press. ISBN 978-0-521-55001-7.

- L.H. Ryder (1996). *Quantum Field Theory* (2nd ed.). Cambridge University Press. p. 62. ISBN 0-52147-8146.

Chapter 21

Special relativity

For history and motivation, see History of special relativity.

In physics, **special relativity** (**SR**, also known as the **special theory of relativity** or **STR**) is the generally accepted and experimentally well confirmed physical theory regarding the relationship between space and time. In Einstein's original pedagogical treatment, it is based on two postulates: (1) that the laws of physics are invariant (i.e. identical) in all inertial systems (non-accelerating frames of reference); and (2) that the speed of light in a vacuum is the same for all observers, regardless of the motion of the light source. It was originally proposed in 1905 by Albert Einstein in the paper "On the Electrodynamics of Moving Bodies".[1] The inconsistency of Newtonian mechanics with Maxwell's equations of electromagnetism and the inability to discover Earth's motion through a luminiferous aether led to the development of special relativity, which corrects mechanics to handle situations involving motions nearing the speed of light. As of today, special relativity is the most accurate model of motion at any speed. Even so, the Newtonian mechanics model is still useful (due to its simplicity and high accuracy) as an approximation at small velocities relative to the speed of light.

Special relativity implies a wide range of consequences, which have been experimentally verified,[2] including length contraction, time dilation, relativistic mass, mass–energy equivalence, a universal speed limit, and relativity of simultaneity. It has replaced the conventional notion of an absolute universal time with the notion of a time that is dependent on reference frame and spatial position. Rather than an invariant time interval between two events, there is an invariant spacetime interval. Combined with other laws of physics, the two postulates of special relativity predict the equivalence of mass and energy, as expressed in the mass–energy equivalence formula $E = mc^2$, where c is the speed of light in vacuum.[3][4]

A defining feature of special relativity is the replacement of the Galilean transformations of Newtonian mechanics with the Lorentz transformations. Time and space cannot be defined separately from each other. Rather space and time are interwoven into a single continuum known as spacetime. Events that occur at the same time for one observer could occur at different times for another.

The theory is "special" in that it only applies in the special case where the curvature of spacetime due to gravity is negligible.[5][6] In order to include gravity, Einstein formulated general relativity in 1915. Special relativity, contrary to some outdated descriptions, is capable of handling accelerated frames of reference.[7]

As Galilean relativity is now considered an approximation of special relativity that is valid for low speeds, special relativity is considered an approximation of general relativity that is valid for weak gravitational fields, i.e. at a sufficiently small scale and in conditions of free fall. Whereas general relativity incorporates noneuclidean geometry in order to represent gravitational effects as the geometric curvature of spacetime, special relativity is restricted to the flat spacetime known as Minkowski space. A locally Lorentz-invariant frame that abides by special relativity can be defined at sufficiently small scales, even in curved spacetime.

Galileo Galilei had already postulated that there is no absolute and well-defined state of rest (no privileged reference frames), a principle now called Galileo's principle of relativity. Einstein extended this principle so that it accounted for the constant speed of light,[8] a phenomenon that had been recently observed in the Michelson–Morley experiment. He also postulated that it holds for all the laws of physics, including both the laws of mechanics and of electrodynamics.[9]

Albert Einstein around 1905, the year his "Annus Mirabilis papers" – which included Zur Elektrodynamik bewegter Körper, *the paper founding special relativity – were published.*

21.1 Postulates

Einstein discerned two fundamental propositions that seemed to be the most assured, regardless of the exact validity of the (then) known laws of either mechanics or electrodynamics. These propositions were the constancy of the speed of light and the independence of physical laws (especially the constancy of the speed of light) from the choice of inertial system. In his initial presentation of special relativity in 1905 he expressed these postulates as:[1]

- The Principle of Relativity – The laws by which the states of physical systems undergo change are not affected, whether these changes of state be referred to the one or the other of two systems in uniform translatory motion relative to each other.[1]

- The Principle of Invariant Light Speed – "... light is always propagated in empty space with a definite velocity [speed] c which is independent of the state of motion of the emitting body" (from the preface).[1] That is, light in vacuum propagates with the speed c (a fixed constant, independent of direction) in at least one system of inertial coordinates (the "stationary system"), regardless of the state of motion of the light source.

The derivation of special relativity depends not only on these two explicit postulates, but also on several tacit assumptions (made in almost all theories of physics), including the isotropy and homogeneity of space and the independence of measuring rods and clocks from their past history.[11]

Following Einstein's original presentation of special relativity in 1905, many different sets of postulates have been proposed in various alternative derivations.[12] However, the most common set of postulates remains those employed by Einstein in his original paper. A more mathematical statement of the Principle of Relativity made later by Einstein, which introduces the concept of simplicity not mentioned above is:

> *Special principle of relativity*: If a system of coordinates K is chosen so that, in relation to it, physical laws hold good in their simplest form, the *same* laws hold good in relation to any other system of coordinates K' moving in uniform translation relatively to K.[13]

Henri Poincaré provided the mathematical framework for relativity theory by proving that Lorentz transformations are a subset of his Poincaré group of symmetry transformations. Einstein later derived these transformations from his axioms.

Many of Einstein's papers present derivations of the Lorentz transformation based upon these two principles.[14]

Einstein consistently based the derivation of Lorentz invariance (the essential core of special relativity) on just the two basic principles of relativity and light-speed invariance. He wrote:

> The insight fundamental for the special theory of relativity is this: The assumptions relativity and light speed invariance are compatible if relations of a new type ("Lorentz transformation") are postulated for the conversion of coordinates and times of events... The universal principle of the special theory of relativity is contained in the postulate: The laws of physics are invariant with respect to Lorentz transformations (for the transition from one inertial system to any other arbitrarily chosen inertial system). This is a restricting principle for natural laws...[10]

Thus many modern treatments of special relativity base it on the single postulate of universal Lorentz covariance, or, equivalently, on the single postulate of Minkowski spacetime.[15][16]

From the principle of relativity alone without assuming the constancy of the speed of light (i.e. using the isotropy of space and the symmetry implied by the principle of special relativity) one can show that the spacetime transformations between inertial frames are either Euclidean, Galilean, or Lorentzian. In the Lorentzian case, one can then obtain relativistic interval conservation and a certain finite limiting speed. Experiments suggest that this speed is the speed of light in vacuum.[17][18]

The constancy of the speed of light was motivated by Maxwell's theory of electromagnetism and the lack of evidence for the luminiferous ether. There is conflicting evidence on the extent to which Einstein was influenced by the null result of the Michelson–Morley experiment.[19][20] In any case, the null result of the Michelson–Morley experiment helped the notion of the constancy of the speed of light gain widespread and rapid acceptance.

21.2   Lack of an absolute reference frame

The principle of relativity, which states that there is no preferred inertial reference frame, dates back to Galileo, and was incorporated into Newtonian physics. However, in the late 19th century, the existence of electromagnetic waves led physicists to suggest that the universe was filled with a substance that they called "aether", which would act as the medium through which these waves, or vibrations travelled. The aether was thought to constitute an absolute reference frame against which speeds could be measured, and could be considered fixed and motionless. Aether supposedly possessed some wonderful properties: it was sufficiently elastic to support electromagnetic waves, and those waves could interact with matter, yet it offered no resistance to bodies passing through it. The results of various experiments, including the Michelson–Morley experiment, led to the theory of special relativity, by showing that there was no aether.[21] Einstein's solution was to discard the notion of an aether and the absolute state of rest. In relativity, any reference frame moving with uniform motion will observe the same laws of physics. In particular, the speed of light in vacuum is always measured to be c, even when measured by multiple systems that are moving at different (but constant) velocities.

21.3   Reference frames, coordinates and the Lorentz transformation

Main article: Lorentz transformation

Reference frames play a crucial role in relativity theory. The term reference frame as used here is an observational

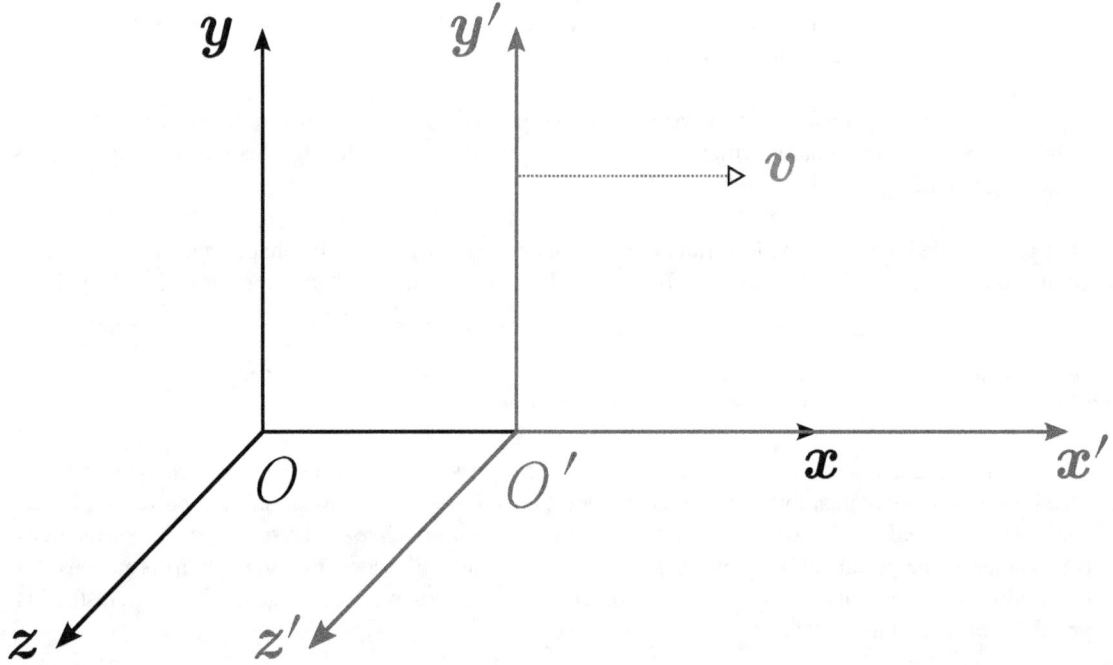

The primed system is in motion relative to the unprimed system with constant velocity v only along the x-axis, from the perspective of an observer stationary in the unprimed system. By the principle of relativity, an observer stationary in the primed system will view a likewise construction except that the velocity they record will be −v. The changing of the speed of propagation of interaction from infinite in non-relativistic mechanics to a finite value will require a modification of the transformation equations mapping events in one frame to another.

perspective in space which is not undergoing any change in motion (acceleration), from which a position can be measured along 3 spatial axes. In addition, a reference frame has the ability to determine measurements of the time of events using a 'clock' (any reference device with uniform periodicity).

An event is an occurrence that can be assigned a single unique time and location in space relative to a reference frame: it is a "point" in spacetime. Since the speed of light is constant in relativity in each and every reference frame, pulses of light can be used to unambiguously measure distances and refer back the times that events occurred to the clock, even though light takes time to reach the clock after the event has transpired.

For example, the explosion of a firecracker may be considered to be an "event". We can completely specify an event by its four spacetime coordinates: The time of occurrence and its 3-dimensional spatial location define a reference point. Let's call this reference frame S.

In relativity theory we often want to calculate the position of a point from a different reference point.

Suppose we have a second reference frame S', whose spatial axes and clock exactly coincide with that of S at time zero, but it is moving at a constant velocity v with respect to S along the x-axis.

Since there is no absolute reference frame in relativity theory, a concept of 'moving' doesn't strictly exist, as everything is always moving with respect to some other reference frame. Instead, any two frames that move at the same speed in the same direction are said to be *comoving*. Therefore, S and S' are not *comoving*.

Define the event to have spacetime coordinates (t,x,y,z) in system S and (t',x',y',z') in S'. Then the Lorentz transformation specifies that these coordinates are related in the following way:

$$t' = \gamma \left(t - vx/c^2 \right)$$
$$x' = \gamma \left(x - vt \right)$$
$$y' = y$$
$$z' = z,$$

where

$$\gamma = \frac{1}{\sqrt{1 - \frac{v^2}{c^2}}}$$

is the Lorentz factor and c is the speed of light in vacuum, and the velocity v of S' is parallel to the x-axis. The y and z coordinates are unaffected; only the x and t coordinates are transformed. These Lorentz transformations form a one-parameter group of linear mappings, that parameter being called rapidity.

There is nothing special about the x-axis, the transformation can apply to the y or z axes, or indeed in any direction, which can be done by directions parallel to the motion (which are warped by the γ factor) and perpendicular; see main article for details.

A quantity invariant under Lorentz transformations is known as a Lorentz scalar.

Writing the Lorentz transformation and its inverse in terms of coordinate differences, where for instance one event has coordinates (x_1, t_1) and (x'_1, t'_1), another event has coordinates (x_2, t_2) and (x'_2, t'_2), and the differences are defined as

$$\Delta x' = x'_2 - x'_1, \quad \Delta x = x_2 - x_1,$$
$$\Delta t' = t'_2 - t'_1, \quad \Delta t = t_2 - t_1,$$

we get

$$\Delta x' = \gamma \left(\Delta x - v\,\Delta t \right), \quad \Delta x = \gamma \left(\Delta x' + v\,\Delta t' \right),$$
$$\Delta t' = \gamma \left(\Delta t - \frac{v\,\Delta x}{c^2} \right), \quad \Delta t = \gamma \left(\Delta t' + \frac{v\,\Delta x'}{c^2} \right).$$

These effects are not merely appearances; they are explicitly related to our way of measuring *time intervals* between events which occur at the same place in a given coordinate system (called "co-local" events). These time intervals will be *different* in another coordinate system moving with respect to the first, unless the events are also simultaneous. Similarly, these effects also relate to our measured distances between separated but simultaneous events in a given coordinate system of choice. If these events are not co-local, but are separated by distance (space), they will *not* occur at the same *spatial distance* from each other when seen from another moving coordinate system. However, the spacetime interval will be the same for all observers. The underlying reality remains the same. Only our perspective changes.

21.4   Consequences derived from the Lorentz transformation

See also: Twin paradox and Relativistic mechanics

The consequences of special relativity can be derived from the Lorentz transformation equations.[22] These transformations, and hence special relativity, lead to different physical predictions than those of Newtonian mechanics when relative velocities become comparable to the speed of light. The speed of light is so much larger than anything humans encounter that some of the effects predicted by relativity are initially counterintuitive.

21.4.1   Relativity of simultaneity

See also: Relativity of simultaneity and Ladder paradox

Two events happening in two different locations that occur simultaneously in the reference frame of one inertial observer, may occur non-simultaneously in the reference frame of another inertial observer (lack of absolute simultaneity).

From the first equation of the Lorentz transformation in terms of coordinate differences

$$\Delta t' = \gamma \left(\Delta t - \frac{v\,\Delta x}{c^2} \right)$$

it is clear that two events that are simultaneous in frame S (satisfying $\Delta t = 0$), are not necessarily simultaneous in another inertial frame S' (satisfying $\Delta t' = 0$). Only if these events are additionally co-local in frame S (satisfying $\Delta x = 0$), will they be simultaneous in another frame S'.

21.4.2   Time dilation

See also: Time dilation

The time lapse between two events is not invariant from one observer to another, but is dependent on the relative speeds of the observers' reference frames (e.g., the twin paradox which concerns a twin who flies off in a spaceship traveling near the speed of light and returns to discover that his or her twin sibling has aged much more).

Suppose a clock is at rest in the unprimed system S. The location of the clock on two different ticks is then characterized by $\Delta x = 0$. To find the relation between the times between these ticks as measured in both systems, the first equation can be used to find:

$$\Delta t' = \gamma\,\Delta t \text{ for events satisfying } \Delta x = 0 \ .$$

This shows that the time ($\Delta t'$) between the two ticks as seen in the frame in which the clock is moving (S'), is *longer* than the time (Δt) between these ticks as measured in the rest frame of the clock (S). Time dilation explains a number of physical phenomena; for example, the decay rate of muons produced by cosmic rays impinging on the Earth's atmosphere.[23]

21.4.3   Length contraction

See also: Lorentz contraction

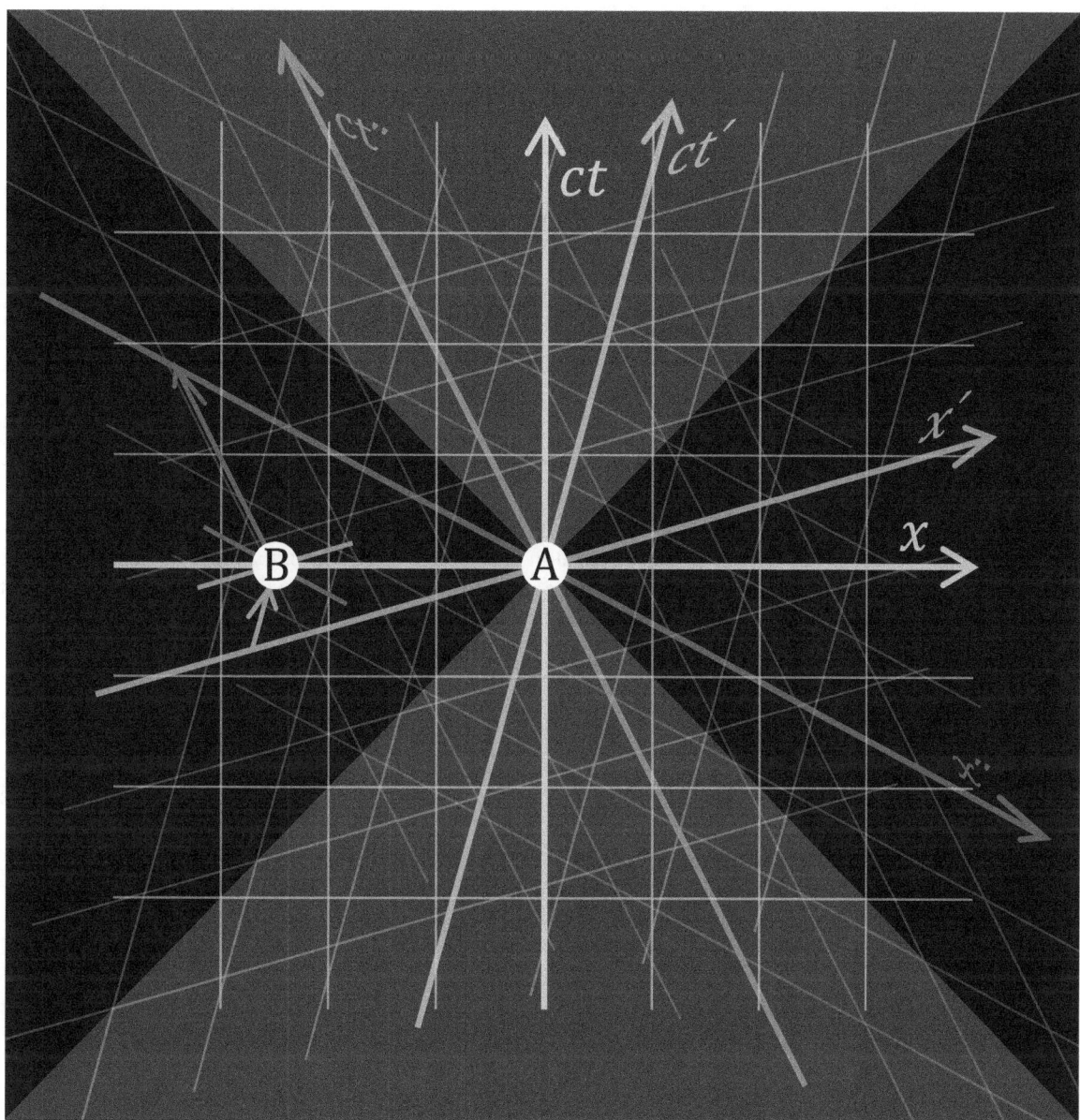

Event B is simultaneous with A in the green reference frame, but it occurs before A in the blue frame, and occurs after A in the red frame.

The dimensions (e.g., length) of an object as measured by one observer may be smaller than the results of measurements of the same object made by another observer (e.g., the ladder paradox involves a long ladder traveling near the speed of light and being contained within a smaller garage).

Similarly, suppose a measuring rod is at rest and aligned along the x-axis in the unprimed system S. In this system, the length of this rod is written as Δx. To measure the length of this rod in the system S', in which the clock is moving, the distances x' to the end points of the rod must be measured simultaneously in that system S'. In other words, the measurement is characterized by $\Delta t' = 0$, which can be combined with the fourth equation to find the relation between the lengths Δx and $\Delta x'$:

$$\Delta x' = \frac{\Delta x}{\gamma} \text{ for events satisfying } \Delta t' = 0 \ .$$

This shows that the length ($\Delta x'$) of the rod as measured in the frame in which it is moving (S'), is *shorter* than its length (Δx) in its own rest frame (S).

21.4.4   Composition of velocities

See also: Velocity-addition formula

Velocities (speeds) do not simply add. If the observer in S measures an object moving along the x axis at velocity u, then the observer in the S' system, a frame of reference moving at velocity v in the x direction with respect to S, will measure the object moving with velocity u' where (from the Lorentz transformations above):

$$u' = \frac{dx'}{dt'} = \frac{\gamma\,(dx - vdt)}{\gamma\,(dt - vdx/c^2)} = \frac{(dx/dt) - v}{1 - (v/c^2)(dx/dt)} = \frac{u - v}{1 - uv/c^2}\,.$$

The other frame S will measure:

$$u = \frac{dx}{dt} = \frac{\gamma\,(dx' + vdt')}{\gamma\,(dt' + vdx'/c^2)} = \frac{(dx'/dt') + v}{1 + (v/c^2)(dx'/dt')} = \frac{u' + v}{1 + u'v/c^2}\,.$$

Notice that if the object were moving at the speed of light in the S system (i.e. $u = c$), then it would also be moving at the speed of light in the S' system. Also, if both u and v are small with respect to the speed of light, we will recover the intuitive Galilean transformation of velocities

$$u' \approx u - v\,.$$

The usual example given is that of a train (frame S' above) traveling due east with a velocity v with respect to the tracks (frame S). A child inside the train throws a baseball due east with a velocity u' with respect to the train. In nonrelativistic physics, an observer at rest on the tracks will measure the velocity of the baseball (due east) as $u = u' + v$, while in special relativity this is no longer true; instead the velocity of the baseball (due east) is given by the second equation: $u = (u' + v)/(1 + u'v/c^2)$. Again, there is nothing special about the x or east directions. This formalism applies to any direction by considering parallel and perpendicular motion to the direction of relative velocity v, see main article for details.

21.5   Other consequences

21.5.1   Thomas rotation

See also: Thomas rotation

The orientation of an object (i.e. the alignment of its axes with the observer's axes) may be different for different observers. Unlike other relativistic effects, this effect becomes quite significant at fairly low velocities as can be seen in the spin of moving particles.

21.5.2   Equivalence of mass and energy

Main article: Mass–energy equivalence

As an object's speed approaches the speed of light from an observer's point of view, its relativistic mass increases thereby making it more and more difficult to accelerate it from within the observer's frame of reference.

The energy content of an object at rest with mass m equals mc^2. Conservation of energy implies that, in any reaction, a decrease of the sum of the masses of particles must be accompanied by an increase in kinetic energies of the particles after the reaction. Similarly, the mass of an object can be increased by taking in kinetic energies.

In addition to the papers referenced above—which give derivations of the Lorentz transformation and describe the foundations of special relativity—Einstein also wrote at least four papers giving heuristic arguments for the equivalence (and transmutability) of mass and energy, for $E = mc^2$.

Mass–energy equivalence is a consequence of special relativity. The energy and momentum, which are separate in Newtonian mechanics, form a four-vector in relativity, and this relates the time component (the energy) to the space components (the momentum) in a nontrivial way. For an object at rest, the energy–momentum four-vector is $(E, 0, 0, 0)$: it has a time component which is the energy, and three space components which are zero. By changing frames with a Lorentz transformation in the x direction with a small value of the velocity v, the energy momentum four-vector becomes $(E, Ev/c^2, 0, 0)$. The momentum is equal to the energy multiplied by the velocity divided by c^2. As such, the Newtonian mass of an object, which is the ratio of the momentum to the velocity for slow velocities, is equal to E/c^2.

The energy and momentum are properties of matter and radiation, and it is impossible to deduce that they form a four-vector just from the two basic postulates of special relativity by themselves, because these don't talk about matter or radiation, they only talk about space and time. The derivation therefore requires some additional physical reasoning. In his 1905 paper, Einstein used the additional principles that Newtonian mechanics should hold for slow velocities, so that there is one energy scalar and one three-vector momentum at slow velocities, and that the conservation law for energy and momentum is exactly true in relativity. Furthermore, he assumed that the energy of light is transformed by the same Doppler-shift factor as its frequency, which he had previously shown to be true based on Maxwell's equations.[1] The first of Einstein's papers on this subject was "Does the Inertia of a Body Depend upon its Energy Content?" in 1905.[24] Although Einstein's argument in this paper is nearly universally accepted by physicists as correct, even self-evident, many authors over the years have suggested that it is wrong.[25] Other authors suggest that the argument was merely inconclusive because it relied on some implicit assumptions.[26]

Einstein acknowledged the controversy over his derivation in his 1907 survey paper on special relativity. There he notes that it is problematic to rely on Maxwell's equations for the heuristic mass–energy argument. The argument in his 1905 paper can be carried out with the emission of any massless particles, but the Maxwell equations are implicitly used to make it obvious that the emission of light in particular can be achieved only by doing work. To emit electromagnetic waves, all you have to do is shake a charged particle, and this is clearly doing work, so that the emission is of energy.[27][28]

21.5.3 How far can one travel from the Earth?

See also: Space travel using constant acceleration

Since one can not travel faster than light, one might conclude that a human can never travel farther from Earth than 40 light years if the traveler is active between the age of 20 and 60. One would easily think that a traveler would never be able to reach more than the very few solar systems which exist within the limit of 20–40 light years from the earth. But that would be a mistaken conclusion. Because of time dilation, a hypothetical spaceship can travel thousands of light years during the pilot's 40 active years. If a spaceship could be built that accelerates at a constant 1 g, it will after a little less than a year be traveling at almost the speed of light as seen from Earth. Time dilation will increase his life span as seen from the reference system of the Earth, but his lifespan measured by a clock traveling with him will not thereby change. During his journey, people on Earth will experience more time than he does. A 5-year round trip for him will take 6½ Earth years and cover a distance of over 6 light-years. A 20-year round trip for him (5 years accelerating, 5 decelerating, twice each) will land him back on Earth having traveled for 335 Earth years and a distance of 331 light years.[29] A full 40-year trip at 1 g will appear on Earth to last 58,000 years and cover a distance of 55,000 light years. A 40-year trip at 1.1 g will take 148,000 Earth years and cover about 140,000 light years. A one-way 28 year (14 years accelerating, 14 decelerating as measured with the cosmonaut's clock) trip at 1 g acceleration could reach 2,000,000 light-years to the Andromeda Galaxy.[30] This same time dilation is why a muon traveling close to c is observed to travel much further than c times its half-life (when at rest).[31]

21.6   Causality and prohibition of motion faster than light

See also: Causality (physics) and Tachyonic antitelephone

In diagram 2 the interval AB is 'time-like'; i.e., there is a frame of reference in which events A and B occur at the same location in space, separated only by occurring at different times. If A precedes B in that frame, then A precedes B in all frames. It is hypothetically possible for matter (or information) to travel from A to B, so there can be a causal relationship (with A the cause and B the effect).

The interval AC in the diagram is 'space-like'; i.e., there is a frame of reference in which events A and C occur simultaneously, separated only in space. There are also frames in which A precedes C (as shown) and frames in which C precedes A. If it were possible for a cause-and-effect relationship to exist between events A and C, then paradoxes of causality would result. For example, if A was the cause, and C the effect, then there would be frames of reference in which the effect preceded the cause. Although this in itself won't give rise to a paradox, one can show[32][33] that faster than light signals can be sent back into one's own past. A causal paradox can then be constructed by sending the signal if and only if no signal was received previously.

Therefore, if causality is to be preserved, one of the consequences of special relativity is that no information signal or material object can travel faster than light in vacuum. However, some "things" can still move faster than light. For example, the location where the beam of a search light hits the bottom of a cloud can move faster than light when the search light is turned rapidly.[34]

Even without considerations of causality, there are other strong reasons why faster-than-light travel is forbidden by special relativity. For example, if a constant force is applied to an object for a limitless amount of time, then integrating $F = dp/dt$ gives a momentum that grows without bound, but this is simply because $p = m\gamma v$ approaches infinity as v approaches c. To an observer who is not accelerating, it appears as though the object's inertia is increasing, so as to produce a smaller acceleration in response to the same force. This behavior is observed in particle accelerators, where each charged particle is accelerated by the electromagnetic force.

21.7   Geometry of spacetime

Main article: Minkowski space

21.7.1   Comparison between flat Euclidean space and Minkowski space

See also: line element

Special relativity uses a 'flat' 4-dimensional Minkowski space – an example of a spacetime. Minkowski spacetime appears to be very similar to the standard 3-dimensional Euclidean space, but there is a crucial difference with respect to time.

In 3D space, the differential of distance (line element) ds is defined by

$$ds^2 = d\mathbf{x} \cdot d\mathbf{x} = dx_1^2 + dx_2^2 + dx_3^2,$$

where $d\mathbf{x} = (dx_1, dx_2, dx_3)$ are the differentials of the three spatial dimensions. In Minkowski geometry, there is an extra dimension with coordinate X^0 derived from time, such that the distance differential fulfills

$$ds^2 = -dX_0^2 + dX_1^2 + dX_2^2 + dX_3^2,$$

where $d\mathbf{X} = (dX_0, dX_1, dX_2, dX_3)$ are the differentials of the four spacetime dimensions. This suggests a deep theoretical insight: special relativity is simply a rotational symmetry of our spacetime, analogous to the rotational symmetry of Euclidean space (see image right).[36] Just as Euclidean space uses a Euclidean metric, so spacetime uses a Minkowski metric. Basically, special relativity can be stated as the *invariance of any spacetime interval* (that is the 4D distance between any two events) when viewed from *any inertial reference frame.* All equations and effects of special relativity can be derived from this rotational symmetry (the Poincaré group) of Minkowski spacetime.

The actual form of ds above depends on the metric and on the choices for the X^0 coordinate. To make the time coordinate look like the space coordinates, it can be treated as imaginary: $X_0 = ict$ (this is called a Wick rotation). According to Misner, Thorne and Wheeler (1971, §2.3), ultimately the deeper understanding of both special and general relativity will come from the study of the Minkowski metric (described below) and to take $X^0 = ct$, rather than a "disguised" Euclidean metric using ict as the time coordinate.

Some authors use $X^0 = t$, with factors of c elsewhere to compensate; for instance, spatial coordinates are divided by c or factors of $c^{\pm 2}$ are included in the metric tensor.[37] These numerous conventions can be superseded by using natural units where $c = 1$. Then space and time have equivalent units, and no factors of c appear anywhere.

21.7.2 3D spacetime

If we reduce the spatial dimensions to 2, so that we can represent the physics in a 3D space

$$ds^2 = dx_1^2 + dx_2^2 - c^2 dt^2,$$

we see that the null geodesics lie along a dual-cone (see image right) defined by the equation;

$$ds^2 = 0 = dx_1^2 + dx_2^2 - c^2 dt^2$$

or simply

$$dx_1^2 + dx_2^2 = c^2 dt^2,$$

which is the equation of a circle of radius $c \, dt$.

21.7.3 4D spacetime

If we extend this to three spatial dimensions, the null geodesics are the 4-dimensional cone:

$$ds^2 = 0 = dx_1^2 + dx_2^2 + dx_3^2 - c^2 dt^2$$

so

$$dx_1^2 + dx_2^2 + dx_3^2 = c^2 dt^2.$$

This null dual-cone represents the "line of sight" of a point in space. That is, when we look at the stars and say "The light from that star which I am receiving is X years old", we are looking down this line of sight: a null geodesic. We are looking at an event a distance $d = \sqrt{x_1^2 + x_2^2 + x_3^2}$ away and a time d/c in the past. For this reason the null dual cone is also known as the 'light cone'. (The point in the lower left of the picture above right represents the star, the origin represents the observer, and the line represents the null geodesic "line of sight".)

The cone in the $-t$ region is the information that the point is 'receiving', while the cone in the $+t$ section is the information that the point is 'sending'.

The geometry of Minkowski space can be depicted using Minkowski diagrams, which are useful also in understanding many of the thought-experiments in special relativity.

Note that, in 4d spacetime, the concept of the center of mass becomes more complicated, see center of mass (relativistic).

21.8   Physics in spacetime

21.8.1   Transformations of physical quantities between reference frames

Above, the Lorentz transformation for the time coordinate and three space coordinates illustrates that they are intertwined. This is true more generally: certain pairs of "timelike" and "spacelike" quantities naturally combine on equal footing under the same Lorentz transformation.

The Lorentz transformation in standard configuration above, i.e. for a boost in the x direction, can be recast into matrix form as follows:

$$
\begin{pmatrix} ct' \\ x' \\ y' \\ z' \end{pmatrix} = \begin{pmatrix} \gamma & -\beta\gamma & 0 & 0 \\ -\beta\gamma & \gamma & 0 & 0 \\ 0 & 0 & 1 & 0 \\ 0 & 0 & 0 & 1 \end{pmatrix} \begin{pmatrix} ct \\ x \\ y \\ z \end{pmatrix} = \begin{pmatrix} \gamma ct - \gamma\beta x \\ \gamma x - \beta\gamma ct \\ y \\ z \end{pmatrix}.
$$

In Newtonian mechanics, quantities which have magnitude and direction are mathematically described as 3d vectors in Euclidean space, and in general they are parametrized by time. In special relativity, this notion is extended by adding the appropriate timelike quantity to a spacelike vector quantity, and we have 4d vectors, or "four vectors", in Minkowski spacetime. The components of vectors are written using tensor index notation, as this has numerous advantages. The notation makes it clear the equations are manifestly covariant under the Poincaré group, thus bypassing the tedious calculations to check this fact. In constructing such equations, we often find that equations previously thought to be unrelated are, in fact, closely connected being part of the same tensor equation. Recognizing other physical quantities as tensors simplifies their transformation laws. Throughout, upper indices (superscripts) are contravariant indices rather than exponents except when they indicate a square (this is should be clear from the context), and lower indices (subscripts) are covariant indices. For simplicity and consistency with the earlier equations, Cartesian coordinates will be used.

The simplest example of a four-vector is the position of an event in spacetime, which constitutes a timelike component ct and spacelike component $\mathbf{x} = (x, y, z)$, in a contravariant position four vector with components:

$$
X^\nu = (X^0, X^1, X^2, X^3) = (ct, x, y, z) = (ct, \mathbf{x}).
$$

where we define $X^0 = ct$ so that the time coordinate has the same dimension of distance as the other spatial dimensions; so that space and time are treated equally.[38][39][40] Now the transformation of the contravariant components of the position 4-vector can be compactly written as:

$$
X^{\mu'} = \Lambda^{\mu'}{}_\nu X^\nu
$$

where there is an implied summation on ν from 0 to 3, and $\Lambda^{\mu'}{}_\nu$ is a matrix.

More generally, all contravariant components of a four-vector T^ν transform from one frame to another frame by a Lorentz transformation:

$$
T^{\mu'} = \Lambda^{\mu'}{}_\nu T^\nu
$$

Examples of other 4-vectors include the four-velocity U^μ, defined as the derivative of the position 4-vector with respect to proper time:

$$
U^\mu = \frac{dX^\mu}{d\tau} = \gamma(v)(c, v_x, v_y, v_z) = \gamma(v)(c, \mathbf{v}).
$$

where the Lorentz factor is:

$$\gamma(v) = \frac{1}{\sqrt{1 - (v/c)^2}}, \quad v^2 = v_x^2 + v_y^2 + v_z^2.$$

The relativistic energy $E = \gamma(v)mc^2$ and relativistic momentum $\mathbf{p} = \gamma(v)m\mathbf{v}$ of an object are respectively the timelike and spacelike components of a contravariant four momentum vector:

$$P^\mu = mU^\mu = m\gamma(v)(c, v_x, v_y, v_z) = (E/c, p_x, p_y, p_z) = (E/c, \mathbf{p}).$$

where m is the invariant mass.

The four-acceleration is the proper time derivative of 4-velocity:

$$A^\mu = \frac{dU^\mu}{d\tau}.$$

The transformation rules for *three*-dimensional velocities and accelerations are very awkward; even above in standard configuration the velocity equations are quite complicated owing to their non-linearity. On the other hand, the transformation of *four*-velocity and *four*-acceleration are simpler by means of the Lorentz transformation matrix.

The four-gradient of a scalar field φ transforms covariantly rather than contravariantly:

$$\begin{pmatrix} \frac{1}{c}\frac{\partial\phi}{\partial t'} & \frac{\partial\phi}{\partial x'} & \frac{\partial\phi}{\partial y'} & \frac{\partial\phi}{\partial z'} \end{pmatrix} = \begin{pmatrix} \frac{1}{c}\frac{\partial\phi}{\partial t} & \frac{\partial\phi}{\partial x} & \frac{\partial\phi}{\partial y} & \frac{\partial\phi}{\partial z} \end{pmatrix} \begin{pmatrix} \gamma & -\beta\gamma & 0 & 0 \\ -\beta\gamma & \gamma & 0 & 0 \\ 0 & 0 & 1 & 0 \\ 0 & 0 & 0 & 1 \end{pmatrix}.$$

that is:

$$(\partial_{\mu'}\phi) = \Lambda_{\mu'}{}^\nu(\partial_\nu\phi), \quad \partial_\mu \equiv \frac{\partial}{\partial x^\mu}.$$

only in Cartesian coordinates. It's the covariant derivative which transforms in manifest covariance, in Cartesian coordinates this happens to reduce to the partial derivatives, but not in other coordinates.

More generally, the *co*variant components of a 4-vector transform according to the *inverse* Lorentz transformation:

$$\Lambda_{\mu'}{}^\nu T^{\mu'} = T^\nu$$

where $\Lambda_{\mu'}{}^\nu$ is the reciprocal matrix of $\Lambda^{\mu'}{}_\nu$.

The postulates of special relativity constrain the exact form the Lorentz transformation matrices take.

More generally, most physical quantities are best described as (components of) tensors. So to transform from one frame to another, we use the well-known tensor transformation law[41]

$$T^{\alpha'\beta'\cdots\zeta'}_{\theta'\iota'\cdots\kappa'} = \Lambda^{\alpha'}{}_\mu\Lambda^{\beta'}{}_\nu\cdots\Lambda^{\zeta'}{}_\rho\Lambda_{\theta'}{}^\sigma\Lambda_{\iota'}{}^\upsilon\cdots\Lambda_{\kappa'}{}^\phi T^{\mu\nu\cdots\rho}_{\sigma\upsilon\cdots\phi}$$

where $\Lambda_{\chi'}{}^\psi$ is the reciprocal matrix of $\Lambda^{\chi'}{}_\psi$. All tensors transform by this rule.

An example of a four dimensional second order antisymmetric tensor is the relativistic angular momentum, which has six components: three are the classical angular momentum, and the other three are related to the boost of the center of mass of the system. The derivative of the relativistic angular momentum with respect to proper time is the relativistic torque, also second order antisymmetric tensor.

The electromagnetic field tensor is another second order antisymmetric tensor field, with six components: three for the electric field and another three for the magnetic field. There is also the stress–energy tensor for the electromagnetic field, namely the electromagnetic stress–energy tensor.

21.8.2   Metric

The metric tensor allows one to define the inner product of two vectors, which in turn allows one to assign a magnitude to the vector. Given the four-dimensional nature of spacetime the Minkowski metric η has components (valid in any inertial reference frame) which can be arranged in a 4×4 matrix:

$$\eta_{\alpha\beta} = \begin{pmatrix} -1 & 0 & 0 & 0 \\ 0 & 1 & 0 & 0 \\ 0 & 0 & 1 & 0 \\ 0 & 0 & 0 & 1 \end{pmatrix}$$

which is equal to its reciprocal, $\eta^{\alpha\beta}$, in those frames. Throughout we use the signs as above, different authors use different conventions – see Minkowski metric alternative signs.

The Poincaré group is the most general group of transformations which preserves the Minkowski metric:

$$\eta_{\alpha\beta} = \eta_{\mu'\nu'} \Lambda^{\mu'}{}_{\alpha} \Lambda^{\nu'}{}_{\beta}$$

and this is the physical symmetry underlying special relativity.

The metric can be used for raising and lowering indices on vectors and tensors. Invariants can be constructed using the metric, the inner product of a 4-vector T with another 4-vector S is:

$$T^{\alpha} S_{\alpha} = T^{\alpha} \eta_{\alpha\beta} S^{\beta} = T_{\alpha} \eta^{\alpha\beta} S_{\beta} = \text{scalar invariant}$$

Invariant means that it takes the same value in all inertial frames, because it is a scalar (0 rank tensor), and so no Λ appears in its trivial transformation. The magnitude of the 4-vector T is the positive square root of the inner product with itself:

$$|\mathbf{T}| = \sqrt{T^{\alpha} T_{\alpha}}$$

One can extend this idea to tensors of higher order, for a second order tensor we can form the invariants:

$$T^{\alpha}{}_{\alpha}, T^{\alpha}{}_{\beta} T^{\beta}{}_{\alpha}, T^{\alpha}{}_{\beta} T^{\beta}{}_{\gamma} T^{\gamma}{}_{\alpha} = \text{scalars invariant},$$

similarly for higher order tensors. Invariant expressions, particularly inner products of 4-vectors with themselves, provide equations that are useful for calculations, because one doesn't need to perform Lorentz transformations to determine the invariants.

21.8.3   Relativistic kinematics and invariance

The coordinate differentials transform also contravariantly:

$$dX^{\mu'} = \Lambda^{\mu'}{}_{\nu} dX^{\nu}$$

so the squared length of the differential of the position four-vector dX^{μ} constructed using

$$d\mathbf{X}^2 = dX^{\mu} dX_{\mu} = \eta_{\mu\nu} dX^{\mu} dX^{\nu} = -(cdt)^2 + (dx)^2 + (dy)^2 + (dz)^2$$

is an invariant. Notice that when the line element $d\mathbf{X}^2$ is negative that $\sqrt{-d\mathbf{X}^2}$ is the differential of proper time, while when $d\mathbf{X}^2$ is positive, $\sqrt{d\mathbf{X}^2}$ is differential of the proper distance.

The 4-velocity U^μ has an invariant form:

$$\mathbf{U}^2 = \eta_{\nu\mu}U^\nu U^\mu = -c^2\,,$$

which means all velocity four-vectors have a magnitude of c. This is an expression of the fact that there is no such thing as being at coordinate rest in relativity: at the least, you are always moving forward through time. Differentiating the above equation by τ produces:

$$2\eta_{\mu\nu}A^\mu U^\nu = 0.$$

So in special relativity, the acceleration four-vector and the velocity four-vector are orthogonal.

21.8.4 Relativistic dynamics and invariance

The invariant magnitude of the momentum 4-vector generates the energy–momentum relation:

$$\mathbf{P}^2 = \eta^{\mu\nu}P_\mu P_\nu = -(E/c)^2 + p^2.$$

We can work out what this invariant is by first arguing that, since it is a scalar, it doesn't matter in which reference frame we calculate it, and then by transforming to a frame where the total momentum is zero.

$$\mathbf{P}^2 = -(E_{\text{rest}}/c)^2 = -(mc)^2.$$

We see that the rest energy is an independent invariant. A rest energy can be calculated even for particles and systems in motion, by translating to a frame in which momentum is zero.

The rest energy is related to the mass according to the celebrated equation discussed above:

$$E_{\text{rest}} = mc^2.$$

Note that the mass of systems measured in their center of momentum frame (where total momentum is zero) is given by the total energy of the system in this frame. It may not be equal to the sum of individual system masses measured in other frames.

To use Newton's third law of motion, both forces must be defined as the rate of change of momentum with respect to the same time coordinate. That is, it requires the 3D force defined above. Unfortunately, there is no tensor in 4D which contains the components of the 3D force vector among its components.

If a particle is not traveling at c, one can transform the 3D force from the particle's co-moving reference frame into the observer's reference frame. This yields a 4-vector called the four-force. It is the rate of change of the above energy momentum four-vector with respect to proper time. The covariant version of the four-force is:

$$F_\nu = \frac{dP_\nu}{d\tau} = mA_\nu$$

In the rest frame of the object, the time component of the four force is zero unless the "invariant mass" of the object is changing (this requires a non-closed system in which energy/mass is being directly added or removed from the object) in which case it is the negative of that rate of change of mass, times c. In general, though, the components of the four force are not equal to the components of the three-force, because the three force is defined by the rate of change of momentum with respect to coordinate time, i.e. dp/dt while the four force is defined by the rate of change of momentum with respect to proper time, i.e. $dp/d\tau$.

In a continuous medium, the 3D *density of force* combines with the *density of power* to form a covariant 4-vector. The spatial part is the result of dividing the force on a small cell (in 3-space) by the volume of that cell. The time component is $-1/c$ times the power transferred to that cell divided by the volume of the cell. This will be used below in the section on electromagnetism.

21.9   Relativity and unifying electromagnetism

Main articles: Classical electromagnetism and special relativity and Covariant formulation of classical electromagnetism

Theoretical investigation in classical electromagnetism led to the discovery of wave propagation. Equations generalizing the electromagnetic effects found that finite propagation speed of the **E** and **B** fields required certain behaviors on charged particles. The general study of moving charges forms the Liénard–Wiechert potential, which is a step towards special relativity.

The Lorentz transformation of the electric field of a moving charge into a non-moving observer's reference frame results in the appearance of a mathematical term commonly called the magnetic field. Conversely, the *magnetic* field generated by a moving charge disappears and becomes a purely *electrostatic* field in a comoving frame of reference. Maxwell's equations are thus simply an empirical fit to special relativistic effects in a classical model of the Universe. As electric and magnetic fields are reference frame dependent and thus intertwined, one speaks of *electromagnetic* fields. Special relativity provides the transformation rules for how an electromagnetic field in one inertial frame appears in another inertial frame.

Maxwell's equations in the 3D form are already consistent with the physical content of special relativity, although they are easier to manipulate in a manifestly covariant form, i.e. in the language of tensor calculus.[42] See main links for more detail.

21.10   Status

Main articles: Tests of special relativity and Criticism of relativity theory

Special relativity in its Minkowski spacetime is accurate only when the absolute value of the gravitational potential is much less than c^2 in the region of interest.[43] In a strong gravitational field, one must use general relativity. General relativity becomes special relativity at the limit of weak field. At very small scales, such as at the Planck length and below, quantum effects must be taken into consideration resulting in quantum gravity. However, at macroscopic scales and in the absence of strong gravitational fields, special relativity is experimentally tested to extremely high degree of accuracy (10^{-20})[44] and thus accepted by the physics community. Experimental results which appear to contradict it are not reproducible and are thus widely believed to be due to experimental errors.

Special relativity is mathematically self-consistent, and it is an organic part of all modern physical theories, most notably quantum field theory, string theory, and general relativity (in the limiting case of negligible gravitational fields).

Newtonian mechanics mathematically follows from special relativity at small velocities (compared to the speed of light) – thus Newtonian mechanics can be considered as a special relativity of slow moving bodies. See classical mechanics for a more detailed discussion.

Several experiments predating Einstein's 1905 paper are now interpreted as evidence for relativity. Of these it is known Einstein was aware of the Fizeau experiment before 1905,[45] and historians have concluded that Einstein was at least aware of the Michelson–Morley experiment as early as 1899 despite claims he made in his later years that it played no role in his development of the theory.[20]

- The Fizeau experiment (1851, repeated by Michelson and Morley in 1886) measured the speed of light in moving media, with results that are consistent with relativistic addition of colinear velocities.

- The famous Michelson–Morley experiment (1881, 1887) gave further support to the postulate that detecting an absolute reference velocity was not achievable. It should be stated here that, contrary to many alternative claims, it said little about the invariance of the speed of light with respect to the source and observer's velocity, as both source and observer were travelling together at the same velocity at all times.

- The Trouton–Noble experiment (1903) showed that the torque on a capacitor is independent of position and inertial reference frame.

- The Experiments of Rayleigh and Brace (1902, 1904) showed that length contraction doesn't lead to birefringence for a co-moving observer, in accordance with the relativity principle.

Particle accelerators routinely accelerate and measure the properties of particles moving at near the speed of light, where their behavior is completely consistent with relativity theory and inconsistent with the earlier Newtonian mechanics. These machines would simply not work if they were not engineered according to relativistic principles. In addition, a considerable number of modern experiments have been conducted to test special relativity. Some examples:

- Tests of relativistic energy and momentum – testing the limiting speed of particles

- Ives–Stilwell experiment – testing relativistic Doppler effect and time dilation

- Time dilation of moving particles – relativistic effects on a fast-moving particle's half-life

- Kennedy–Thorndike experiment – time dilation in accordance with Lorentz transformations

- Hughes–Drever experiment – testing isotropy of space and mass

- Modern searches for Lorentz violation – various modern tests

- Experiments to test emission theory demonstrated that the speed of light is independent of the speed of the emitter.

- Experiments to test the aether drag hypothesis – no "aether flow obstruction".

21.11 Theories of relativity and quantum mechanics

Special relativity can be combined with quantum mechanics to form relativistic quantum mechanics. It is an unsolved problem in physics how *general* relativity and quantum mechanics can be unified; quantum gravity and a "theory of everything", which require such a unification, are active and ongoing areas in theoretical research.

The early Bohr–Sommerfeld atomic model explained the fine structure of alkali metal atoms using both special relativity and the preliminary knowledge on quantum mechanics of the time.[46]

In 1928, Paul Dirac constructed an influential relativistic wave equation, now known as the Dirac equation in his honour,[47] that is fully compatible both with special relativity and with the final version of quantum theory existing after 1926. This equation explained not only the intrinsic angular momentum of the electrons called *spin*, it also led to the prediction of the antiparticle of the electron (the positron),[47][48] and fine structure could only be fully explained with special relativity. It was the first foundation of *relativistic quantum mechanics*. In non-relativistic quantum mechanics, spin is phenomenological and cannot be explained.

On the other hand, the existence of antiparticles leads to the conclusion that relativistic quantum mechanics is not enough for a more accurate and complete theory of particle interactions. Instead, a theory of particles interpreted as quantized fields, called *quantum field theory*, becomes necessary; in which particles can be created and destroyed throughout space and time.

21.12   See also

People: Hendrik Lorentz | Henri Poincaré | Albert Einstein | Max Planck | Hermann Minkowski | Max von Laue | Arnold Sommerfeld | Max Born | Gustav Herglotz | Richard C. Tolman

Relativity: Theory of relativity | History of special relativity | Principle of relativity | General relativity | Frame of reference | Inertial frame of reference | Lorentz transformations | Bondi k-calculus | Einstein synchronisation | Rietdijk–Putnam argument | Special relativity (alternative formulations) | Criticism of relativity theory | Relativity priority dispute

Physics: Newtonian Mechanics | spacetime | speed of light | simultaneity | center of mass (relativistic) | physical cosmology | Doppler effect | relativistic Euler equations | Aether drag hypothesis | Lorentz ether theory | Moving magnet and conductor problem | Shape waves | Relativistic heat conduction | Relativistic disk | Thomas precession | Born rigidity | Born coordinates

Mathematics: Derivations of the Lorentz transformations | Minkowski space | four-vector | world line | light cone | Lorentz group | Poincaré group | geometry | tensors | split-complex number | Relativity in the APS formalism

Philosophy: actualism | conventionalism | formalism

Paradoxes: Twin paradox | Ehrenfest paradox | Ladder paradox | Bell's spaceship paradox | Velocity composition paradox

21.13   References

[1] Albert Einstein (1905) "*Zur Elektrodynamik bewegter Körper*", *Annalen der Physik* 17: 891; English translation On the Electrodynamics of Moving Bodies by George Barker Jeffery and Wilfrid Perrett (1923); Another English translation On the Electrodynamics of Moving Bodies by Megh Nad Saha (1920).

[2] Tom Roberts and Siegmar Schleif (October 2007). "What is the experimental basis of Special Relativity?". *Usenet Physics FAQ*. Retrieved 2008-09-17.

[3] Albert Einstein (2001). *Relativity: The Special and the General Theory* (Reprint of 1920 translation by Robert W. Lawson ed.). Routledge. p. 48. ISBN 0-415-25384-5.

[4] Richard Phillips Feynman (1998). *Six Not-so-easy Pieces: Einstein's relativity, symmetry, and space–time* (Reprint of 1995 ed.). Basic Books. p. 68. ISBN 0-201-32842-9.

[5] Sean Carroll, Lecture Notes on General Relativity, ch. 1, "Special relativity and flat spacetime," http://ned.ipac.caltech.edu/level5/March01/Carroll3/Carroll1.html

[6] Wald, General Relativity, p. 60: "...the special theory of relativity asserts that spacetime is the manifold $\mathbb{R}^4$ with a flat metric of Lorentz signature defined on it. Conversely, the entire content of special relativity ... is contained in this statement ..."

[7] Rindler, W., 1969, Essential Relativity: Special, General, and Cosmological

[8] Edwin F. Taylor and John Archibald Wheeler (1992). *Spacetime Physics: Introduction to Special Relativity*. W. H. Freeman. ISBN 0-7167-2327-1.

[9] Wolfgang Rindler (1977). *Essential Relativity*. Birkhäuser. p. §1,11 p. 7. ISBN 3-540-07970-X.

[10] Einstein, Autobiographical Notes, 1949.

[11] Einstein, "Fundamental Ideas and Methods of the Theory of Relativity", 1920

[12] For a survey of such derivations, see Lucas and Hodgson, Spacetime and Electromagnetism, 1990

[13] Einstein, A., Lorentz, H. A., Minkowski, H., & Weyl, H. (1952). *The Principle of Relativity: a collection of original memoirs on the special and general theory of relativity*. Courier Dover Publications. p. 111. ISBN 0-486-60081-5.

[14] Einstein, On the Relativity Principle and the Conclusions Drawn from It, 1907; "The Principle of Relativity and Its Consequences in Modern Physics", 1910; "The Theory of Relativity", 1911; Manuscript on the Special Theory of Relativity, 1912; Theory of Relativity, 1913; Einstein, Relativity, the Special and General Theory, 1916; The Principle Ideas of the Theory of Relativity, 1916; What Is The Theory of Relativity?, 1919; The Principle of Relativity (Princeton Lectures), 1921; Physics and Reality, 1936; The Theory of Relativity, 1949.

[15] Das, A. (1993) *The Special Theory of Relativity, A Mathematical Exposition*, Springer, ISBN 0-387-94042-1.

[16] Schutz, J. (1997) Independent Axioms for Minkowski Spacetime, Addison Wesley Longman Limited, ISBN 0-582-31760-6.

[17] Yaakov Friedman (2004). *Physical Applications of Homogeneous Balls*. Progress in Mathematical Physics **40**. pp. 1–21. ISBN 0-8176-3339-1.

[18] David Morin (2007) *Introduction to Classical Mechanics*, Cambridge University Press, Cambridge, chapter 11, Appendix I, ISBN 1-139-46837-5.

[19] Michael Polanyi (1974) *Personal Knowledge: Towards a Post-Critical Philosophy*, ISBN 0-226-67288-3, footnote page 10–11: Einstein reports, via Dr N Balzas in response to Polanyi's query, that "The Michelson–Morley experiment had no role in the foundation of the theory." and "..the theory of relativity was not founded to explain its outcome at all."

[20] Jeroen van Dongen (2009). "On the role of the Michelson–Morley experiment: Einstein in Chicago" (PDF). *Eprint arXiv: 0908.1545* **0908**: 1545. arXiv:0908.1545. Bibcode:2009arXiv0908.1545V.

[21] Staley, Richard (2009), "Albert Michelson, the Velocity of Light, and the Ether Drift", *Einstein's generation. The origins of the relativity revolution,* Chicago: University of Chicago Press, ISBN 0-226-77057-5

[22] Robert Resnick (1968). *Introduction to special relativity*. Wiley. pp. 62–63.

[23] Daniel Kleppner and David Kolenkow (1973). *An Introduction to Mechanics*. pp. 468–70. ISBN 0-07-035048-5.

[24] Does the inertia of a body depend upon its energy content? A. Einstein, *Annalen der Physik*. **18**:639, 1905 (English translation by W. Perrett and G.B. Jeffery)

[25] Max Jammer (1997). *Concepts of Mass in Classical and Modern Physics*. Courier Dover Publications. pp. 177–178. ISBN 0-486-29998-8.

[26] John J. Stachel (2002). *Einstein from B to Z*. Springer. p. 221. ISBN 0-8176-4143-2.

[27] *On the Inertia of Energy Required by the Relativity Principle*, A. Einstein, Annalen der Physik 23 (1907): 371–384

[28] In a letter to Carl Seelig in 1955, Einstein wrote "I had already previously found that Maxwell's theory did not account for the micro-structure of radiation and could therefore have no general validity.", Einstein letter to Carl Seelig, 1955.

[29] Philip Gibbs and Don Koks. "The Relativistic Rocket". Retrieved 30 August 2012.

[30] Philip Gibbs and Don Koks. "The Relativistic Rocket". Retrieved 13 October 2013.

[31] The special theory of relativity shows that time and space are affected by motion. Library.thinkquest.org. Retrieved on 2013-04-24.

[32] R. C. Tolman, *The theory of the Relativity of Motion*, (Berkeley 1917), p. 54

[33] G. A. Benford, D. L. Book, and W. A. Newcomb (1970). "The Tachyonic Antitelephone". *Physical Review D* **2** (2): 263. Bibcode:1970PhRvD...2..263B. doi:10.1103/PhysRevD.2.263.

[34] Wesley C. Salmon (2006). *Four Decades of Scientific Explanation*. University of Pittsburgh. p. 107. ISBN 0-8229-5926-7., Section 3.7 page 107

[35] J.A. Wheeler, C. Misner, K.S. Thorne (1973). *Gravitation*. W.H. Freeman & Co. p. 58. ISBN 0-7167-0344-0.

[36] J.R. Forshaw, A.G. Smith (2009). *Dynamics and Relativity*. Wiley. p. 247. ISBN 978-0-470-01460-8.

[37] R. Penrose (2007). *The Road to Reality*. Vintage books. ISBN 0-679-77631-1.

[38] Jean-Bernard Zuber & Claude Itzykson, *Quantum Field Theory*, pg 5, ISBN 0-07-032071-3

[39] Charles W. Misner, Kip S. Thorne & John A. Wheeler, *Gravitation*, pg 51, ISBN 0-7167-0344-0

[40] George Sterman, *An Introduction to Quantum Field Theory*, pg 4 , ISBN 0-521-31132-2

[41] Sean M. Carroll (2004). *Spacetime and Geometry: An Introduction to General Relativity*. Addison Wesley. p. 22. ISBN 0-8053-8732-3.

[42] E. J. Post (1962). *Formal Structure of Electromagnetics: General Covariance and Electromagnetics*. Dover Publications Inc. ISBN 0-486-65427-3.

[43] Øyvind Grøn and Sigbjørn Hervik (2007). *Einstein's general theory of relativity: with modern applications in cosmology*. Springer. p. 195. ISBN 0-387-69199-5., Extract of page 195 (with units where c=1)

[44] The number of works is vast, see as example:
      Sidney Coleman, Sheldon L. Glashow (1997). "Cosmic Ray and Neutrino Tests of Special Relativity". *Phys. Lett.* **B405** (3–4): 249–252. arXiv:hep-ph/9703240. Bibcode:1997PhLB..405..249C. doi:10.1016/S0370-2693(97)00638-2.
      An overview can be found on this page

[45] John D. Norton, John D. (2004). "Einstein's Investigations of Galilean Covariant Electrodynamics prior to 1905". *Archive for History of Exact Sciences* **59**: 45–105. Bibcode:2004AHES...59...45N. doi:10.1007/s00407-004-0085-6.

[46] R. Resnick, R. Eisberg (1985). *Quantum Physics of Atoms, Molecules, Solids, Nuclei and Particles* (2nd ed.). John Wiley & Sons. pp. 114–116. ISBN 978-0-471-87373-0.

[47] P.A.M. Dirac (1930). "A Theory of Electrons and Protons". *Proceedings o f the Royal Society* **A126** (801): 360. Bibcode: doi:10.1098/rspa.1930.0013. JSTOR 95359.

[48] C.D. Anderson (1933). "The Positiv e Electron". *Phys. Rev.* **43** (6): 491–494. Bibcode:1933PhRv...43..491A. doi:10.1103/

21.13.1   Textbooks

- Einstein, Albert (1920). Relativity: The Special and General Theory.

- Einstein, Albert (1996). *The Meaning of Relativity*. Fine Communications. ISBN 1-56731-136-9

- Logunov, Anatoly A. (2005) Henri Poincaré and the Relativity Theory (transl. from Russian by G. Pontocorvo and V. O. Soleviev, edited by V. A. Petrov) Nauka, Moscow.

- Charles Misner, Kip Thorne, and John Archibald Wheeler (1971) *Gravitation*. W. H. Freeman & Co. ISBN 0-7167-0334-3

- Post, E.J., 1997 (1962) *Formal Structure of Electromagnetics: General Covariance and Electromagnetics*. Dover Publications.

- Wolfgang Rindler (1991). Introduction to Special Relativity (2nd ed.), Oxford University Press. ISBN 978-0-19-853952-0; ISBN 0-19-853952-5

- Harvey R. Brown (2005). Physical relativity: space–time structure from a dynamical perspective, Oxford University Press, ISBN 0-19-927583-1; ISBN 978-0-19-927583-0

- Qadir, Asghar (1989). *Relativity: An Introduction to the Special Theory*. Singapore: World Scientific Publications. p. 128. ISBN 9971-5-0612-2.

- Silberstein, Ludwik (1914) The Theory of Relativity.

- Lawrence Sklar (1977). *Space, Time and Spacetime*. University of California Press. ISBN 0-520-03174-1.

- Lawrence Sklar (1992). *Philosophy of Physics*. Westview Press. ISBN 0-8133-0625-6.

- Taylor, Edwin, and John Archibald Wheeler (1992) *Spacetime Physics* (2nd ed.). W.H. Freeman & Co. ISBN 0-7167-2327-1

- Tipler, Paul, and Llewellyn, Ralph (2002). *Modern Physics* (4th ed.). W. H. Freeman & Co. ISBN 0-7167-4345-0

21.13.2 Journal articles

- Alvager, T.; Farley, F. J. M.; Kjellman, J.; Wallin, L.; et al. (1964). "Test of the Second Postulate of Special Relativity in the GeV region". *Physics Letters* **12** (3): 260. Bibcode:1964PhL....12..260A. doi:10.1016/0031-9163(64)91095-9.

- Darrigol, Olivier (2004). "The Mystery of the Poincaré–Einstein Connection". *Isis* **95** (4): 614–26. doi:10.1086 PMID 16011297.

- Wolf, Peter; Petit, Gerard (1997). "Satellite test of Special Relativity using the Global Positioning System". *Physical Review A* **56** (6): 4405–09. Bibcode:1997PhRvA..56.4405W. doi:10.1103/PhysRevA.56.4405.

- Special Relativity Scholarpedia

- Special relativity: Kinematics Wolfgang Rindler, Scholarpedia, 6(2):8520. doi:10.4249/scholarpedia.8520

21.14 External links

21.14.1 Original works

- *Zur Elektrodynamik bewegter Körper* Einstein's original work in German, Annalen der Physik, Bern 1905

- *On the Electrodynamics of Moving Bodies* English Translation as published in the 1923 book *The Principle of Relativity*.

21.14.2 Special relativity for a general audience (no mathematical knowledge required)

- Einstein Light An award-winning, non-technical introduction (film clips and demonstrations) supported by dozens of pages of further explanations and animations, at levels with or without mathematics.

- Einstein Online Introduction to relativity theory, from the Max Planck Institute for Gravitational Physics.

- Audio: Cain/Gay (2006) – Astronomy Cast. Einstein's Theory of Special Relativity

21.14.3 Special relativity explained (using simple or more advanced mathematics)

- Greg Egan's *Foundations*.

- The Hogg Notes on Special Relativity A good introduction to special relativity at the undergraduate level, using calculus.

- Relativity Calculator: Special Relativity – An algebraic and integral calculus derivation for $E = mc^2$.

- MathPages – Reflections on Relativity A complete online book on relativity with an extensive bibliography.

- Relativity An introduction to special relativity at the undergraduate level, without calculus.

-

- *Relativity: the Special and General Theory* at Project Gutenberg, by Albert Einstein

- Special Relativity Lecture Notes is a standard introduction to special relativity containing illustrative explanations based on drawings and spacetime diagrams from Virginia Polytechnic Institute and State University.

- Understanding Special Relativity The theory of special relativity in an easily understandable way.

- An Introduction to the Special Theory of Relativity (1964) by Robert Katz, "an introduction ... that is accessible to any student who has had an introduction to general physics and some slight acquaintance with the calculus" (130 pp; pdf format).

- Lecture Notes on Special Relativity by J D Cresser Department of Physics Macquarie University.

- SpecialRelativity.net - An overview with visualizations and minimal mathematics.

21.14.4   Visualization

- Raytracing Special Relativity Software visualizing several scenarios under the influence of special relativity.

- Real Time Relativity The Australian National University. Relativistic visual effects experienced through an interactive program.

- Spacetime travel A variety of visualizations of relativistic effects, from relativistic motion to black holes.

- Through Einstein's Eyes The Australian National University. Relativistic visual effects explained with movies and images.

- Warp Special Relativity Simulator A computer program to show the effects of traveling close to the speed of light.

- Animation clip on YouTube visualizing the Lorentz transformation.

- Original interactive FLASH Animations from John de Pillis illustrating Lorentz and Galilean frames, Train and Tunnel Paradox, the Twin Paradox, Wave Propagation, Clock Synchronization, etc.

- Relativistic Optics at the ANU

- lightspeed An OpenGL-based program developed to illustrate the effects of special relativity on the appearance of moving objects.

- Animation showing the stars near Earth, as seen from a spacecraft accelerating rapidly to light speed.

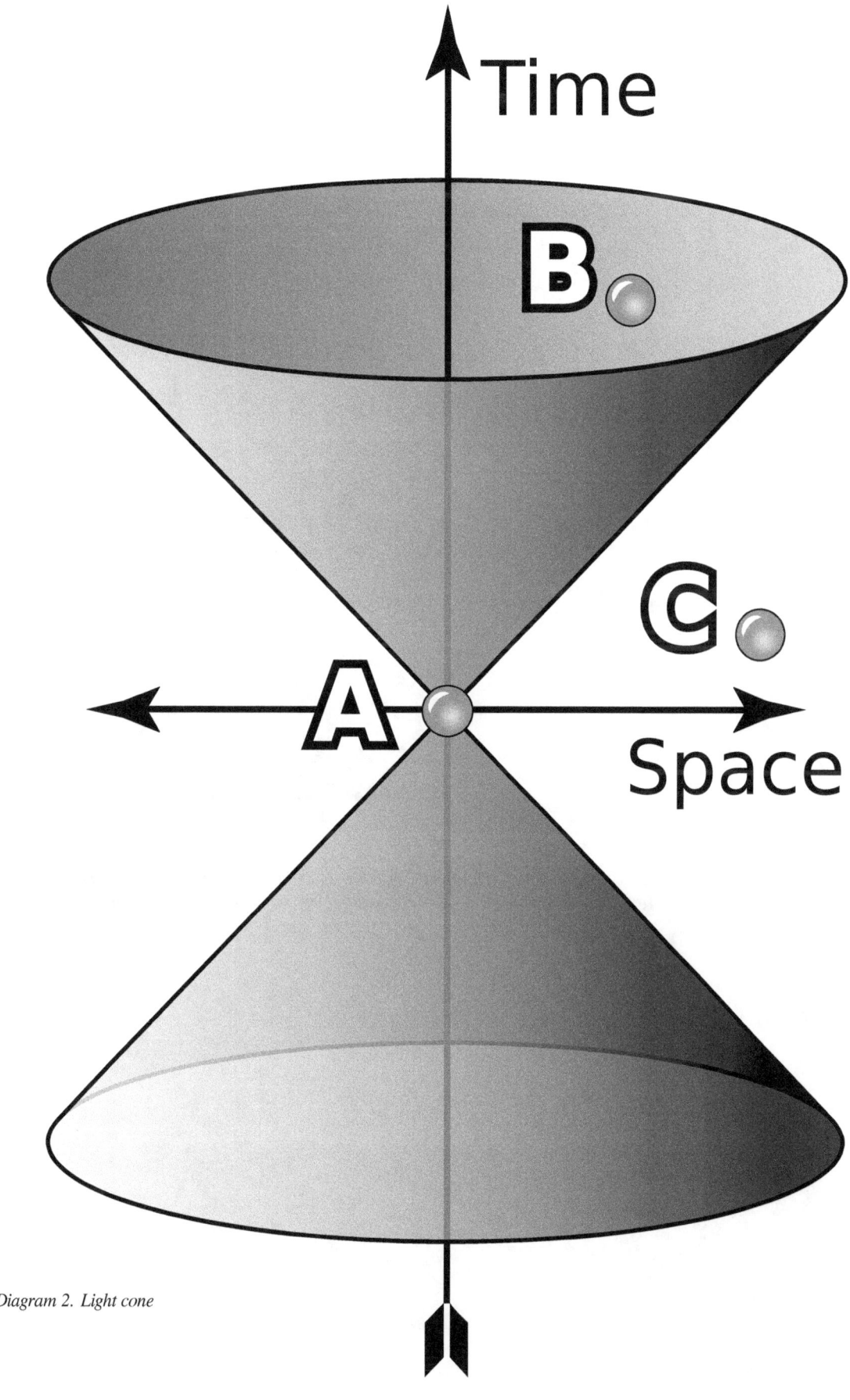

Diagram 2. Light cone

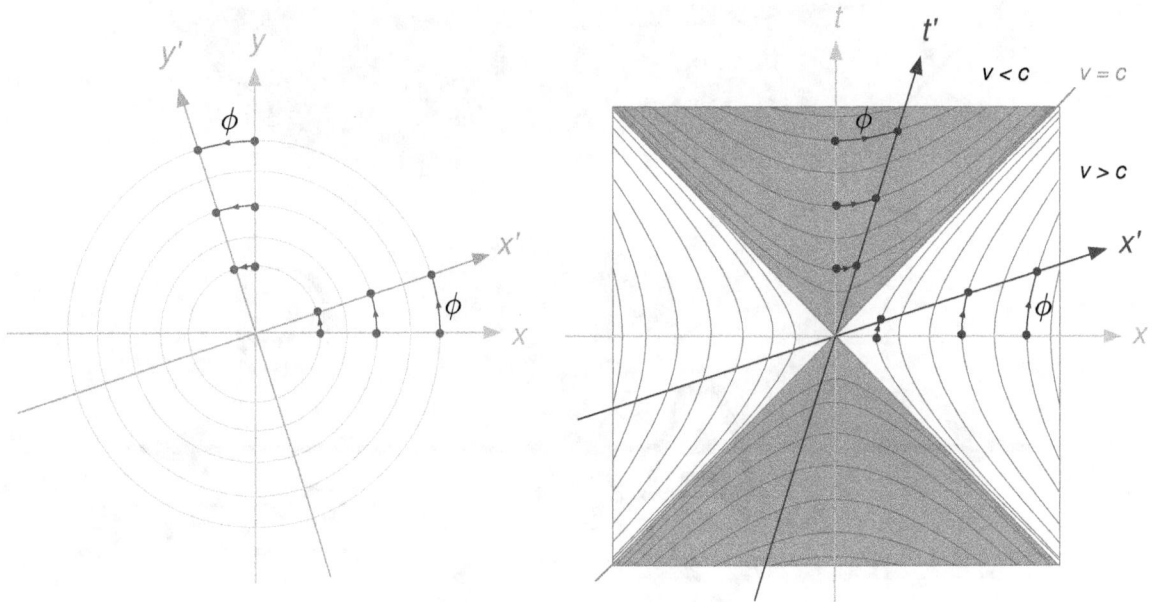

Orthogonality and rotation of coordinate systems compared between **left:** *Euclidean space through circular angle* φ, **right:** *in Minkowski spacetime through hyperbolic angle* φ *(red lines labelled* c *denote the worldlines of a light signal, a vector is orthogonal to itself if it lies on this line).*[35]

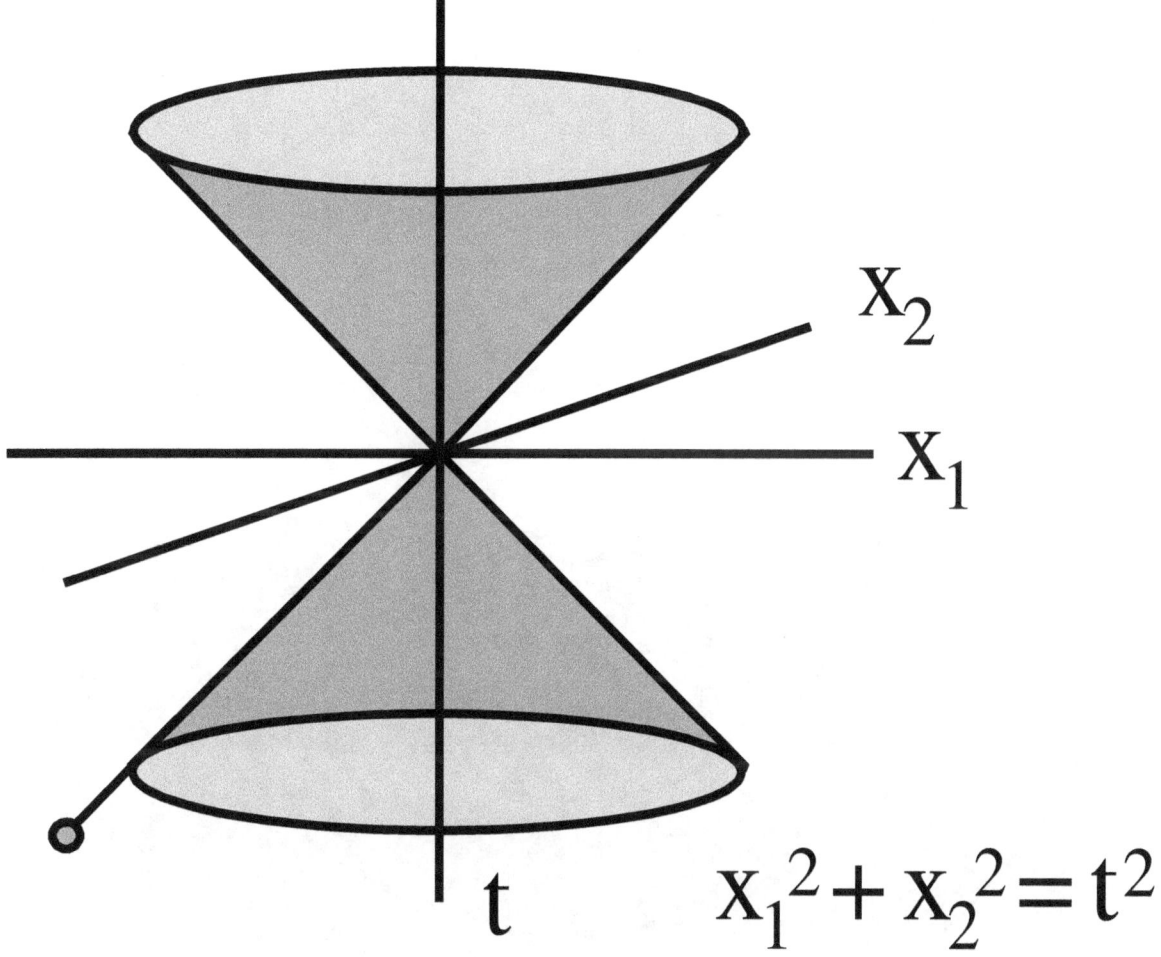

$$x_1^2 + x_2^2 = t^2$$

Three-dimensional dual-cone.

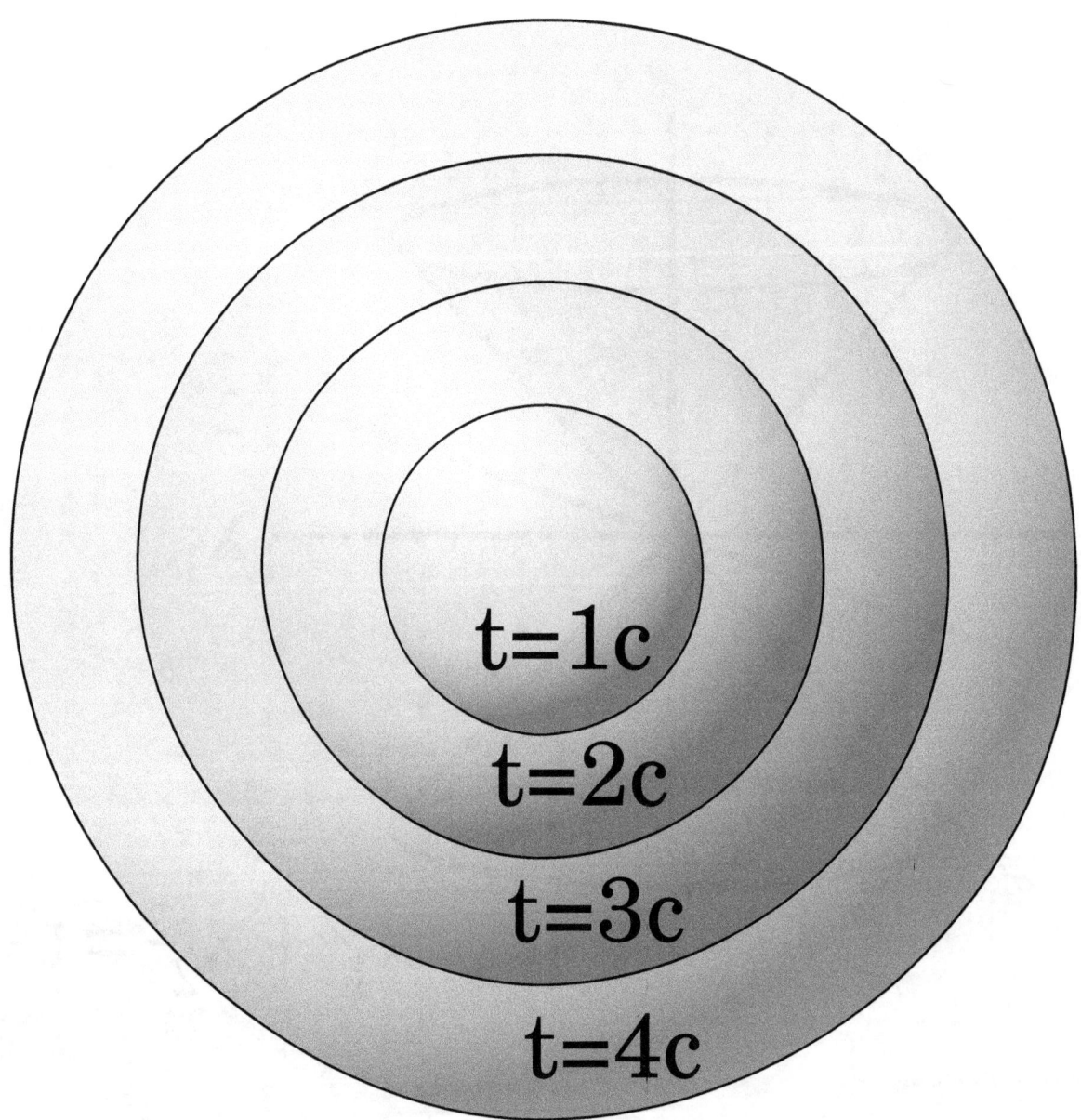

Null spherical space.

Chapter 22

Global symmetry

In physics, a **global symmetry** is a symmetry that holds at all points in the spacetime under consideration, as opposed to a local symmetry which varies from point to point.

Global symmetries require conservation laws, but not forces, in physics.

An example of a global symmetry is the action of the $U(1) = e^{iq\theta}$ (for θ a constant - making it a global transformation) group on the Dirac Lagrangian:

$$\mathcal{L}_D = \bar{\psi} \left(i\gamma^\mu \partial_\mu - m \right) \psi$$

Under this transformation the wavefunction changes as $\psi \to e^{iq\theta}\psi$ and $\bar{\psi} \to e^{-iq\theta}\bar{\psi}$ and so:

$$\mathcal{L} \to \bar{\mathcal{L}} = e^{-iq\theta}\bar{\psi} \left(i\gamma^\mu \partial_\mu - m \right) e^{iq\theta}\psi = e^{-iq\theta}e^{iq\theta}\bar{\psi} \left(i\gamma^\mu \partial_\mu - m \right) \psi = \mathcal{L}$$

22.1 See also

- Field (physics)

- Global spacetime structure

- Local spacetime structure

Chapter 23

Local symmetry

In physics, a **local symmetry** is symmetry of some physical quantity, which smoothly depends on the point of the base manifold. Such quantities can be for example an observable, a tensor or the Lagrangian of a theory. If a symmetry is local in this sense, then one can apply a local transformation (resp. local gauge transformation), which means that the representation of the symmetry group is a function of the manifold and can thus be taken to act differently on different points of spacetime.

The diffeomorphism group is a local symmetry and thus every geometrical or generally covariant theory (i.e. a theory whose equations are tensor equations, for example general relativity) has local symmetries.

Often the term local symmetry is specifically associated with local gauge symmetries in Yang–Mills theory (see also standard model) where the Lagrangian is locally symmetric under some compact Lie group. Local gauge symmetries always come together with some bosonic gauge fields, like the photon or gluon field, which induce a force in addition to requiring conservation laws.[1]

23.1 Examples

- General relativity has a local symmetry (general covariance, diffeomorphisms) which can be seen as generating the gravitational force.[2] Special relativity only has a global symmetry (Lorentz symmetry or more generally Poincaré symmetry)

- There are many global symmetries (such as SU(2) of isospin symmetry) and local symmetries (like SU(2) of weak interactions) in particle physics. The standard model of particle physics consists of Yang-Mills Theories

- The symmetry group of Supergravity is a local symmetry, whereas supersymmetry is a global symmetry.

23.2 See also

- Field (physics)

- Global spacetime structure

- Local spacetime structure

- Gauge theory

- Gravitation (book)

23.3 References

[1] Kaku, Michio (1993). *Quantum Field Theory: A Modern Introduction*. New York: Oxford University Press. ISBN 0-19-507652-4.

[2] Misner, Charles W.; Thorne, Kip S.; Wheeler, John Archibald (1973-09-15). "Gravitation". San Francisco: W. H. Freeman. ISBN 978-0-7167-0344-0.

Chapter 24

Gauge theory

For a more accessible and less technical introduction to this topic, see Introduction to gauge theory.

In physics, a **gauge theory** is a type of field theory in which the Lagrangian is invariant under a continuous group of local transformations.

The term *gauge* refers to redundant degrees of freedom in the Lagrangian. The transformations between possible gauges, called *gauge transformations*, form a Lie group—referred to as the *symmetry group* or the *gauge group* of the theory. Associated with any Lie group is the Lie algebra of group generators. For each group generator there necessarily arises a corresponding vector field called the *gauge field*. Gauge fields are included in the Lagrangian to ensure its invariance under the local group transformations (called *gauge invariance*). When such a theory is quantized, the quanta of the gauge fields are called *gauge bosons*. If the symmetry group is non-commutative, the gauge theory is referred to as *non-abelian*, the usual example being the Yang–Mills theory.

Many powerful theories in physics are described by Lagrangians that are invariant under some symmetry transformation groups. When they are invariant under a transformation identically performed at *every* point in the space in which the physical processes occur, they are said to have a global symmetry. The requirement of local symmetry, the cornerstone of gauge theories, is a stricter constraint. In fact, a global symmetry is just a local symmetry whose group's parameters are fixed in space-time.

Gauge theories are important as the successful field theories explaining the dynamics of elementary particles. Quantum electrodynamics is an abelian gauge theory with the symmetry group U(1) and has one gauge field, the electromagnetic four-potential, with the photon being the gauge boson. The Standard Model is a non-abelian gauge theory with the symmetry group U(1)×SU(2)×SU(3) and has a total of twelve gauge bosons: the photon, three weak bosons and eight gluons.

Gauge theories are also important in explaining gravitation in the theory of general relativity. Its case is somewhat unique in that the gauge field is a tensor, the Lanczos tensor. Theories of quantum gravity, beginning with gauge gravitation theory, also postulate the existence of a gauge boson known as the graviton. Gauge symmetries can be viewed as analogues of the principle of general covariance of general relativity in which the coordinate system can be chosen freely under arbitrary diffeomorphisms of spacetime. Both gauge invariance and diffeomorphism invariance reflect a redundancy in the description of the system. An alternative theory of gravitation, gauge theory gravity, replaces the principle of general covariance with a true gauge principle with new gauge fields.

Historically, these ideas were first stated in the context of classical electromagnetism and later in general relativity. However, the modern importance of gauge symmetries appeared first in the relativistic quantum mechanics of electrons – quantum electrodynamics, elaborated on below. Today, gauge theories are useful in condensed matter, nuclear and high energy physics among other subfields.

24.1 History and importance

The earliest field theory having a gauge symmetry was Maxwell's formulation, in 1864–65, of electrodynamics ("A Dynamical Theory of the Electromagnetic Field"). The importance of this symmetry remained unnoticed in the earliest formulations. Similarly unnoticed, Hilbert had derived the Einstein field equations by postulating the invariance of the action under a general coordinate transformation. Later Hermann Weyl, in an attempt to unify general relativity and electromagnetism, conjectured that *Eichinvarianz* or invariance under the change of scale (or "gauge") might also be a local symmetry of general relativity. After the development of quantum mechanics, Weyl, Vladimir Fock and Fritz London modified gauge by replacing the scale factor with a complex quantity and turned the scale transformation into a change of phase, which is a U(1) gauge symmetry. This explained the electromagnetic field effect on the wave function of a charged quantum mechanical particle. This was the first widely recognised gauge theory, popularised by Pauli in the 1940s.[1]

In 1954, attempting to resolve some of the great confusion in elementary particle physics, Chen Ning Yang and Robert Mills introduced **non-abelian gauge theories** as models to understand the strong interaction holding together nucleons in atomic nuclei. (Ronald Shaw, working under Abdus Salam, independently introduced the same notion in his doctoral thesis.) Generalizing the gauge invariance of electromagnetism, they attempted to construct a theory based on the action of the (non-abelian) SU(2) symmetry group on the isospin doublet of protons and neutrons. This is similar to the action of the U(1) group on the spinor fields of quantum electrodynamics. In particle physics the emphasis was on using **quantized gauge theories**.

This idea later found application in the quantum field theory of the weak force, and its unification with electromagnetism in the electroweak theory. Gauge theories became even more attractive when it was realized that non-abelian gauge theories reproduced a feature called asymptotic freedom. Asymptotic freedom was believed to be an important characteristic of strong interactions. This motivated searching for a strong force gauge theory. This theory, now known as quantum chromodynamics, is a gauge theory with the action of the SU(3) group on the color triplet of quarks. The Standard Model unifies the description of electromagnetism, weak interactions and strong interactions in the language of gauge theory.

In the 1970s, Sir Michael Atiyah began studying the mathematics of solutions to the classical Yang–Mills equations. In 1983, Atiyah's student Simon Donaldson built on this work to show that the differentiable classification of smooth 4-manifolds is very different from their classification up to homeomorphism. Michael Freedman used Donaldson's work to exhibit exotic $\mathbf{R}^4$s, that is, exotic differentiable structures on Euclidean 4-dimensional space. This led to an increasing interest in gauge theory for its own sake, independent of its successes in fundamental physics. In 1994, Edward Witten and Nathan Seiberg invented gauge-theoretic techniques based on supersymmetry that enabled the calculation of certain topological invariants (the Seiberg–Witten invariants). These contributions to mathematics from gauge theory have led to a renewed interest in this area.

The importance of gauge theories in physics is exemplified in the tremendous success of the mathematical formalism in providing a unified framework to describe the quantum field theories of electromagnetism, the weak force and the strong force. This theory, known as the Standard Model, accurately describes experimental predictions regarding three of the four fundamental forces of nature, and is a gauge theory with the gauge group SU(3) × SU(2) × U(1). Modern theories like string theory, as well as general relativity, are, in one way or another, gauge theories.

See Pickering[2] for more about the history of gauge and quantum field theories.

24.2 Description

24.2.1 Global and local symmetries

In physics, the mathematical description of any physical situation usually contains excess degrees of freedom; the same physical situation is equally well described by many equivalent mathematical configurations. For instance, in Newtonian dynamics, if two configurations are related by a Galilean transformation (an inertial change of reference frame) they represent the same physical situation. These transformations form a group of "symmetries" of the theory, and a physical situation corresponds not to an individual mathematical configuration but to a class of configurations related to one another by this symmetry group.

This idea can be generalized to include local as well as global symmetries, analogous to much more abstract "changes of coordinates" in a situation where there is no preferred "inertial" coordinate system that covers the entire physical system. A gauge theory is a mathematical model that has symmetries of this kind, together with a set of techniques for making physical predictions consistent with the symmetries of the model.

24.2.2 Example of global symmetry

When a quantity occurring in the mathematical configuration is not just a number but has some geometrical significance, such as a velocity or an axis of rotation, its representation as numbers arranged in a vector or matrix is also changed by a coordinate transformation. For instance, if one description of a pattern of fluid flow states that the fluid velocity in the neighborhood of (x=1, y=0) is 1 m/s in the positive x direction, then a description of the same situation in which the coordinate system has been rotated clockwise by 90 degrees states that the fluid velocity in the neighborhood of (x=0, y=1) is 1 m/s in the positive y direction. The coordinate transformation has affected both the coordinate system used to identify the *location* of the measurement and the basis in which its *value* is expressed. As long as this transformation is performed globally (affecting the coordinate basis in the same way at every point), the effect on values that represent the *rate of change* of some quantity along some path in space and time as it passes through point P is the same as the effect on values that are truly local to P.

24.2.3 Use of fiber bundles to describe local symmetries

In order to adequately describe physical situations in more complex theories, it is often necessary to introduce a "coordinate basis" for some of the objects of the theory that do not have this simple relationship to the coordinates used to label points in space and time. (In mathematical terms, the theory involves a fiber bundle in which the fiber at each point of the base space consists of possible coordinate bases for use when describing the values of objects at that point.) In order to spell out a mathematical configuration, one must choose a particular coordinate basis at each point (a *local section* of the fiber bundle) and express the values of the objects of the theory (usually "fields" in the physicist's sense) using this basis. Two such mathematical configurations are equivalent (describe the same physical situation) if they are related by a transformation of this abstract coordinate basis (a change of local section, or *gauge transformation*).

In most gauge theories, the set of possible transformations of the abstract gauge basis at an individual point in space and time is a finite-dimensional Lie group. The simplest such group is U(1), which appears in the modern formulation of quantum electrodynamics (QED) via its use of complex numbers. QED is generally regarded as the first, and simplest, physical gauge theory. The set of possible gauge transformations of the entire configuration of a given gauge theory also forms a group, the *gauge group* of the theory. An element of the gauge group can be parameterized by a smoothly varying function from the points of spacetime to the (finite-dimensional) Lie group, such that the value of the function and its derivatives at each point represents the action of the gauge transformation on the fiber over that point.

A gauge transformation with constant parameter at every point in space and time is analogous to a rigid rotation of the geometric coordinate system; it represents a global symmetry of the gauge representation. As in the case of a rigid rotation, this gauge transformation affects expressions that represent the rate of change along a path of some gauge-dependent quantity in the same way as those that represent a truly local quantity. A gauge transformation whose parameter is *not* a constant function is referred to as a local symmetry; its effect on expressions that involve a derivative is qualitatively different from that on expressions that don't. (This is analogous to a non-inertial change of reference frame, which can produce a Coriolis effect.)

24.2.4 Gauge fields

The "gauge covariant" version of a gauge theory accounts for this effect by introducing a gauge field (in mathematical language, an Ehresmann connection) and formulating all rates of change in terms of the covariant derivative with respect to this connection. The gauge field becomes an essential part of the description of a mathematical configuration. A configuration in which the gauge field can be eliminated by a gauge transformation has the property that its field strength (in mathematical language, its curvature) is zero everywhere; a gauge theory is *not* limited to these configurations. In other words, the distinguishing characteristic of a gauge theory is that the gauge field does not merely compensate for a poor choice of coordinate system; there is generally no gauge transformation that makes the gauge field vanish.

When analyzing the dynamics of a gauge theory, the gauge field must be treated as a dynamical variable, similarly to other objects in the description of a physical situation. In addition to its interaction with other objects via the covariant derivative, the gauge field typically contributes energy in the form of a "self-energy" term. One can obtain the equations for the gauge theory by:

- starting from a naïve ansatz without the gauge field (in which the derivatives appear in a "bare" form);

- listing those global symmetries of the theory that can be characterized by a continuous parameter (generally an abstract equivalent of a rotation angle);

- computing the correction terms that result from allowing the symmetry parameter to vary from place to place; and

- reinterpreting these correction terms as couplings to one or more gauge fields, and giving these fields appropriate self-energy terms and dynamical behavior.

This is the sense in which a gauge theory "extends" a global symmetry to a local symmetry, and closely resembles the historical development of the gauge theory of gravity known as general relativity.

24.2.5 Physical experiments

Gauge theories are used to model the results of physical experiments, essentially by:

- limiting the universe of possible configurations to those consistent with the information used to set up the experiment, and then

- computing the probability distribution of the possible outcomes that the experiment is designed to measure.

The mathematical descriptions of the "setup information" and the "possible measurement outcomes" (loosely speaking, the "boundary conditions" of the experiment) are generally not expressible without reference to a particular coordinate system, including a choice of gauge. (If nothing else, one assumes that the experiment has been adequately isolated from "external" influence, which is itself a gauge-dependent statement.) Mishandling gauge dependence in boundary conditions is a frequent source of anomalies in gauge theory calculations, and gauge theories can be broadly classified by their approaches to anomaly avoidance.

24.2.6 Continuum theories

The two gauge theories mentioned above (continuum electrodynamics and general relativity) are examples of continuum field theories. The techniques of calculation in a continuum theory implicitly assume that:

- given a completely fixed choice of gauge, the boundary conditions of an individual configuration can in principle be completely described;

- given a completely fixed gauge and a complete set of boundary conditions, the principle of least action determines a unique mathematical configuration (and therefore a unique physical situation) consistent with these bounds;

- the likelihood of possible measurement outcomes can be determined by:

  - establishing a probability distribution over all physical situations determined by boundary conditions that are consistent with the setup information,

  - establishing a probability distribution of measurement outcomes for each possible physical situation, and

- convolving these two probability distributions to get a distribution of possible measurement outcomes consistent with the setup information; and

- fixing the gauge introduces no anomalies in the calculation, due either to gauge dependence in describing partial information about boundary conditions or to incompleteness of the theory.

These assumptions are close enough to be valid across a wide range of energy scales and experimental conditions, to allow these theories to make accurate predictions about almost all of the phenomena encountered in daily life, from light, heat, and electricity to eclipses and spaceflight. They fail only at the smallest and largest scales (due to omissions in the theories themselves) and when the mathematical techniques themselves break down (most notably in the case of turbulence and other chaotic phenomena).

24.2.7   Quantum field theories

Other than these classical continuum field theories, the most widely known gauge theories are quantum field theories, including quantum electrodynamics and the Standard Model of elementary particle physics. The starting point of a quantum field theory is much like that of its continuum analog: a gauge-covariant action integral that characterizes "allowable" physical situations according to the principle of least action. However, continuum and quantum theories differ significantly in how they handle the excess degrees of freedom represented by gauge transformations. Continuum theories, and most pedagogical treatments of the simplest quantum field theories, use a gauge fixing prescription to reduce the orbit of mathematical configurations that represent a given physical situation to a smaller orbit related by a smaller gauge group (the global symmetry group, or perhaps even the trivial group).

More sophisticated quantum field theories, in particular those that involve a non-abelian gauge group, break the gauge symmetry within the techniques of perturbation theory by introducing additional fields (the Faddeev–Popov ghosts) and counterterms motivated by anomaly cancellation, in an approach known as BRST quantization. While these concerns are in one sense highly technical, they are also closely related to the nature of measurement, the limits on knowledge of a physical situation, and the interactions between incompletely specified experimental conditions and incompletely understood physical theory . The mathematical techniques that have been developed in order to make gauge theories tractable have found many other applications, from solid-state physics and crystallography to low-dimensional topology.

24.3   Classical gauge theory

24.3.1   Classical electromagnetism

Historically, the first example of gauge symmetry discovered was classical electromagnetism. In electrostatics, one can either discuss the electric field, $\mathbf{E}$, or its corresponding electric potential, V. Knowledge of one makes it possible to find the other, except that potentials differing by a constant, $V \to V + C$, correspond to the same electric field. This is because the electric field relates to *changes* in the potential from one point in space to another, and the constant C would cancel out when subtracting to find the change in potential. In terms of vector calculus, the electric field is the gradient of the potential, $\mathbf{E} = -\nabla V$. Generalizing from static electricity to electromagnetism, we have a second potential, the vector potential $\mathbf{A}$, with

$$\mathbf{E} = -\nabla V - \frac{\partial \mathbf{A}}{\partial t}$$
$$\mathbf{B} = \nabla \times \mathbf{A}$$

The general gauge transformations now become not just $V \to V + C$ but

$$\mathbf{A} \to \mathbf{A} + \nabla f$$
$$V \to V - \frac{\partial f}{\partial t}$$

where f is any function that depends on position and time. The fields remain the same under the gauge transformation, and therefore Maxwell's equations are still satisfied. That is, Maxwell's equations have a gauge symmetry.

24.3.2 An example: Scalar O(n) gauge theory

The remainder of this section requires some familiarity with classical or quantum field theory, and the use of Lagrangians.

Definitions in this section: gauge group, gauge field, interaction Lagrangian, gauge boson.

The following illustrates how local gauge invariance can be "motivated" heuristically starting from global symmetry properties, and how it leads to an interaction between originally non-interacting fields.

Consider a set of n non-interacting real scalar fields, with equal masses m. This system is described by an action that is the sum of the (usual) action for each scalar field φ_i

$$S = \int \mathrm{d}^4x \sum_{i=1}^{n} \left[\frac{1}{2} \partial_\mu \varphi_i \partial^\mu \varphi_i - \frac{1}{2} m^2 \varphi_i^2 \right]$$

The Lagrangian (density) can be compactly written as

$$\mathcal{L} = \frac{1}{2} (\partial_\mu \Phi)^T \partial^\mu \Phi - \frac{1}{2} m^2 \Phi^T \Phi$$

by introducing a vector of fields

$$\Phi = (\varphi_1, \varphi_2, \ldots, \varphi_n)^T$$

The term ∂_μ is Einstein notation for the partial derivative of Φ in each of the four dimensions.

It is now transparent that the Lagrangian is invariant under the transformation

$$\Phi \mapsto \Phi' = G\Phi$$

whenever G is a *constant* matrix belonging to the n-by-n orthogonal group O(n). This is seen to preserve the Lagrangian, since the derivative of Φ transforms identically to Φ and both quantities appear inside dot products in the Lagrangian (orthogonal transformations preserve the dot product).

$$(\partial_\mu \Phi) \mapsto (\partial_\mu \Phi)' = G\partial_\mu \Phi$$

This characterizes the *global* symmetry of this particular Lagrangian, and the symmetry group is often called the **gauge group**; the mathematical term is **structure group**, especially in the theory of G-structures. Incidentally, Noether's theorem implies that invariance under this group of transformations leads to the conservation of the *currents*

$$J_\mu^a = i\partial_\mu \Phi^T T^a \Phi$$

where the T^a matrices are generators of the SO(n) group. There is one conserved current for every generator.

Now, demanding that this Lagrangian should have *local* O(n)-invariance requires that the G matrices (which were earlier constant) should be allowed to become functions of the space-time coordinates x.

170   CHAPTER 24. GAUGE THEORY

In this case, the G matrices do not "pass through" the derivatives, when $G = G(x)$,

$$\partial_\mu(G\Phi) \neq G(\partial_\mu\Phi)$$

The failure of the derivative to commute with "G" introduces an additional term (in keeping with the product rule), which spoils the invariance of the Lagrangian. In order to rectify this we define a new derivative operator such that the derivative of Φ again transforms identically with Φ

$$(D_\mu\Phi)' = GD_\mu\Phi$$

This new "derivative" is called a (gauge) covariant derivative and takes the form

$$D_\mu = \partial_\mu + igA_\mu$$

Where g is called the coupling constant; a quantity defining the strength of an interaction. After a simple calculation we can see that the **gauge field** $A(x)$ must transform as follows

$$A'_\mu = GA_\mu G^{-1} + \frac{i}{g}(\partial_\mu G)G^{-1}$$

The gauge field is an element of the Lie algebra, and can therefore be expanded as

$$A_\mu = \sum_a A_\mu^a T^a$$

There are therefore as many gauge fields as there are generators of the Lie algebra.

Finally, we now have a *locally gauge invariant* Lagrangian

$$\mathcal{L}_{\text{loc}} = \frac{1}{2}(D_\mu\Phi)^T D^\mu\Phi - \frac{1}{2}m^2\Phi^T\Phi$$

Pauli uses the term *gauge transformation of the first type* to mean the transformation of Φ, while the compensating transformation in A is called a *gauge transformation of the second type*.

The difference between this Lagrangian and the original *globally gauge-invariant* Lagrangian is seen to be the **interaction Lagrangian**

$$\mathcal{L}_{\text{int}} = i\frac{g}{2}\Phi^T A_\mu^T \partial^\mu\Phi + i\frac{g}{2}(\partial_\mu\Phi)^T A^\mu\Phi - \frac{g^2}{2}(A_\mu\Phi)^T A^\mu\Phi$$

This term introduces interactions between the n scalar fields just as a consequence of the demand for local gauge invariance. However, to make this interaction physical and not completely arbitrary, the mediator $A(x)$ needs to propagate in space. That is dealt with in the next section by adding yet another term, $\mathcal{L}_{\text{gf}}$, to the Lagrangian. In the quantized version of the obtained classical field theory, the quanta of the gauge field $A(x)$ are called gauge bosons. The interpretation of the interaction Lagrangian in quantum field theory is of scalar bosons interacting by the exchange of these gauge bosons.

24.3.3   The Yang–Mills Lagrangian for the gauge field

Main article: Yang–Mills theory

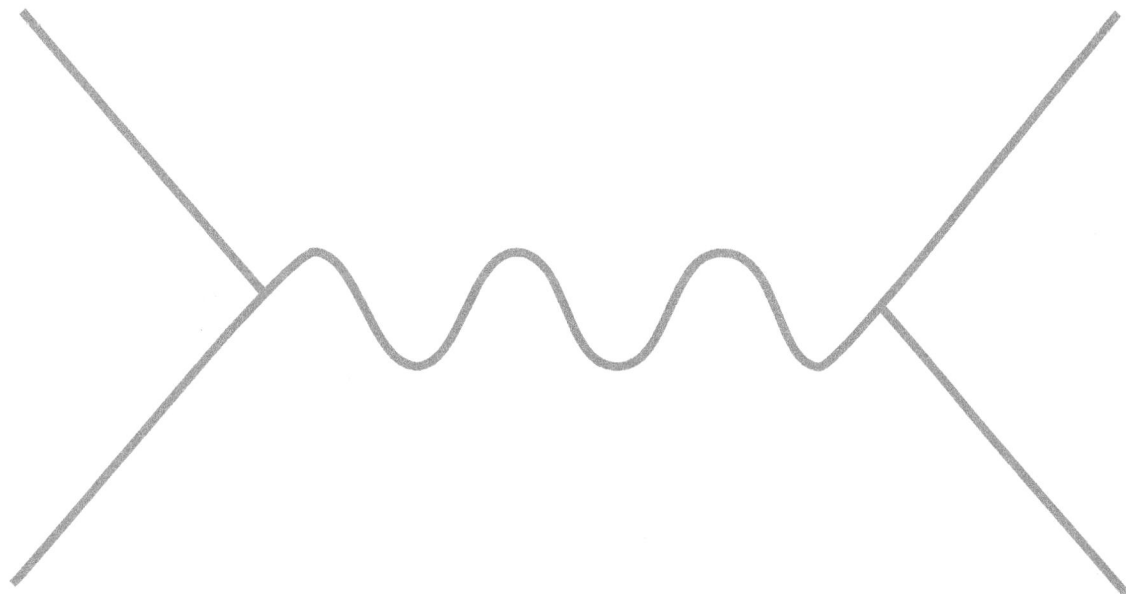

Feynman diagram of scalar bosons interacting via a gauge boson

The picture of a classical gauge theory developed in the previous section is almost complete, except for the fact that to define the covariant derivatives D, one needs to know the value of the gauge field $A(x)$ at all space-time points. Instead of manually specifying the values of this field, it can be given as the solution to a field equation. Further requiring that the Lagrangian that generates this field equation is locally gauge invariant as well, one possible form for the gauge field Lagrangian is (conventionally) written as

$$\mathcal{L}_{\text{gf}} = -\frac{1}{2}\,\text{Tr}(F^{\mu\nu}F_{\mu\nu})$$

with

$$F_{\mu\nu} = \frac{1}{ig}[D_\mu, D_\nu]$$

and the trace being taken over the vector space of the fields. This is called the **Yang–Mills action**. Other gauge invariant actions also exist (e.g., nonlinear electrodynamics, Born–Infeld action, Chern–Simons model, theta term, etc.).

Note that in this Lagrangian term there is no field whose transformation counterweighs the one of A. Invariance of this term under gauge transformations is a particular case of *a priori* classical (geometrical) symmetry. This symmetry must be restricted in order to perform quantization, the procedure being denominated gauge fixing, but even after restriction, gauge transformations may be possible.[3]

The complete Lagrangian for the gauge theory is now

$$\mathcal{L} = \mathcal{L}_{\text{loc}} + \mathcal{L}_{\text{gf}} = \mathcal{L}_{\text{global}} + \mathcal{L}_{\text{int}} + \mathcal{L}_{\text{gf}}$$

24.3.4 An example: Electrodynamics

As a simple application of the formalism developed in the previous sections, consider the case of electrodynamics, with only the electron field. The bare-bones action that generates the electron field's Dirac equation is

$$\mathcal{S} = \int \bar{\psi}(i\hbar c\,\gamma^\mu \partial_\mu - mc^2)\psi\,\mathrm{d}^4x$$

The global symmetry for this system is

$$\psi \mapsto e^{i\theta}\psi$$

The gauge group here is U(1), just rotations of the phase angle of the field, with the particular rotation determined by the constant θ.

"Localising" this symmetry implies the replacement of θ by $\theta(x)$. An appropriate covariant derivative is then

$$D_\mu = \partial_\mu - i\frac{e}{\hbar}A_\mu$$

Identifying the "charge" e (not to be confused with the mathematical constant e in the symmetry description) with the usual electric charge (this is the origin of the usage of the term in gauge theories), and the gauge field $A(x)$ with the four-vector potential of electromagnetic field results in an interaction Lagrangian

$$\mathcal{L}_{\text{int}} = \frac{e}{\hbar}\bar{\psi}(x)\gamma^\mu \psi(x)A_\mu(x) = J^\mu(x)A_\mu(x)$$

where $J^\mu(x)$ is the usual four vector electric current density. The gauge principle is therefore seen to naturally introduce the so-called minimal coupling of the electromagnetic field to the electron field.

Adding a Lagrangian for the gauge field $A_\mu(x)$ in terms of the field strength tensor exactly as in electrodynamics, one obtains the Lagrangian used as the starting point in quantum electrodynamics.

$$\mathcal{L}_{\text{QED}} = \bar{\psi}(i\hbar c\,\gamma^\mu D_\mu - mc^2)\psi - \frac{1}{4\mu_0}F_{\mu\nu}F^{\mu\nu}$$

> *See also: Dirac equation, Maxwell's equations, Quantum electrodynamics*

24.4   Mathematical formalism

Gauge theories are usually discussed in the language of differential geometry. Mathematically, a *gauge* is just a choice of a (local) section of some principal bundle. A **gauge transformation** is just a transformation between two such sections.

Although gauge theory is dominated by the study of connections (primarily because it's mainly studied by high-energy physicists), the idea of a connection is not central to gauge theory in general. In fact, a result in general gauge theory shows that affine representations (i.e., affine modules) of the gauge transformations can be classified as sections of a jet bundle satisfying certain properties. There are representations that transform covariantly pointwise (called by physicists gauge transformations of the first kind), representations that transform as a connection form (called by physicists gauge transformations of the second kind, an affine representation)—and other more general representations, such as the B field in BF theory. There are more general nonlinear representations (realizations), but these are extremely complicated. Still, nonlinear sigma models transform nonlinearly, so there are applications.

If there is a principal bundle P whose base space is space or spacetime and structure group is a Lie group, then the sections of P form a principal homogeneous space of the group of gauge transformations.

Connections (gauge connection) define this principal bundle, yielding a covariant derivative ∇ in each associated vector bundle. If a local frame is chosen (a local basis of sections), then this covariant derivative is represented by the connection form A, a Lie algebra-valued 1-form, which is called the **gauge potential** in physics. This is evidently not an intrinsic but a frame-dependent quantity. The curvature form Γ, a Lie algebra-valued 2-form that is an intrinsic quantity, is constructed from a connection form by

$$\mathbf{F} = d\mathbf{A} + \mathbf{A} \wedge \mathbf{A}$$

where d stands for the exterior derivative and $\wedge$ stands for the wedge product. ($\mathbf{A}$ is an element of the vector space spanned by the generators T^a, and so the components of $\mathbf{A}$ do not commute with one another. Hence the wedge product $\mathbf{A} \wedge \mathbf{A}$ does not vanish.)

Infinitesimal gauge transformations form a Lie algebra, which is characterized by a smooth Lie-algebra-valued scalar, ε. Under such an infinitesimal gauge transformation,

$$\delta_\varepsilon \mathbf{A} = [\varepsilon, \mathbf{A}] - d\varepsilon$$

where $[\cdot, \cdot]$ is the Lie bracket.

One nice thing is that if $\delta_\varepsilon X = \varepsilon X$, then $\delta_\varepsilon DX = \varepsilon DX$ where D is the covariant derivative

$$DX \stackrel{\text{def}}{=} dX + \mathbf{A}X$$

Also, $\delta_\varepsilon \mathbf{F} = \varepsilon \mathbf{F}$, which means $\mathbf{F}$ transforms covariantly.

Not all gauge transformations can be generated by infinitesimal gauge transformations in general. An example is when the base manifold is a compact manifold without boundary such that the homotopy class of mappings from that manifold to the Lie group is nontrivial. See instanton for an example.

The *Yang–Mills action* is now given by

$$\frac{1}{4g^2} \int \text{Tr}[*F \wedge F]$$

where * stands for the Hodge dual and the integral is defined as in differential geometry.

A quantity which is **gauge-invariant** (i.e., invariant under gauge transformations) is the Wilson loop, which is defined over any closed path, γ, as follows:

$$\chi^{(\rho)} \left(\mathcal{P} \left\{ e^{\int_\gamma A} \right\} \right)$$

where χ is the character of a complex representation ρ and $\mathcal{P}$ represents the path-ordered operator.

24.5 Quantization of gauge theories

Main article: Quantum gauge theory

Gauge theories may be quantized by specialization of methods which are applicable to any quantum field theory. However, because of the subtleties imposed by the gauge constraints (see section on Mathematical formalism, above) there are many technical problems to be solved which do not arise in other field theories. At the same time, the richer structure of gauge theories allows simplification of some computations: for example Ward identities connect different renormalization constants.

24.5.1 Methods and aims

The first gauge theory quantized was quantum electrodynamics (QED). The first methods developed for this involved gauge fixing and then applying canonical quantization. The Gupta–Bleuler method was also developed to handle this problem. Non-abelian gauge theories are now handled by a variety of means. Methods for quantization are covered in the article on quantization.

The main point to quantization is to be able to compute quantum amplitudes for various processes allowed by the theory. Technically, they reduce to the computations of certain correlation functions in the vacuum state. This involves a renormalization of the theory.

When the running coupling of the theory is small enough, then all required quantities may be computed in perturbation theory. Quantization schemes intended to simplify such computations (such as canonical quantization) may be called **perturbative quantization schemes**. At present some of these methods lead to the most precise experimental tests of gauge theories.

However, in most gauge theories, there are many interesting questions which are non-perturbative. Quantization schemes suited to these problems (such as lattice gauge theory) may be called **non-perturbative quantization schemes**. Precise computations in such schemes often require supercomputing, and are therefore less well-developed currently than other schemes.

24.5.2 Anomalies

Some of the symmetries of the classical theory are then seen not to hold in the quantum theory; a phenomenon called an **anomaly**. Among the most well known are:

- The scale anomaly, which gives rise to a *running coupling constant*. In QED this gives rise to the phenomenon of the Landau pole. In Quantum Chromodynamics (QCD) this leads to asymptotic freedom.

- The chiral anomaly in either chiral or vector field theories with fermions. This has close connection with topology through the notion of instantons. In QCD this anomaly causes the decay of a pion to two photons.

- The gauge anomaly, which must cancel in any consistent physical theory. In the electroweak theory this cancellation requires an equal number of quarks and leptons.

24.6 Pure gauge

A pure gauge is the set of field configurations obtained by a gauge transformation on the null-field configuration, i.e., a gauge-transform of zero. So it is a particular "gauge orbit" in the field configuration's space.

Thus, in the abelian case, where $A_\mu(x) \to A'_\mu(x) = A_\mu(x) + \partial_\mu f(x)$, the pure gauge is just the set of field configurations $A'_\mu(x) = \partial_\mu f(x)$ for all $f(x)$.

24.7 See also

24.8 References

[1] Wolfgang Pauli (1941) "Relativistic Field Theories of Elementary Particles," *Rev. Mod. Phys.* **13**: 203–32.

[2] Pickering, A. (1984). *Constructing Quarks*. University of Chicago Press. ISBN 0-226-66799-5.

[3] Sakurai, *Advanced Quantum Mechanics*, sect 1–4

24.9 Bibliography

General readers

- Schumm, Bruce (2004) *Deep Down Things*. Johns Hopkins University Press. Esp. chpt. 8. A serious attempt by a physicist to explain gauge theory and the Standard Model with little formal mathematics.

Texts

- Bromley, D.A. (2000). *Gauge Theory of Weak Interactions*. Springer. ISBN 3-540-67672-4.

- Cheng, T.-P.; Li, L.-F. (1983). *Gauge Theory of Elementary Particle Physics*. Oxford University Press. ISBN 0-19-851961-3.

- Frampton, P. (2008). *Gauge Field Theories* (3rd ed.). Wiley-VCH.

- Kane, G.L. (1987). *Modern Elementary Particle Physics*. Perseus Books. ISBN 0-201-11749-5.

Articles

- Becchi, C. (1997). "Introduction to Gauge Theories": 5211. arXiv:hep-ph/9705211. Bibcode:1997hep.ph....5211B.

- Gross, D. (1992). "Gauge theory – Past, Present and Future" (PDF). Retrieved 2009-04-23.

- Jackson, J.D. (2002). "From Lorenz to Coulomb and other explicit gauge transformations". *Am.J.Phys* **70** (9): 917–928. arXiv:physics/0204034. Bibcode:2002AmJPh..70..917J. doi:10.1119/1.1491265.

- Svetlichny, George (1999). "Preparation for Gauge Theory": 2027. arXiv:math-ph/9902027. Bibcode:1999

24.10 External links

- Hazewinkel, Michiel, ed. (2001), "Gauge transformation", *Encyclopedia of Mathematics*, Springer, ISBN 978-1-55608-010-4

- Yang–Mills equations on DispersiveWiki

- Gauge theories on Scholarpedia

Chapter 25

Continuous symmetry

In mathematics, **continuous symmetry** is an intuitive idea corresponding to the concept of viewing some symmetries as motions, as opposed to e.g. reflection symmetry, which is invariance under a kind of flip from one state to another.

25.1 Formalization

The notion of continuous symmetry has largely and successfully been formalised in the mathematical notions of topological group, Lie group and group action. For most practical purposes continuous symmetry is modelled by a *group action* of a topological group.

25.1.1 One-parameter subgroups

The simplest motions follow a one-parameter subgroup of a Lie group, such as the Euclidean group of three-dimensional space. For example translation parallel to the x-axis by u units, as u varies, is a one-parameter group of motions. Rotation around the z-axis is also a one-parameter group.

25.2 Noether's theorem

Continuous symmetry has a basic role in Noether's theorem in theoretical physics, in the derivation of conservation laws from symmetry principles, specifically for continuous symmetries. The search for continuous symmetries only intensified with the further developments of quantum field theory.

25.3 See also

- Goldstone's theorem

- Infinitesimal transformation

- Noether's theorem

- Sophus Lie

25.4 References

- William H. Barker, Roger Howe (2007), *Continuous Symmetry: from Euclid to Klein*

Chapter 26

Continuous function

In mathematics, a **continuous function** is, roughly speaking, a function for which small changes in the input result in small changes in the output. Otherwise, a function is said to be a *discontinuous* function. A continuous function with a continuous inverse function is called a homeomorphism.

Continuity of functions is one of the core concepts of topology, which is treated in full generality below. The introductory portion of this article focuses on the special case where the inputs and outputs of functions are real numbers. In addition, this article discusses the definition for the more general case of functions between two metric spaces. In order theory, especially in domain theory, one considers a notion of continuity known as Scott continuity. Other forms of continuity do exist but they are not discussed in this article.

As an example, consider the function $h(t)$, which describes the height of a growing flower at time t. This function is continuous. By contrast, if $M(t)$ denotes the amount of money in a bank account at time t, then the function jumps whenever money is deposited or withdrawn, so the function $M(t)$ is discontinuous.

26.1 History

A form of this epsilon-delta definition of continuity was first given by Bernard Bolzano in 1817. Augustin-Louis Cauchy defined continuity of $y = f(x)$ as follows: an infinitely small increment α of the independent variable x always produces an infinitely small change $f(x + \alpha) - f(x)$ of the dependent variable y (see e.g., *Cours d'Analyse*, p. 34). Cauchy defined infinitely small quantities in terms of variable quantities, and his definition of continuity closely parallels the infinitesimal definition used today (see microcontinuity). The formal definition and the distinction between pointwise continuity and uniform continuity were first given by Bolzano in the 1830s but the work wasn't published until the 1930s. Eduard Heine provided the first published definition of uniform continuity in 1872, but based these ideas on lectures given by Peter Gustav Lejeune Dirichlet in 1854.[1]

26.2 Real-valued continuous functions

26.2.1 Definition

A function from the set of real numbers to the real numbers can be represented by a graph in the Cartesian plane; such a function is continuous if, roughly speaking, the graph is a single unbroken curve with no "holes" or "jumps".

A function is *continuous at a point* if it does not have a hole or jump. A "hole" or "jump" in the graph of a function occurs if the value of the function at a point c differs from its limiting value along points that are nearby. Such a point is called a *discontinuity*. A function is then *continuous* if it has no holes or jumps: that is, if it is continuous at every point of its domain. Otherwise, a function is *discontinuous*, at the points where the value of the function differs from its limiting value (if any).

There are several ways to make this definition mathematically rigorous. These definitions are equivalent to one another, so the most convenient definition can be used to determine whether a given function is continuous or not. In the definitions below,

$$f : I \to \mathbf{R}.$$

is a function defined on a subset I of the set $\mathbf{R}$ of real numbers. This subset I is referred to as the domain of f. Some possible choices include $I=\mathbf{R}$, the whole set of real numbers, an open interval

$$I = (a, b) = \{x \in \mathbf{R} \mid a < x < b\},$$

or a closed interval

$$I = [a, b] = \{x \in \mathbf{R} \mid a \le x \le b\}.$$

Here, a and b are real numbers.

Definition in terms of limits of functions

The function f is *continuous at some point* c of its domain if the limit of $f(x)$ as x approaches c through the domain of f exists and is equal to $f(c)$.[2] In mathematical notation, this is written as

$$\lim_{x \to c} f(x) = f(c).$$

In detail this means three conditions: first, f has to be defined at c. Second, the limit on the left hand side of that equation has to exist. Third, the value of this limit must equal $f(c)$.

(We have here assumed that the domain of f does not have any isolated points. For example, an interval or union of intervals has no isolated points.)

Definition in terms of neighborhoods

A neighborhood of a point c is a set that contains all points of the domain within some fixed distance of c. Intuitively, a function is continuous at a point c if the range of the restriction of f to a neighborhood of c shrinks to a single point $f(c)$ as the width of the neighborhood shrinks to zero. More precisely, a function f is continuous at a point c of its domain if, for any neighborhood $N_1(f(c))$ there is a neighborhood $N_2(c)$ such that $f(x) \in N_1(f(c))$ whenever $x \in N_2(c)$.

This definition does not require any assumption on the nature of the domain. For instance, the function f is automatically continuous at every isolated point of its domain. As a specific example, every real valued function on the set of integers is continuous.

Definition in terms of limits of sequences

One can instead require that for any sequence $(x_n)_{n \in \mathbb{N}}$ of points in the domain which converges to c, the corresponding sequence $(f(x_n))_{n \in \mathbb{N}}$ converges to $f(c)$. In mathematical notation, $\forall (x_n)_{n \in \mathbb{N}} \subset I : \lim_{n \to \infty} x_n = c \Rightarrow \lim_{n \to \infty} f(x_n) = f(c)$.

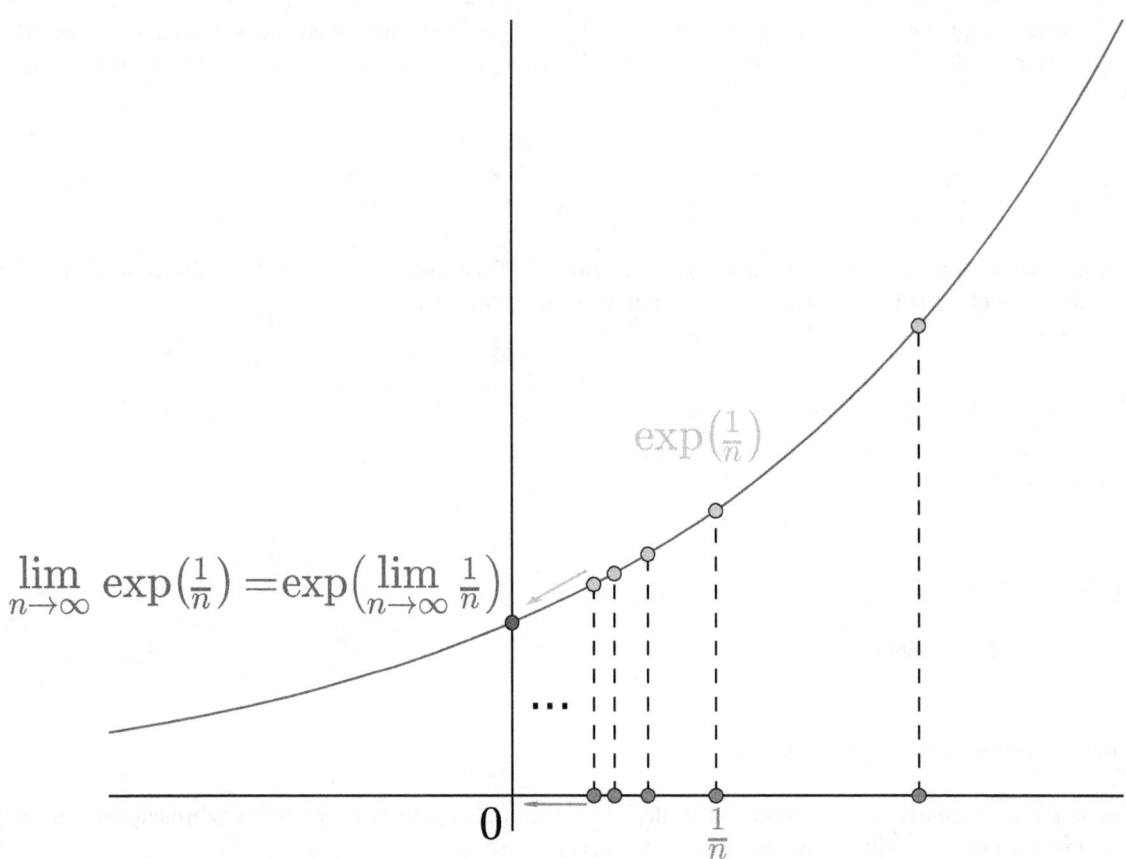

The sequence exp(1/n) converges to exp(0)

Weierstrass definition (epsilon–delta) of continuous functions

Explicitly including the definition of the limit of a function, we obtain a self-contained definition: Given a function f as above and an element c of the domain I, f is said to be continuous at the point c if the following holds: For any number $\varepsilon > 0$, however small, there exists some number $\delta > 0$ such that for all x in the domain of f with $c - \delta < x < c + \delta$, the value of $f(x)$ satisfies

$$f(c) - \varepsilon < f(x) < f(c) + \varepsilon.$$

Alternatively written, continuity of $f : I \to R$ at $c \in I$ means that for every $\varepsilon > 0$ there exists a $\delta > 0$ such that for all $x \in I$,:

$$|x - c| < \delta \Rightarrow |f(x) - f(c)| < \varepsilon.$$

More intuitively, we can say that if we want to get all the $f(x)$ values to stay in some small neighborhood around $f(c)$, we simply need to choose a small enough neighborhood for the x values around c, and we can do that no matter how small the $f(x)$ neighborhood is; f is then continuous at c.

In modern terms, this is generalized by the definition of continuity of a function with respect to a basis for the topology, here the metric topology.

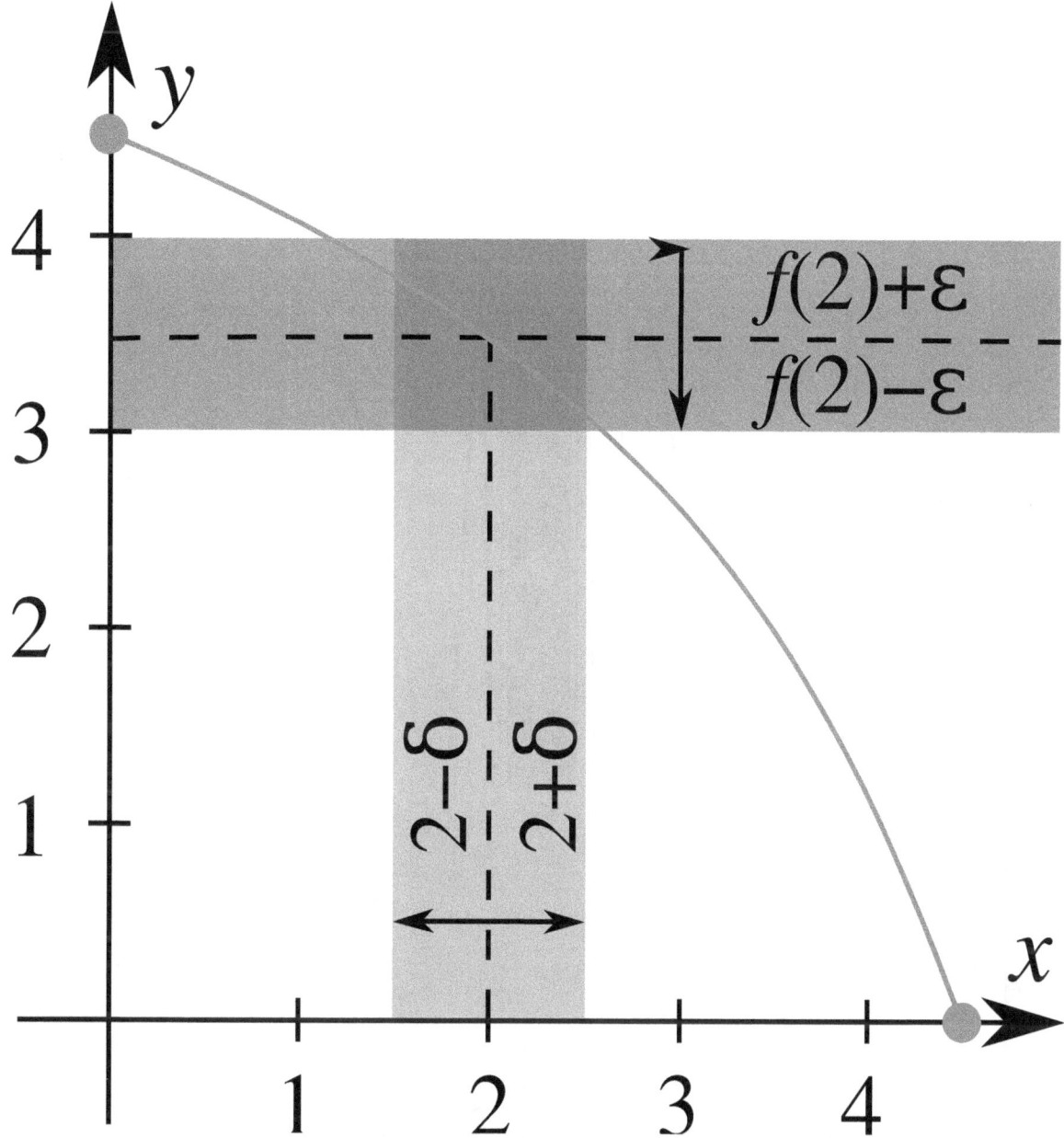

Illustration of the ε-δ-definition: for ε=0.5, c=2, the value δ=0.5 satisfies the condition of the definition.

Definition using oscillation

Continuity can also be defined in terms of oscillation: a function f is continuous at a point x_0 if and only if its oscillation at that point is zero;[3] in symbols, $\omega_f(x_0) = 0$. A benefit of this definition is that it *quantifies* discontinuity: the oscillation gives how *much* the function is discontinuous at a point.

This definition is useful in descriptive set theory to study the set of discontinuities and continuous points – the continuous points are the intersection of the sets where the oscillation is less than ε (hence a Gδ set) – and gives a very quick proof of one direction of the Lebesgue integrability condition.[4]

The oscillation is equivalent to the ε-δ definition by a simple re-arrangement, and by using a limit (lim sup, lim inf) to define oscillation: if (at a given point) for a given ε_0 there is no δ that satisfies the ε-δ definition, then the oscillation is at least ε_0, and conversely if for every ε there is a desired δ, the oscillation is 0. The oscillation definition can be naturally

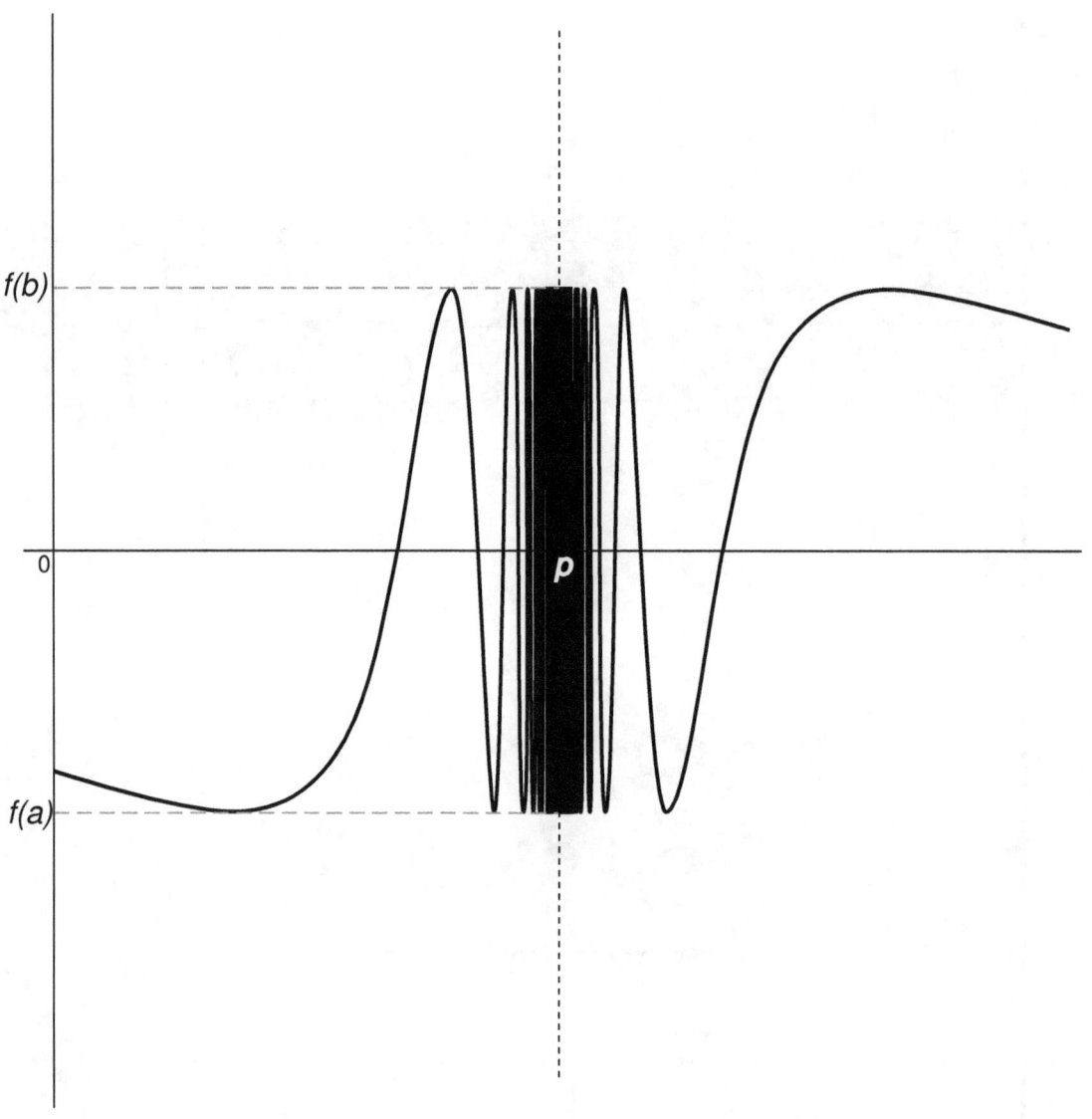

The failure of a function to be continuous at a point is quantified by its oscillation.

generalized to maps from a topological space to a metric space.

Definition using the hyperreals

Cauchy defined continuity of a function in the following intuitive terms: an infinitesimal change in the independent variable corresponds to an infinitesimal change of the dependent variable (see *Cours d'analyse*, page 34). Non-standard analysis is a way of making this mathematically rigorous. The real line is augmented by the addition of infinite and infinitesimal numbers to form the hyperreal numbers. In nonstandard analysis, continuity can be defined as follows.

> A real-valued function f is continuous at x if its natural extension to the hyperreals has the property that for all infinitesimal dx, $f(x+dx) - f(x)$ is infinitesimal[5]

(see microcontinuity). In other words, an infinitesimal increment of the independent variable always produces to an infinitesimal change of the dependent variable, giving a modern expression to Augustin-Louis Cauchy's definition of continuity.

26.2.2 Examples

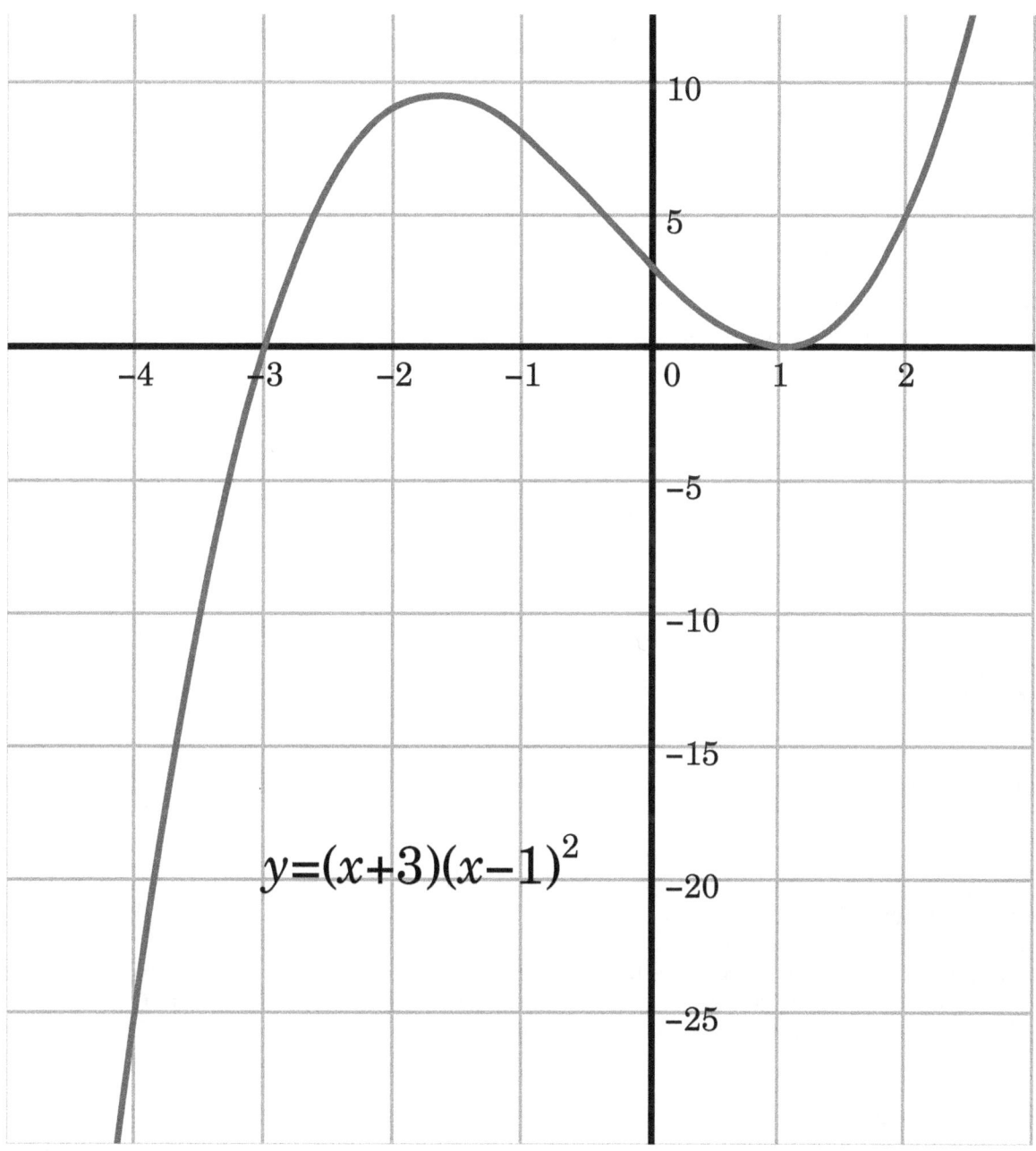

$$y=(x+3)(x-1)^2$$

The graph of a cubic function has no jumps or holes. The function is continuous.

All polynomial functions, such as $f(x) = x^3 + x^2 - 5x + 3$ (pictured), are continuous. This is a consequence of the fact that, given two continuous functions

$$f, g: I \to \mathbf{R}$$

defined on the same domain I, then the sum $f + g$ and the product fg of the two functions are continuous (on the same domain I). Moreover, the function

$$\frac{f}{g}\colon \{x \in I | g(x) \neq 0\} \to \mathbf{R}, x \mapsto \frac{f(x)}{g(x)}$$

is continuous. (The points where $g(x)$ is zero are discarded, as they are not in the domain of f/g.) For example, the function (pictured)

$$f(x) = \frac{2x - 1}{x + 2}$$

is defined for all real numbers $x \neq -2$ and is continuous at every such point. Thus it is a continuous function. The question of continuity at $x = -2$ does not arise, since $x = -2$ is not in the domain of f. There is no continuous function $F\colon \mathbf{R} \to \mathbf{R}$ that agrees with $f(x)$ for all $x \neq -2$. The sinc function $g(x) = (\sin x)/x$, defined for all $x\neq0$ is continuous at these points. Thus it is a continuous function, too. However, unlike the one of the previous example, this one *can* be extended to a continuous function on all real numbers, namely

$$G(x) = \begin{cases} \frac{\sin(x)}{x} & \text{if } x \neq 0 \\ 1 & \text{if } x = 0, \end{cases}$$

since the limit of $g(x)$, when x approaches 0, is 1. Therefore, the point $x=0$ is called a removable singularity of g.

Given two continuous functions

$$f\colon I \to J(\subset \mathbf{R}), g\colon J \to \mathbf{R},$$

the composition

$$g \circ f\colon I \to \mathbf{R}, x \mapsto g(f(x))$$

is continuous.

26.2.3   Non-examples

An example of a discontinuous function is the function f defined by $f(x) = 1$ if $x > 0$, $f(x) = 0$ if $x \leq 0$. Pick for instance $\varepsilon = 1/2$. There is no δ-neighborhood around $x = 0$ that will force all the $f(x)$ values to be within ε of $f(0)$. Intuitively we can think of this type of discontinuity as a sudden jump in function values. Similarly, the signum or sign function

$$\text{sgn}(x) = \begin{cases} 1 & \text{if } x > 0 \\ 0 & \text{if } x = 0 \\ -1 & \text{if } x < 0 \end{cases}$$

is discontinuous at $x = 0$ but continuous everywhere else. Yet another example: the function

$$f(x) = \begin{cases} \sin\left(\frac{1}{x^2}\right) & \text{if } x \neq 0 \\ 0 & \text{if } x = 0 \end{cases}$$

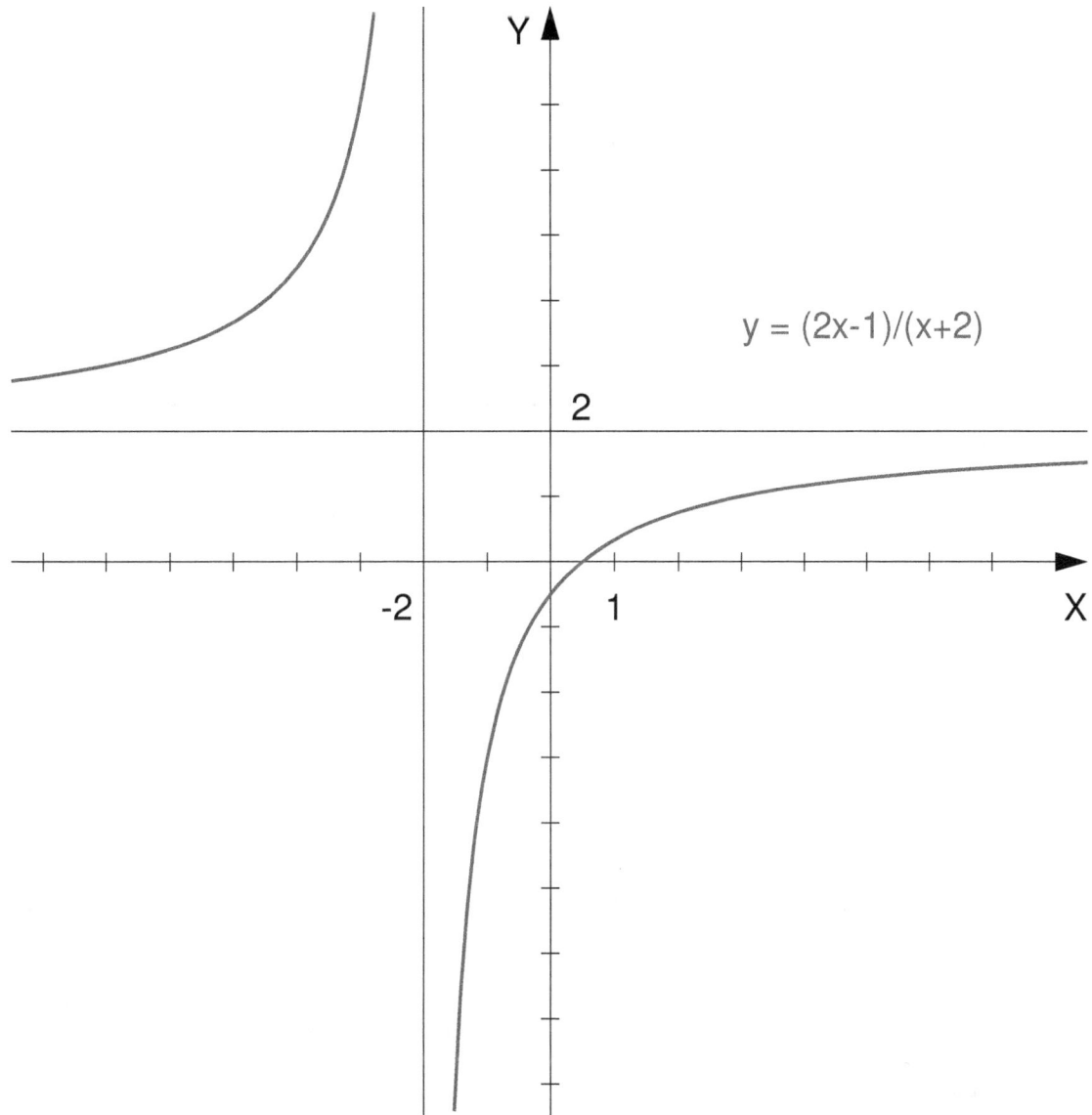

$y = (2x-1)/(x+2)$

The graph of a continuous rational function. The function is not defined for x=−2. The vertical and horizontal lines are asymptotes.

is continuous everywhere apart from $x = 0$.

Thomae's function,

$$f(x) = \begin{cases} 1 \text{ if } x = 0 \\ \frac{1}{q} \text{ if } x = \frac{p}{q}\text{number rational a is terms) lowest (in} \\ 0 \text{ if } x\text{irrational is .} \end{cases}$$

is continuous at all irrational numbers and discontinuous at all rational numbers. In a similar vein, Dirichlet's function

$$D(x) = \begin{cases} 0 \text{ if } x \text{ irrational is } (\in \mathbb{R} \setminus \mathbb{Q}) \\ 1 \text{ if } x \text{ rational is } (\in \mathbb{Q}) \end{cases}$$

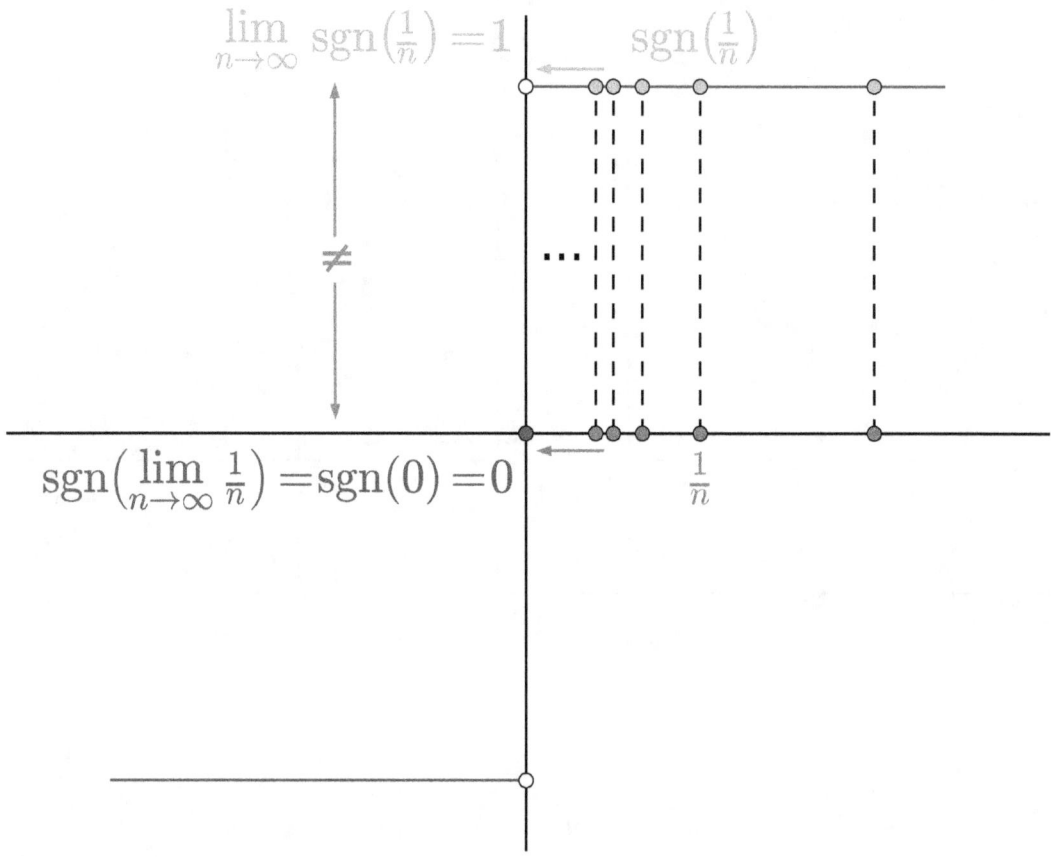

Plot of the signum function. It shows that $\lim_{n\to\infty} \mathrm{sgn}\left(\frac{1}{n}\right) \neq \mathrm{sgn}\left(\lim_{n\to\infty}\frac{1}{n}\right)$. Thus, the signum function is not continuous at the point 0.

is nowhere continuous.

26.2.4   Properties

Intermediate value theorem

The intermediate value theorem is an existence theorem, based on the real number property of completeness, and states:

> If the real-valued function f is continuous on the closed interval $[a, b]$ and k is some number between $f(a)$ and $f(b)$, then there is some number c in $[a, b]$ such that $f(c) = k$.

For example, if a child grows from 1 m to 1.5 m between the ages of two and six years, then, at some time between two and six years of age, the child's height must have been 1.25 m.

As a consequence, if f is continuous on $[a, b]$ and $f(a)$ and $f(b)$ differ in sign, then, at some point c in $[a, b]$, $f(c)$ must equal zero.

Extreme value theorem

The extreme value theorem states that if a function f is defined on a closed interval $[a,b]$ (or any closed and bounded set) and is continuous there, then the function attains its maximum, i.e. there exists $c \in [a,b]$ with $f(c) \geq f(x)$ for all $x \in [a,b]$.

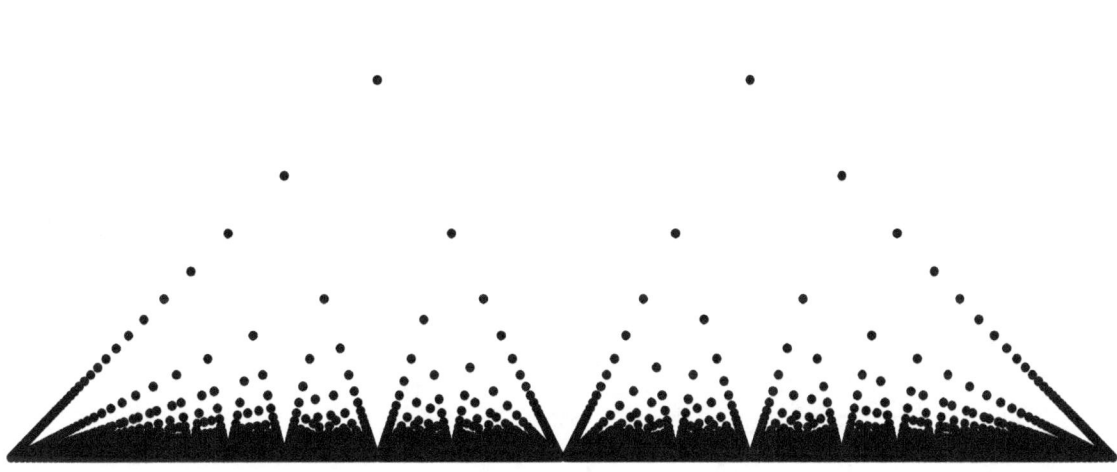

Plot of Thomae's function for the domain 0<x<1.

The same is true of the minimum of f. These statements are not, in general, true if the function is defined on an open interval (a,b) (or any set that is not both closed and bounded), as, for example, the continuous function $f(x) = 1/x$, defined on the open interval $(0,1)$, does not attain a maximum, being unbounded above.

Relation to differentiability and integrability

Every differentiable function

$$f : (a, b) \to \mathbf{R}$$

is continuous, as can be shown. The converse does not hold: for example, the absolute value function

$$f(x) = |x| = \begin{cases} x \text{ if } x \geq 0 \\ -x \text{ if } x < 0 \end{cases}$$

is everywhere continuous. However, it is not differentiable at $x = 0$ (but is so everywhere else). Weierstrass's function is also everywhere continuous but nowhere differentiable.

The derivative $f'(x)$ of a differentiable function $f(x)$ need not be continuous. If $f'(x)$ is continuous, $f(x)$ is said to be continuously differentiable. The set of such functions is denoted $C^1((a, b))$. More generally, the set of functions

$$f : \Omega \to \mathbf{R}$$

(from an open interval (or open subset of $\mathbf{R}$) Ω to the reals) such that f is n times differentiable and such that the n-th derivative of f is continuous is denoted $C^n(\Omega)$. See differentiability class. In the field of computer graphics, these three levels are sometimes called G^0 (continuity of position), G^1 (continuity of tangency), and G^2 (continuity of curvature).

Every continuous function

$f \colon [a, b] \to \mathbf{R}$

is integrable (for example in the sense of the Riemann integral). The converse does not hold, as the (integrable, but discontinuous) sign function shows.

Pointwise and uniform limits

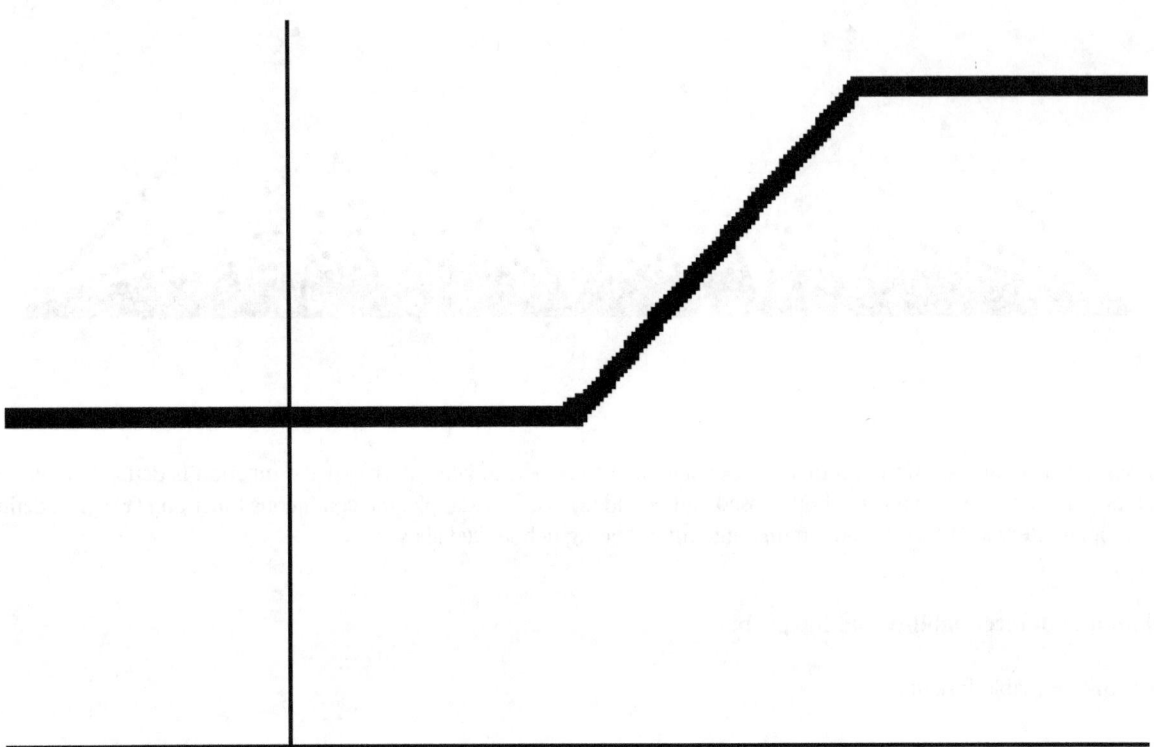

A sequence of continuous functions $f_n(x)$ *whose (pointwise) limit function* $f(x)$ *is discontinuous. The convergence is not uniform.*

Given a sequence

$f_1, f_2, \ldots \colon I \to \mathbf{R}$

of functions such that the limit

$$f(x) := \lim_{n \to \infty} f_n(x)$$

exists for all x in I, the resulting function $f(x)$ is referred to as the pointwise limit of the sequence of functions $(fn)n \in \mathbf{N}$. The pointwise limit function need not be continuous, even if all functions fn are continuous, as the animation at the right shows. However, f is continuous when the sequence converges uniformly, by the uniform convergence theorem. This theorem can be used to show that the exponential functions, logarithms, square root function, trigonometric functions are continuous.

26.2.5   Directional and semi-continuity

- A right-continuous function

- A left-continuous function

Discontinuous functions may be discontinuous in a restricted way, giving rise to the concept of directional continuity (or right and left continuous functions) and semi-continuity. Roughly speaking, a function is *right-continuous* if no jump occurs when the limit point is approached from the right. Formally, f is said to be right-continuous at the point c if the following holds: For any number $\varepsilon > 0$ however small, there exists some number $\delta > 0$ such that for all x in the domain with $c < x < c + \delta$, the value of $f(x)$ will satisfy

$$|f(x) - f(c)| < \varepsilon.$$

This is the same condition as for continuous functions, except that it is required to hold for x strictly larger than c only. Requiring it instead for all x with $c - \delta < x < c$ yields the notion of *left-continuous* functions. A function is continuous if and only if it is both right-continuous and left-continuous.

A function f is *lower semi-continuous* if, roughly, any jumps that might occur only go down, but not up. That is, for any $\varepsilon > 0$, there exists some number $\delta > 0$ such that for all x in the domain with $|x - c| < \delta$, the value of $f(x)$ satisfies

$$f(x) \geq f(c) - \epsilon.$$

The reverse condition is *upper semi-continuity*.

26.3 Continuous functions between metric spaces

The concept of continuous real-valued functions can be generalized to functions between metric spaces. A metric space is a set X equipped with a function (called metric) dX, that can be thought of as a measurement of the distance of any two elements in X. Formally, the metric is a function

$$d_X \colon X \times X \to \mathbf{R}$$

that satisfies a number of requirements, notably the triangle inequality. Given two metric spaces (X, dX) and (Y, dY) and a function

$$f \colon X \to Y$$

then f is continuous at the point c in X (with respect to the given metrics) if for any positive real number ε, there exists a positive real number δ such that all x in X satisfying $dX(x, c) < \delta$ will also satisfy $dY(f(x), f(c)) < \varepsilon$. As in the case of real functions above, this is equivalent to the condition that for every sequence (xn) in X with limit $\lim xn = c$, we have $\lim f(xn) = f(c)$. The latter condition can be weakened as follows: f is continuous at the point c if and only if for every convergent sequence (xn) in X with limit c, the sequence $(f(xn))$ is a Cauchy sequence, and c is in the domain of f.

The set of points at which a function between metric spaces is continuous is a Gδ set – this follows from the ε-δ definition of continuity.

This notion of continuity is applied, for example, in functional analysis. A key statement in this area says that a linear operator

$$T \colon V \to W$$

between normed vector spaces V and W (which are vector spaces equipped with a compatible norm, denoted $\|x\|$) is continuous if and only if it is bounded, that is, there is a constant K such that

$$\|T(x)\| \le K\|x\|$$

for all x in V.

26.3.1   Uniform, Hölder and Lipschitz continuity

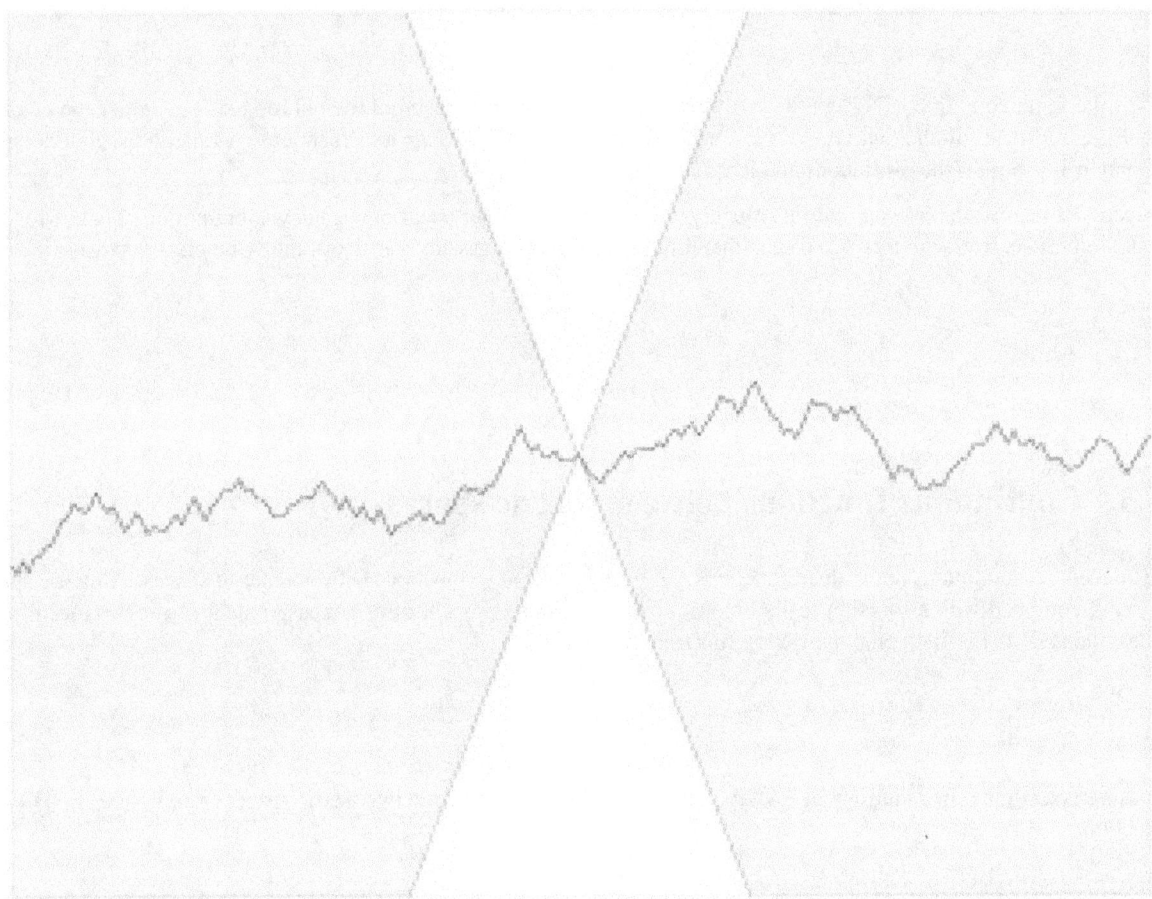

For a Lipschitz continuous function, there is a double cone (shown in white) whose vertex can be translated along the graph, so that the graph always remains entirely outside the cone.

The concept of continuity for functions between metric spaces can be strengthened in various ways by limiting the way δ depends on ε and c in the definition above. Intuitively, a function f as above is uniformly continuous if the δ does not depend on the point c. More precisely, it is required that for every real number $\varepsilon > 0$ there exists $\delta > 0$ such that for every $c, b \in X$ with $dX(b, c) < \delta$, we have that $dY(f(b), f(c)) < \varepsilon$. Thus, any uniformly continuous function is continuous. The converse does not hold in general, but holds when the domain space X is compact. Uniformly continuous maps can be defined in the more general situation of uniform spaces.[6]

A function is Hölder continuous with exponent α (a real number) if there is a constant K such that for all b and c in X, the inequality

$$d_Y(f(b), f(c)) \le K \cdot (d_X(b, c))^{\alpha}$$

holds. Any Hölder continuous function is uniformly continuous. The particular case $\alpha = 1$ is referred to as Lipschitz continuity. That is, a function is Lipschitz continuous if there is a constant K such that the inequality

$d_Y(f(b), f(c)) \leq K \cdot d_X(b,c)$

holds any b, c in X.[7] The Lipschitz condition occurs, for example, in the Picard–Lindelöf theorem concerning the solutions of ordinary differential equations.

26.4   Continuous functions between topological spaces

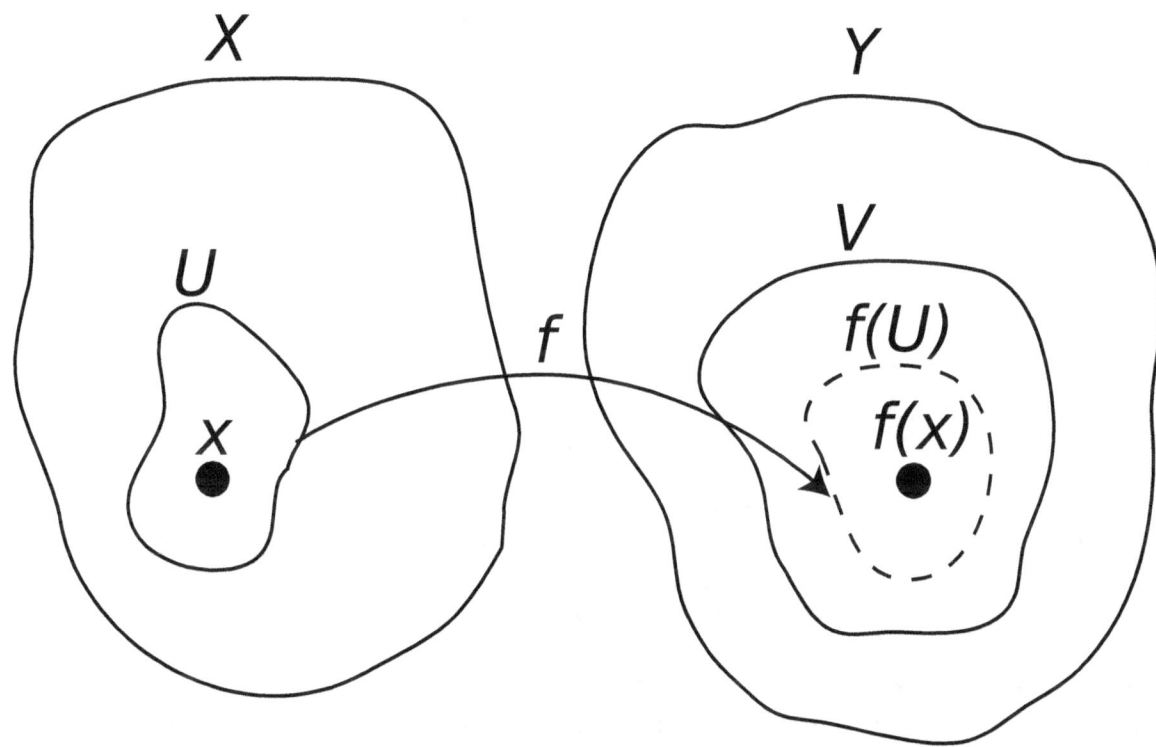

Continuity of a function at a point.

Another, more abstract, notion of continuity is continuity of functions between topological spaces in which there generally is no formal notion of distance, as there is in the case of metric spaces. A topological space is a set X together with a topology on X, which is a set of subsets of X satisfying a few requirements with respect to their unions and intersections that generalize the properties of the open balls in metric spaces while still allowing to talk about the neighbourhoods of a given point. The elements of a topology are called open subsets of X (with respect to the topology).

A function

$f: X \to Y$

between two topological spaces X and Y is continuous if for every open set $V \subseteq Y$, the inverse image

$f^{-1}(V) = \{x \in X \mid f(x) \in V\}$

is an open subset of X. That is, f is a function between the sets X and Y (not on the elements of the topology TX), but the continuity of f depends on the topologies used on X and Y.

This is equivalent to the condition that the preimages of the closed sets (which are the complements of the open subsets) in Y are closed in X.

An extreme example: if a set X is given the discrete topology (in which every subset is open), all functions

$$f\colon X \to T$$

to any topological space T are continuous. On the other hand, if X is equipped with the indiscrete topology (in which the only open subsets are the empty set and X) and the space T set is at least T_0, then the only continuous functions are the constant functions. Conversely, any function whose range is indiscrete is continuous.

26.4.1 Alternative definitions

Several equivalent definitions for a topological structure exist and thus there are several equivalent ways to define a continuous function.

Neighborhood definition

Neighborhoods continuity for functions between topological spaces $(X, \mathcal{T}_X)$ and $(Y, \mathcal{T}_Y)$ at a point may be defined: A function $f : X \to Y$ is continuous at a point $x \in X$ iff for any neighborhood of its image $f(x) \in Y$ the preimage is again a neighborhood of that point: $\forall N \in \mathcal{N}_{f(x)} : f^{-1}(N) \in \mathcal{M}_x$

According to the property that neighborhood systems being upper sets this can be restated as follows:
$\forall N \in \mathcal{N}_{f(x)} \exists M \in \mathcal{M}_x : M \subseteq f^{-1}(N)$
$\forall N \in \mathcal{N}_{f(x)} \exists M \in \mathcal{M}_x : f(M) \subseteq N$
The second one being a restatement involving the image rather than the preimage.
Literally, this means no matter how small the neighborhood is chosen one can always find a neighborhood mapped into it.

Besides, there's a simplification involving only open neighborhoods. In fact, they're equivalent:
$\forall V \in \mathcal{T}_Y, f(x) \in V \exists U \in \mathcal{T}_X, x \in U : U \subseteq f^{-1}(V)$
$\forall V \in \mathcal{T}_Y, f(x) \in V \exists U \in \mathcal{T}_X, x \in U : f(U) \subseteq V$
The second one again being a restatement using images rather than preimages.

If X and Y are metric spaces, it is equivalent to consider the neighborhood system of open balls centered at x and $f(x)$ instead of all neighborhoods. This gives back the above δ-ε definition of continuity in the context of metric spaces. However, in general topological spaces, there is no notion of nearness or distance.

Note, however, that if the target space is Hausdorff, it is still true that f is continuous at a if and only if the limit of f as x approaches a is $f(a)$. At an isolated point, every function is continuous.

Sequences and nets

In several contexts, the topology of a space is conveniently specified in terms of limit points. In many instances, this is accomplished by specifying when a point is the limit of a sequence, but for some spaces that are too large in some sense, one specifies also when a point is the limit of more general sets of points indexed by a directed set, known as nets. A function is (Heine-)continuous only if it takes limits of sequences to limits of sequences. In the former case, preservation of limits is also sufficient; in the latter, a function may preserve all limits of sequences yet still fail to be continuous, and preservation of nets is a necessary and sufficient condition.

In detail, a function $f\colon X \to Y$ is **sequentially continuous** if whenever a sequence (xn) in X converges to a limit x, the sequence $(f(xn))$ converges to $f(x)$. Thus sequentially continuous functions "preserve sequential limits". Every continuous function is sequentially continuous. If X is a first-countable space and countable choice holds, then the converse also holds: any function preserving sequential limits is continuous. In particular, if X is a metric space, sequential continuity and continuity are equivalent. For non first-countable spaces, sequential continuity might be strictly weaker than continuity. (The spaces for which the two properties are equivalent are called sequential spaces.) This motivates the consideration of nets instead of sequences in general topological spaces. Continuous functions preserve limits of nets, and in fact this property characterizes continuous functions.

Closure operator definition

Instead of specifying the open subsets of a topological space, the topology can also be determined by a closure operator (denoted cl) which assigns to any subset $A \subseteq X$ its closure, or an interior operator (denoted int), which assigns to any subset A of X its interior. In these terms, a function

$$f \colon (X, \mathrm{cl}) \to (X', \mathrm{cl}')$$

between topological spaces is continuous in the sense above if and only if for all subsets A of X

$$f(\mathrm{cl}(A)) \subseteq \mathrm{cl}'(f(A)).$$

That is to say, given any element x of X that is in the closure of any subset A, $f(x)$ belongs to the closure of $f(A)$. This is equivalent to the requirement that for all subsets A' of X'

$$f^{-1}(\mathrm{cl}'(A')) \supseteq \mathrm{cl}(f^{-1}(A')).$$

Moreover,

$$f \colon (X, \mathrm{int}) \to (X', \mathrm{int}')$$

is continuous if and only if

$$f^{-1}(\mathrm{int}'(A')) \subseteq \mathrm{int}(f^{-1}(A'))$$

for any subset A' of Y.

26.4.2   Properties

If $f \colon X \to Y$ and $g \colon Y \to Z$ are continuous, then so is the composition $g \circ f \colon X \to Z$. If $f \colon X \to Y$ is continuous and

- X is compact, then $f(X)$ is compact.
- X is connected, then $f(X)$ is connected.
- X is path-connected, then $f(X)$ is path-connected.
- X is Lindelöf, then $f(X)$ is Lindelöf.
- X is separable, then $f(X)$ is separable.

The possible topologies on a fixed set X are partially ordered: a topology τ_1 is said to be coarser than another topology τ_2 (notation: $\tau_1 \subseteq \tau_2$) if every open subset with respect to τ_1 is also open with respect to τ_2. Then, the identity map

$$\mathrm{id}X \colon (X, \tau_2) \to (X, \tau_1)$$

is continuous if and only if $\tau_1 \subseteq \tau_2$ (see also comparison of topologies). More generally, a continuous function

$$(X, \tau_X) \to (Y, \tau_Y)$$

stays continuous if the topology τY is replaced by a coarser topology and/or τX is replaced by a finer topology.

26.4.3   Homeomorphisms

Symmetric to the concept of a continuous map is an open map, for which *images* of open sets are open. In fact, if an open map f has an inverse function, that inverse is continuous, and if a continuous map g has an inverse, that inverse is open. Given a bijective function f between two topological spaces, the inverse function f^{-1} need not be continuous. A bijective continuous function with continuous inverse function is called a *homeomorphism*.

If a continuous bijection has as its domain a compact space and its codomain is Hausdorff, then it is a homeomorphism.

26.4.4   Defining topologies via continuous functions

Given a function

$$f \colon X \to S,$$

where X is a topological space and S is a set (without a specified topology), the final topology on S is defined by letting the open sets of S be those subsets A of S for which $f^{-1}(A)$ is open in X. If S has an existing topology, f is continuous with respect to this topology if and only if the existing topology is coarser than the final topology on S. Thus the final topology can be characterized as the finest topology on S that makes f continuous. If f is surjective, this topology is canonically identified with the quotient topology under the equivalence relation defined by f.

Dually, for a function f from a set S to a topological space, the initial topology on S has as open subsets A of S those subsets for which $f(A)$ is open in X. If S has an existing topology, f is continuous with respect to this topology if and only if the existing topology is finer than the initial topology on S. Thus the initial topology can be characterized as the coarsest topology on S that makes f continuous. If f is injective, this topology is canonically identified with the subspace topology of S, viewed as a subset of X.

More generally, given a set S, specifying the set of continuous functions

$$S \to X$$

into all topological spaces X defines a topology. Dually, a similar idea can be applied to maps

$$X \to S.$$

This is an instance of a universal property.

26.5   Related notions

Various other mathematical domains use the concept of continuity in different, but related meanings. For example, in order theory, an order-preserving function $f \colon X \to Y$ between particular types of partially ordered sets X and Y is continuous if for each directed subset A of X, we have $\sup(f(A)) = f(\sup(A))$. Here sup is the supremum with respect to the orderings in X and Y, respectively. This notion of continuity is the same as topological continuity when the partially ordered sets are given the Scott topology.[8][9]

In category theory, a functor

$$F \colon \mathcal{C} \to \mathcal{D}$$

between two categories is called *continuous*, if it commutes with small limits. That is to say,

$$\varprojlim_{i \in I} F(C_i) \cong F(\varprojlim_{i \in I} C_i)$$

for any small (i.e., indexed by a set I, as opposed to a class) diagram of objects in $\mathcal{C}$.

A *continuity space* is a generalization of metric spaces and posets,[10][11] which uses the concept of quantales, and that can be used to unify the notions of metric spaces and domains.[12]

26.6   See also

- Absolute continuity
- Classification of discontinuities
- Coarse function
- Continuous stochastic process
- Dini continuity
- Discrete function
- Equicontinuity
- Normal function
- Piecewise
- Symmetrically continuous function

26.7   Notes

[1] Rusnock, P.; Kerr-Lawson, A. (2005), "Bolzano and uniform continuity", *Historia Mathematica* **32** (3): 303–311, doi:10.

[2] Lang, Serge (1997), *Undergraduate analysis*, Undergraduate Texts in Mathematics (2nd ed.), Berlin, New York: Springer-Verlag, ISBN 978-0-387-94841-6, section II.4

[3] *Introduction to Real Analysis,* updated April 2010, William F. Trench, Theorem 3.5.2, p. 172

[4] *Introduction to Real Analysis,* updated April 2010, William F. Trench, 3.5 "A More Advanced Look at the Existence of the Proper Riemann Integral", pp. 171–177

[5] "Elementary Calculus". *wisc.edu.*

[6] Gaal, Steven A. (2009), *Point set topology*, New York: Dover Publications, ISBN 978-0-486-47222-5, section IV.10

[7] Searcóid, Mícheál Ó (2006), *Metric spaces*, Springer undergraduate mathematics series, Berlin, New York: Springer-Verlag, ISBN 978-1-84628-369-7, section 9.4

[8] Goubault-Larrecq, Jean (2013). *Non-Hausdorff Topology and Domain Theory: Selected Topics in Point-Set Topology.* Cambridge University Press. ISBN 1107034132.

[9] Gierz, G.; Hofmann, K. H.; Keimel, K.; Lawson, J. D.; Mislove, M. W.; Scott, D. S. (2003). *Continuous Lattices and Domains.* Encyclopedia of Mathematics and its Applications **93**. Cambridge University Press. ISBN 0521803381.

[10] Flagg, R. C. (1997). "Quantales and continuity spaces". *Algebra Universalis.* CiteSeerX: 10.1.1.48.851.

[11] Kopperman, R. (1988). "All topologies come from generalized metrics". *American Mathematical Monthly* **95** (2): 89–97. doi:10.2307/2323060.

[12] Flagg, B.; Kopperman, R. (1997). "Continuity spaces: Reconciling domains and metric spaces". *Theoretical Computer Science* **177** (1): 111–138. doi:10.1016/S0304-3975(97)00236-3.

26.8   References

- Hazewinkel, Michiel, ed. (2001), "Continuous function", *Encyclopedia of Mathematics*, Springer, ISBN 978-1-55608-010-4

- Visual Calculus by Lawrence S. Husch, University of Tennessee (2001).

Chapter 27

Smoothness

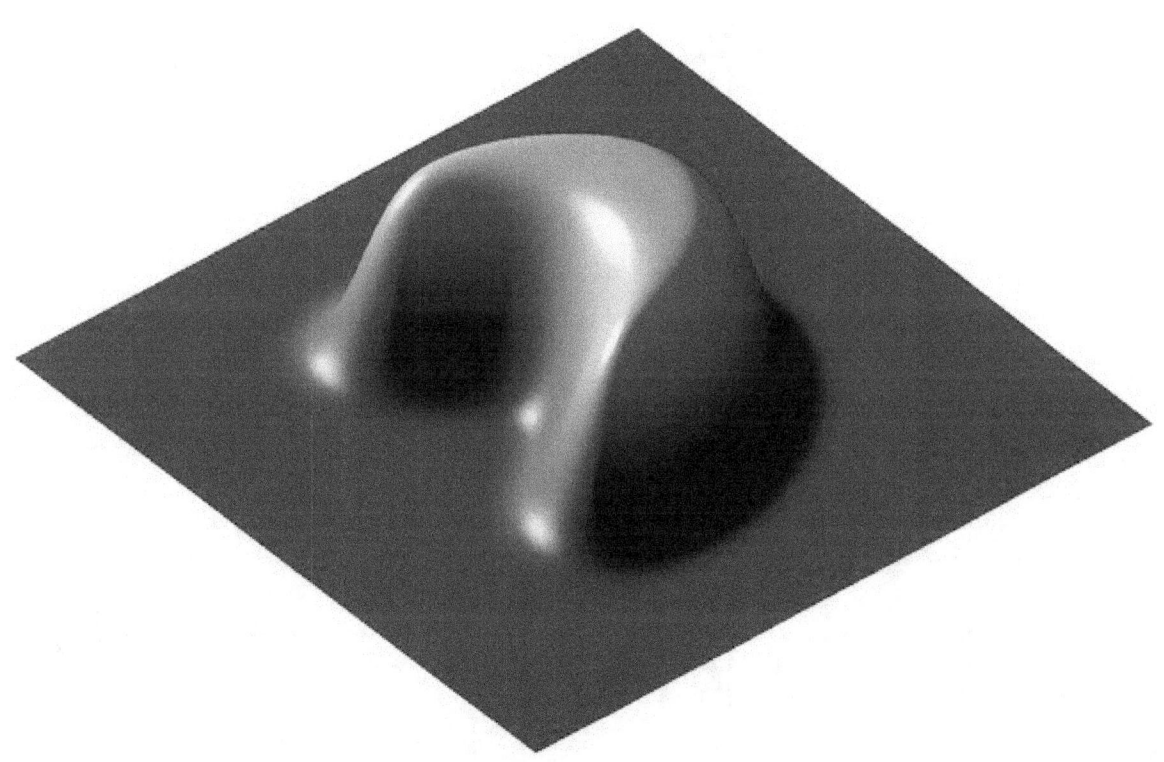

A bump function is a smooth function with compact support.

In mathematical analysis, the **smoothness** of a function is a property measured by the number of derivatives it has which are continuous. A **smooth function** is a function that has derivatives of all orders everywhere in its domain.

27.1 Differentiability classes

Differentiability class is a classification of functions according to the properties of their derivatives. Higher order differentiability classes correspond to the existence of more derivatives.

Consider an open set on the real line and a function f defined on that set with real values. Let k be a non-negative integer. The function f is said to be of (differentiability) **class C^k** if the derivatives f', f'', ..., $f^{(k)}$ exist and are continuous (the continuity is implied by differentiability for all the derivatives except for $f^{(k)}$). The function f is said to be of **class C^∞**, or

smooth, if it has derivatives of all orders.[1] The function f is said to be of **class C^ω**, or **analytic**, if f is smooth *and* if it its Taylor series expansion around any point in its domain converges to the function in some neighborhood of the point. C^ω is thus strictly contained in C^∞. Bump functions are examples of functions in C^∞ but *not* in C^ω.

To put it differently, the class C^0 consists of all continuous functions. The class C^1 consists of all differentiable functions whose derivative is continuous; such functions are called **continuously differentiable**. Thus, a C^1 function is exactly a function whose derivative exists and is of class C^0. In general, the classes C^k can be defined recursively by declaring C^0 to be the set of all continuous functions and declaring C^k for any positive integer k to be the set of all differentiable functions whose derivative is in C^{k-1}. In particular, C^k is contained in C^{k-1} for every k, and there are examples to show that this containment is strict. C^∞, the class of **infinitely differentiable** functions, is the intersection of the sets C^k as k varies over the non-negative integers (i.e. from 0 to ∞).

27.1.1 Examples

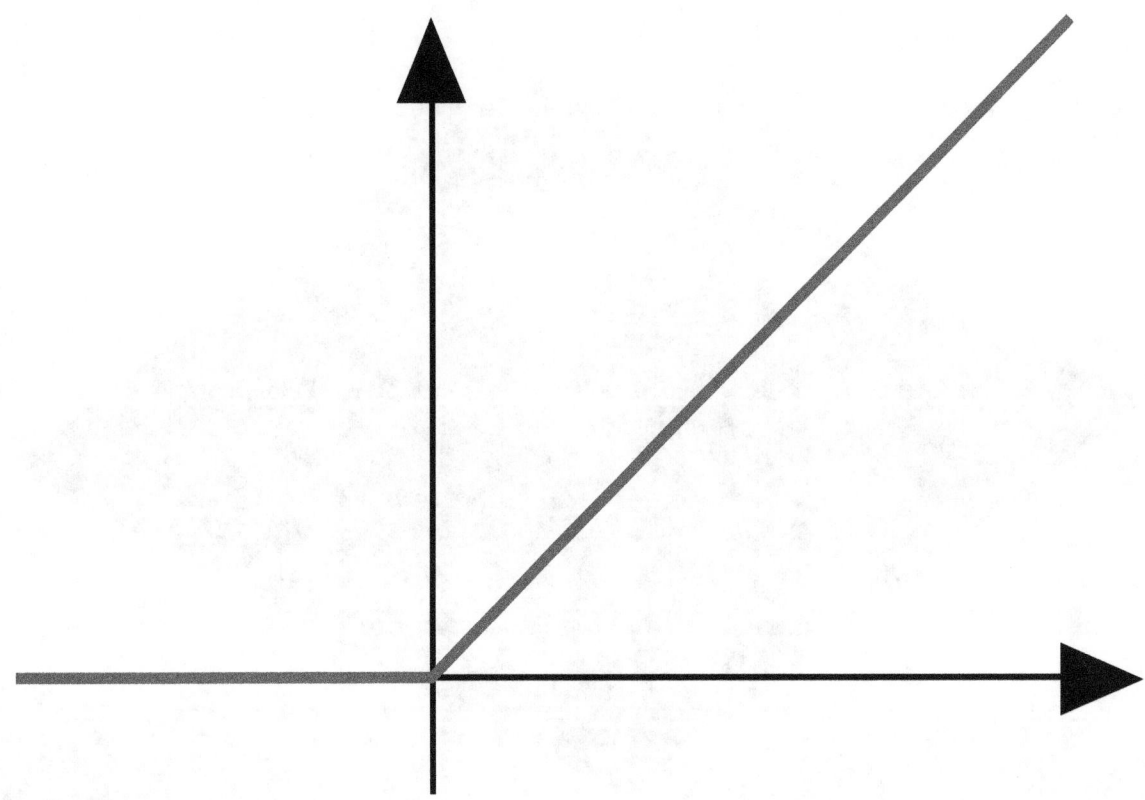

The C^0 function f(x)=x for x≥0 and 0 otherwise.

The function

$$f(x) = \begin{cases} x & \text{if } x \geq 0, \\ 0 & \text{if } x < 0 \end{cases}$$

is continuous, but not differentiable at $x = 0$, so it is of class C^0 but not of class C^1.

The function

$$f(x) = \begin{cases} x^2 \sin\left(\frac{1}{x}\right) & \text{if } x \neq 0, \\ 0 & \text{if } x = 0 \end{cases}$$

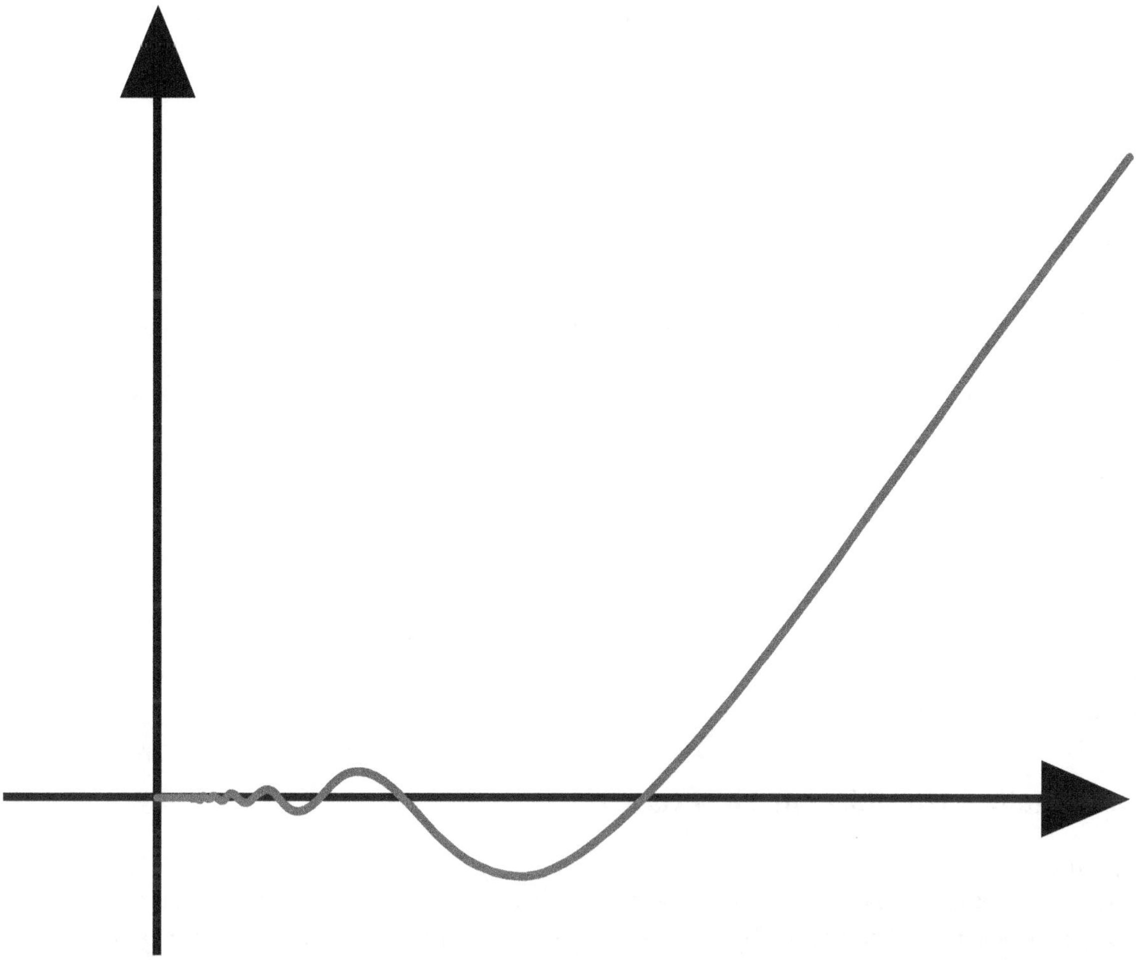

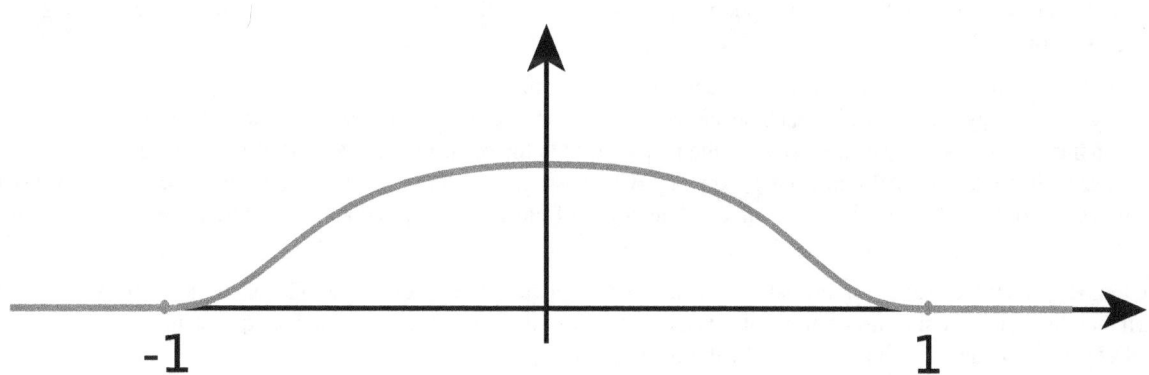

The function f(x)=x^2 *sin(1/x) for* x>0.

A smooth function that is not analytic.

is differentiable, with derivative

$$f'(x) = \begin{cases} -\cos(\frac{1}{x}) + 2x\sin(\frac{1}{x}) & \text{if } x \neq 0, \\ 0 & \text{if } x = 0. \end{cases}$$

Because $\cos(1/x)$ oscillates as $x \to 0$, $f'(x)$ is not continuous at zero. Therefore, this function is differentiable but not of class C^1. Moreover, if one takes $f(x) = x^{4/3}\sin(1/x)$ ($x \neq 0$) in this example, it can be used to show that the derivative function of a differentiable function can be unbounded on a compact set and, therefore, that a differentiable function on a compact set may not be locally Lipschitz continuous.

The functions

$$f(x) = |x|^{k+1}$$

where k is even, are continuous and k times differentiable at all x. But at $x = 0$ they are not $(k+1)$ times differentiable, so they are of class C^k but not of class C^j where $j > k$.

The exponential function is analytic, so, of class C^ω. The trigonometric functions are also analytic wherever they are defined.

The function

$$f(x) = \begin{cases} e^{-\frac{1}{1-x^2}} & \text{if } |x| < 1, \\ 0 & \text{otherwise} \end{cases}$$

is smooth, so of class C^∞, but it is not analytic at $x = \pm 1$, so it is not of class C^ω. The function f is an example of a smooth function with compact support.

27.1.2 Multivariate differentiability classes

Let n and m be some positive integers. If f is a function from an open subset of $\mathbf{R}^n$ with values in $\mathbf{R}^m$, then f has component functions $f_1, ..., fm$. Each of these may or may not have partial derivatives. For a non-negative integer ℓ, we say that f is of **class C^ℓ** if all of the partial derivatives $\frac{\partial^\ell f_i}{\partial x_{i_1}^{\ell_1} \partial x_{i_2}^{\ell_2} \cdots \partial x_{i_k}^{\ell_k}}$ exist and are continuous, where k is a non-negative integer, i is an integer between 1 and m, each of $i_1, i_2, \ldots, i_k$ is an integer between 1 and n, each of $\ell, \ell_1, \ell_2, \ldots, \ell_k$ is an integer between 0 and ℓ, and $\ell_1 + \ell_2 + \cdots + \ell_k = \ell$.[1] The classes C^∞ and C^ω are defined as before.[1]

These criteria of differentiability can be applied to the transition functions of a differential structure. The resulting space is called a C^k manifold.

If one wishes to start with a coordinate-independent definition of the **class C^k**, one may start by considering maps between Banach spaces. A map from one Banach space to another is differentiable at a point if there is an affine map which approximates it at that point. The derivative of the map assigns to the point x the linear part of the affine approximation to the map at x. Since the space of linear maps from one Banach space to another is again a Banach space, we may continue this procedure to define higher order derivatives. A map f is of **class C^k** if it has continuous derivatives up to order k, as before.

Note that $\mathbf{R}^n$ is a Banach space for any value of n, so the coordinate-free approach is applicable in this instance. It can be shown that the definition in terms of partial derivatives and the coordinate-free approach are equivalent; that is, a function f is of **class C^k** by one definition iff it is so by the other definition.

27.1.3 The space of C^k functions

Let D be an open subset of the real line. The set of all C^k functions defined on D and taking real values is a Fréchet vector space with the countable family of seminorms

$$p_{K,m} = \sup_{x \in K} \left| f^{(m)}(x) \right|$$

where K varies over an increasing sequence of compact sets whose union is D, and $m = 0, 1, \ldots, k$.

The set of C^∞ functions over D also forms a Fréchet space. One uses the same seminorms as above, except that m is allowed to range over all non-negative integer values.

The above spaces occur naturally in applications where functions having derivatives of certain orders are necessary; however, particularly in the study of partial differential equations, it can sometimes be more fruitful to work instead with the Sobolev spaces.

27.2 Parametric continuity

Parametric continuity is a concept applied to parametric curves describing the smoothness of the parameter's value with distance along the curve.

27.2.1 Definition

A curve can be said to have C^n continuity if $\dfrac{d^n s}{dt^n}$ is continuous of value throughout the curve.

As an example of a practical application of this concept, a curve describing the motion of an object with a parameter of time, must have C^1 continuity for the object to have finite acceleration. For smoother motion, such as that of a camera's path while making a film, higher orders of parametric continuity are required.

27.2.2 Order of continuity

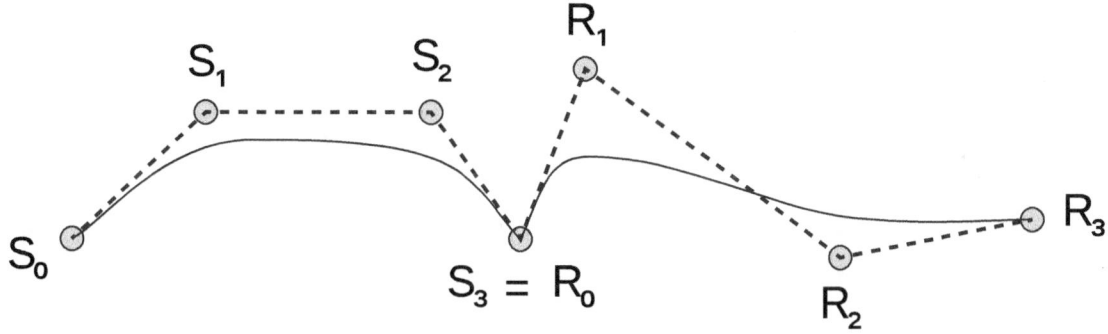

Two curve segments with only C^0 continuity

Two Bézier curve segments attached that is only C^0 continuous.

The various order of parametric continuity can be described as follows:[2]

- C^{-1}: curves include discontinuities
- C^0: curves are joined
- C^1: first derivatives are continuous
- C^2: first and second derivatives are continuous
- C^n: first through n^{th} derivatives are continuous

The term *parametric continuity* was introduced to distinguish it from *geometric continuity* (G^n) which removes restrictions on the speed with which the parameter traces out the curve.[3]

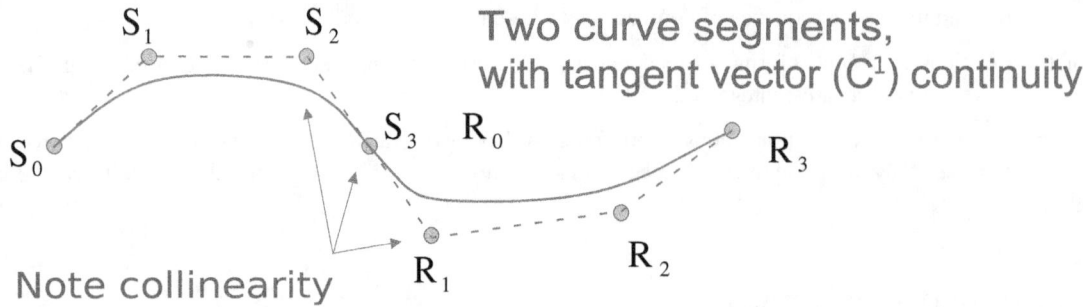

Two Bézier curve segments attached in such a way that they are C^1 continuous.

27.3   Geometric continuity

Geometric continuity is the continuity of the implicit function.

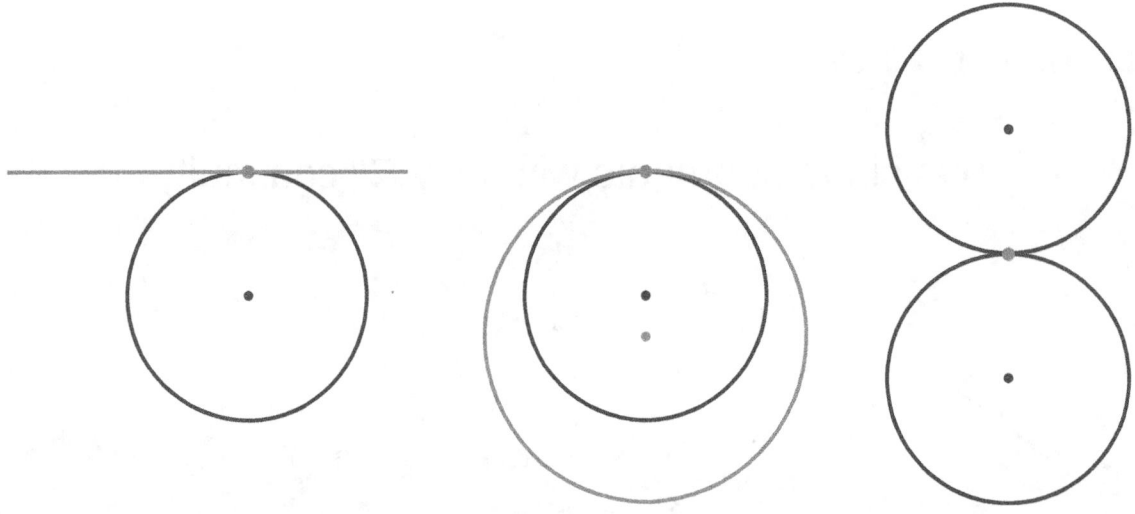

Curves with G^1-contact (circles,line)

The concept of **geometrical** or **geometric continuity** was primarily applied to the conic sections and related shapes by mathematicians such as Leibniz, Kepler, and Poncelet. The concept was an early attempt at describing, through geometry rather than algebra, the concept of continuity as expressed through a parametric function.

The basic idea behind geometric continuity was that the five conic sections were really five different versions of the same shape. An ellipse tends to a circle as the eccentricity approaches zero, or to a parabola as it approaches one; and a hyperbola tends to a parabola as the eccentricity drops toward one; it can also tend to intersecting lines. Thus, there was *continuity* between the conic sections. These ideas led to other concepts of continuity. For instance, if a circle and a straight line were two expressions of the same shape, perhaps a line could be thought of as a circle of infinite radius. For such to be the case, one would have to make the line closed by allowing the point $x = \infty$ to be a point on the circle, and for $x = +\infty$ and $x = -\infty$ to be identical. Such ideas were useful in crafting the modern, algebraically defined, idea of the continuity of a function and of ∞.

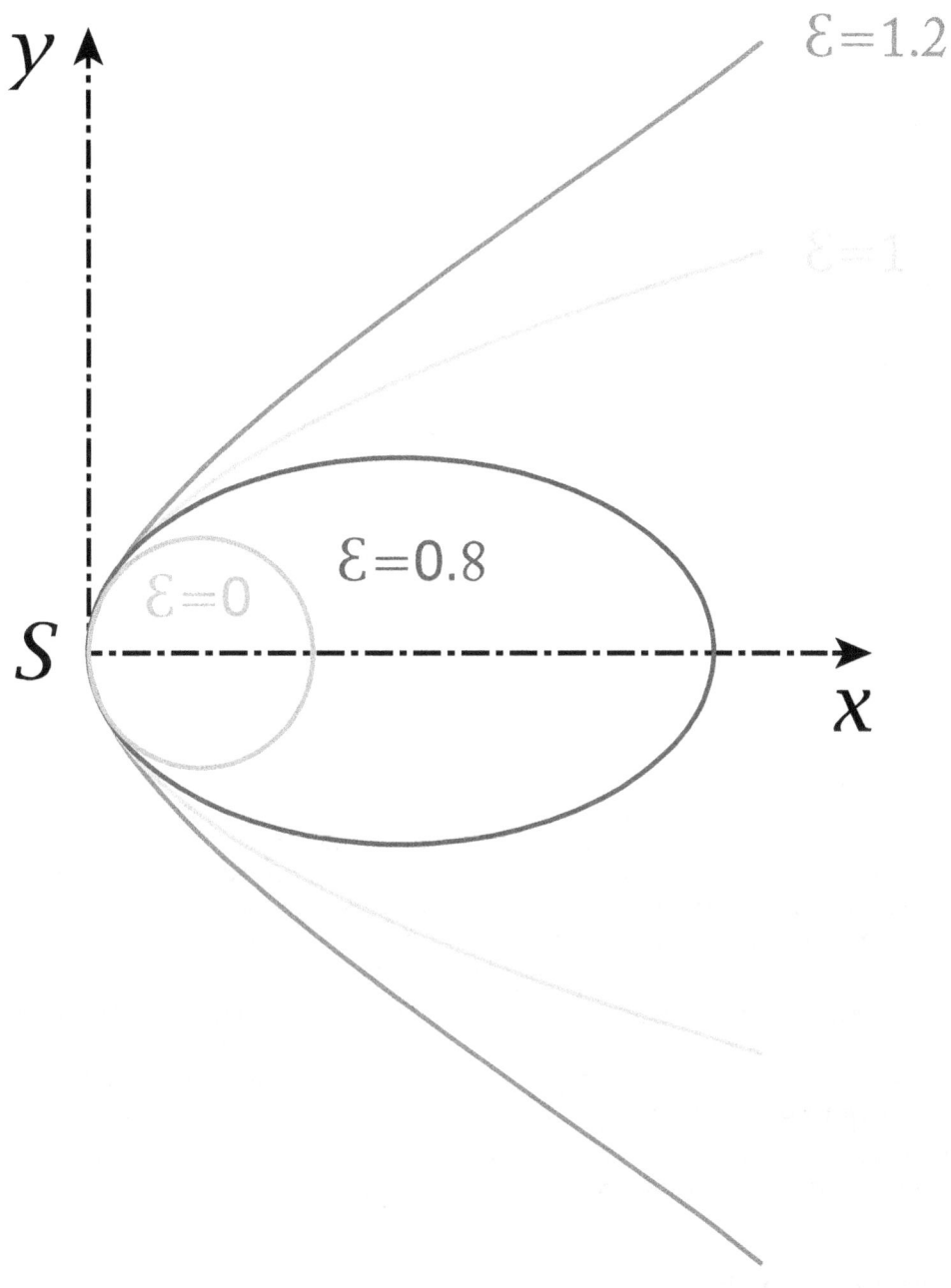

$(1 - \varepsilon^2)x^2 - 2px + y^2 = 0, \ p > 0 \ , \varepsilon \geq 0$
pencil of conic sections with $\mathbf{G}^2$*-contact: p fix, ε variable*
($\varepsilon = 0$: circle, $\varepsilon = 0.8$: ellipse, $\varepsilon = 1$: parabola, $\varepsilon = 1.2$: hyperbola)

27.3.1 Smoothness of curves and surfaces

A curve or surface can be described as having G^n continuity, n being the increasing measure of smoothness. Consider the segments either side of a point on a curve:

- G^0: The curves touch at the join point.

- G^1: The curves also share a common tangent direction at the join point.

- G^2: The curves also share a common center of curvature at the join point.

In general, G^n continuity exists if the curves can be reparameterized to have C^n (parametric) continuity.[4][5] A reparametrization of the curve is geometrically identical to the original; only the parameter is affected.

Equivalently, two vector functions $f(t)$ and $g(t)$ have G^n continuity if $f^{(n)}(t) \neq 0$ and $f^{(n)}(t) \equiv kg^{(n)}(t)$, for a scalar $k > 0$ (i.e., if the direction, but not necessarily the magnitude, of the two vectors is equal).

While it may be obvious that a curve would require G^1 continuity to appear smooth, for good aesthetics, such as those aspired to in architecture and sports car design, higher levels of geometric continuity are required. For example, reflections in a car body will not appear smooth unless the body has G^2 continuity.

A *rounded rectangle* (with ninety degree circular arcs at the four corners) has G^1 continuity, but does not have G^2 continuity. The same is true for a *rounded cube*, with octants of a sphere at its corners and quarter-cylinders along its edges. If an editable curve with G^2 continuity is required, then cubic splines are typically chosen; these curves are frequently used in industrial design.

27.3.2   Smoothness of piecewise defined curves and surfaces

27.4   Smoothness

27.4.1   Relation to analyticity

While all analytic functions are "smooth" (i.e. have all derivatives continuous) on the set on which they are analytic, examples such as bump functions (mentioned above) show that the converse is not true for functions on the reals: there exist smooth real functions which are not analytic. Simple examples of functions which are smooth but not analytic at any point can be made by means of Fourier series; another example is the Fabius function. Although it might seem that such functions are the exception rather than the rule, it turns out that the analytic functions are scattered very thinly among the smooth ones; more rigorously, the analytic functions form a meagre subset of the smooth functions. Furthermore, for every open subset A of the real line, there exist smooth functions which are analytic on A and nowhere else.

It is useful to compare the situation to that of the ubiquity of transcendental numbers on the real line. Both on the real line and the set of smooth functions, the examples we come up with at first thought (algebraic/rational numbers and analytic functions) are far better behaved than the majority of cases: the transcendental numbers and nowhere analytic functions have full measure (their complements are meagre).

The situation thus described is in marked contrast to complex differentiable functions. If a complex function is differentiable just once on an open set it is both infinitely differentiable and analytic on that set.

27.4.2   Smooth partitions of unity

Smooth functions with given closed support are used in the construction of **smooth partitions of unity** (see *partition of unity* and topology glossary); these are essential in the study of smooth manifolds, for example to show that Riemannian metrics can be defined globally starting from their local existence. A simple case is that of a **bump function** on the real line, that is, a smooth function f that takes the value 0 outside an interval $[a,b]$ and such that

$$f(x) > 0 \quad \text{for} \quad a < x < b.$$

Given a number of overlapping intervals on the line, bump functions can be constructed on each of them, and on semi-infinite intervals $(-\infty, c]$ and $[d, +\infty)$ to cover the whole line, such that the sum of the functions is always 1.

From what has just been said, partitions of unity don't apply to holomorphic functions; their different behavior relative to existence and analytic continuation is one of the roots of sheaf theory. In contrast, sheaves of smooth functions tend not to carry much topological information.

27.4.3 Smooth functions between manifolds

Smooth maps between smooth manifolds may be defined by means of charts, since the idea of smoothness of function is independent of the particular chart used. If F is a map from an m-manifold M to an n-manifold N, then F is smooth if, for every $p \in M$, there is a chart (U, φ) in M containing p and a chart (V, ψ) in N containing $F(p)$ with $F(U) \subset V$, such that $\psi \circ F \circ \varphi^{-1}$ is smooth from $\varphi(U)$ to $\psi(V)$ as a function from $\mathbf{R}^m$ to $\mathbf{R}^n$.

Such a map has a first derivative defined on tangent vectors; it gives a fibre-wise linear mapping on the level of tangent bundles.

27.4.4 Smooth functions between subsets of manifolds

There is a corresponding notion of **smooth map** for arbitrary subsets of manifolds. If $f : X \to Y$ is a function whose domain and range are subsets of manifolds $X \subset M$ and $Y \subset N$ respectively. f is said to be **smooth** if for all $x \in X$ there is an open set $U \subset M$ with $x \in U$ and a smooth function $F : U \to N$ such that $F(p) = f(p)$ for all $p \in U \cap X$.

27.5 See also

- Non-analytic smooth function

- Quasi-analytic function

- Spline

- Smooth number (number theory)

- Sinuosity

27.6 References

[1] Warner (1983), p. 5, Definition 1.2.

[2] Parametric Curves

[3] Richard H. Bartels; John C. Beatty; Brian A. Barsky (1987). *An Introduction to Splines for Use in Computer Graphics and Geometric Modeling*. Morgan Kaufmann. Chapter 13. Parametric vs. Geometric Continuity. ISBN 978-1-55860-400-1.

[4] Brian A. Barsky and Tony D. DeRose, "Geometric Continuity of Parametric Curves: Three Equivalent Characterizations," IEEE Computer Graphics and Applications, 9(6), Nov. 1989, pp. 60–68.

[5] Erich Hartmann:*Geometry and Algorithms for COMPUTER AIDED DESIGN*, page 55

- This article incorporates text from a publication now in the public domain: Chisholm, Hugh, ed. (1911). *Encyclopædia Britannica* (11th ed.). Cambridge University Press.

- Guillemin, Victor; Pollack, Alan (1974). *Differential Topology*. Englewood Cliffs: Prentice-Hall. ISBN 0-13-212605-2.

- Warner, Frank Wilson (1983). *Foundations of differentiable manifolds and Lie groups*. Springer. ISBN 978-0-387-90894-6.

Chapter 28

Discrete symmetry

In mathematics, a **discrete symmetry** is a symmetry that describes non-continuous changes in a system. For example, a square possesses discrete rotational symmetry, as only rotations by multiples of right angles will preserve the square's original appearance. Discrete symmetries sometimes involve some type of 'swapping', these swaps usually being called *reflections* or *interchanges*. In mathematics and theoretical physics, a **discrete symmetry** is a symmetry under the transformations of a discrete group—e.g. a topological group with a discrete topology whose elements form a finite or a countable set.

One of the most prominent discrete symmetries in physics is parity symmetry. It manifests itself in various elementary physical quantum systems, such as quantum harmonic oscillator, electron orbitals of Hydrogen-like atoms by forcing wavefunctions to be even or odd. This in turn gives rise to selection rules that determine which transition lines are visible in atomic absorption spectra.

28.1 References

- Slavik V. Jablan, *Symmetry, Ornament and Modularity*, Volume 30 of K & E Series on Knots and Everything, World Scientific, 2002. ISBN 9812380809

Chapter 29

T-symmetry

In theoretical physics, **T-symmetry** is the theoretical symmetry of physical laws under a **time reversal** transformation:

$$T : t \mapsto -t.$$

Although in restricted contexts one may find this symmetry, the observable universe itself does not show symmetry under time reversal, primarily due to the second law of thermodynamics. Hence time is said to be non-symmetric, or asymmetric, except for equilibrium states when the second law of thermodynamics predicts the time symmetry to hold. However, quantum noninvasive measurements are predicted to violate time symmetry even in equilibrium,[1] contrary to their classical counterparts, although it has not yet been experimentally confirmed.

Time *asymmetries* are generally distinguished as between those intrinsic to the dynamic physical laws, those due to the initial conditions of our universe, and due to measurements

1. The T-asymmetry of the weak force is of the first kind,

2. The T-asymmetry of the second law of thermodynamics is of the second kind, while

3. The T-asymmetry of the noninvasive measurements is of the third kind.

29.1 Invariance

Physicists also discuss the time-reversal invariance of local and/or macroscopic descriptions of physical systems, independent of the invariance of the underlying microscopic physical laws. For example, Maxwell's equations with material absorption or Newtonian mechanics with friction are not time-reversal invariant at the macroscopic level where they are normally applied, even if they are invariant at the microscopic level; when one includes the atomic motions, the "lost" energy is translated into heat.

29.2 Macroscopic phenomena: the second law of thermodynamics

Our daily experience shows that T-symmetry does not hold for the behavior of bulk materials. Of these macroscopic laws, most notable is the second law of thermodynamics. Many other phenomena, such as the relative motion of bodies with friction, or viscous motion of fluids, reduce to this, because the underlying mechanism is the dissipation of usable energy (for example, kinetic energy) into heat.

The question of whether this time-asymmetric dissipation is really inevitable has been considered by many physicists, often in the context of **Maxwell's demon**. The name comes from a thought experiment described by James Clerk Maxwell in which a microscopic demon guards a gate between two halves of a room. It only lets slow molecules into one half, only fast ones into the other. By eventually making one side of the room cooler than before and the other hotter, it seems to reduce the entropy of the room, and reverse the arrow of time. Many analyses have been made of this; all show that when the entropy of room and demon are taken together, this total entropy does increase. Modern analyses of this problem have taken into account Claude E. Shannon's relation between entropy and information. Many interesting results in modern computing are closely related to this problem — reversible computing, quantum computing and physical limits to computing, are examples. These seemingly metaphysical questions are today, in these ways, slowly being converted to the stuff of the physical sciences.

The current consensus hinges upon the Boltzmann-Shannon identification of the logarithm of phase space volume with the negative of Shannon information, and hence to entropy. In this notion, a fixed initial state of a macroscopic system corresponds to relatively low entropy because the coordinates of the molecules of the body are constrained. As the system evolves in the presence of dissipation, the molecular coordinates can move into larger volumes of phase space, becoming more uncertain, and thus leading to increase in entropy.

One can, however, equally well imagine a state of the universe in which the motions of all of the particles at one instant were the reverse (strictly, the CPT reverse). Such a state would then evolve in reverse, so presumably entropy would decrease (Loschmidt's paradox). Why is 'our' state preferred over the other?

One position is to say that the constant increase of entropy we observe happens *only* because of the initial state of our universe. Other possible states of the universe (for example, a universe at heat death equilibrium) would actually result in no increase of entropy. In this view, the apparent T-asymmetry of our universe is a problem in cosmology: why did the universe start with a low entropy? This view, if it remains viable in the light of future cosmological observation, would connect this problem to one of the big open questions beyond the reach of today's physics — the question of *initial conditions* of the universe.

29.3   Macroscopic phenomena: black holes

An object can cross through the event horizon of a black hole from the outside, and then fall rapidly to the central region where our understanding of physics breaks down. Since within a black hole the forward light-cone is directed towards the center and the backward light-cone is directed outward, it is not even possible to define time-reversal in the usual manner. The only way anything can escape from a black hole is as Hawking radiation.

The time reversal of a black hole would be a hypothetical object known as a white hole. From the outside they appear similar. While a black hole has a beginning and is inescapable, a white hole has an ending and cannot be entered. The forward light-cones of a white hole are directed outward; and its backward light-cones are directed towards the center.

The event horizon of a black hole may be thought of as a surface moving outward at the local speed of light and is just on the edge between escaping and falling back. The event horizon of a white hole is a surface moving inward at the local speed of light and is just on the edge between being swept outward and succeeding in reaching the center. They are two different kinds of horizons—the horizon of a white hole is like the horizon of a black hole turned inside-out.

The modern view of black hole irreversibility is to relate it to the second law of thermodynamics, since black holes are viewed as thermodynamic objects. Indeed, according to the Gauge–gravity duality conjecture, all microscopic processes in a black hole are reversible, and only the collective behavior is irreversible, as in any other macroscopic, thermal system.

29.4   Kinetic consequences: detailed balance and Onsager reciprocal relations

In physical and chemical kinetics, T-symmetry of the mechanical microscopic equations implies two important laws: the principle of detailed balance and the Onsager reciprocal relations. T-symmetry of the microscopic description together with its kinetic consequences are called microscopic reversibility.

29.5 Effect of time reversal on some variables of classical physics

29.5.1 Even

Classical variables that do not change upon time reversal include:

$\vec{x}$, Position of a particle in three-space

$\vec{a}$, Acceleration of the particle

$\vec{F}$, Force on the particle

E, Energy of the particle

ϕ, Electric potential (voltage)

$\vec{E}$, Electric field

$\vec{D}$, Electric displacement

ρ, Density of electric charge

$\vec{P}$, Electric polarization

Energy density of the electromagnetic field

Maxwell stress tensor

All masses, charges, coupling constants, and other physical constants, except those associated with the weak force.

29.5.2 Odd

Classical variables that time reversal negates include:

t, The time when an event occurs

$\vec{v}$, Velocity of a particle

$\vec{p}$, Linear momentum of a particle

$\vec{l}$, Angular momentum of a particle (both orbital and spin)

$\vec{A}$, Electromagnetic vector potential

$\vec{B}$, Magnetic induction

$\vec{H}$, Magnetic field

$\vec{j}$, Density of electric current

$\vec{M}$, Magnetization

$\vec{S}$, Poynting vector

Power (rate of work done).

29.6 Microscopic phenomena: time reversal invariance

Since most systems are asymmetric under time reversal, it is interesting to ask whether there are phenomena that do have this symmetry. In classical mechanics, a velocity v reverses under the operation of T, but an acceleration does not. Therefore, one models dissipative phenomena through terms that are odd in v. However, delicate experiments in which known sources of dissipation are removed reveal that the laws of mechanics are time reversal invariant. Dissipation itself is originated in the second law of thermodynamics.

The motion of a charged body in a magnetic field, B involves the velocity through the Lorentz force term $v \times B$, and might seem at first to be asymmetric under T. A closer look assures us that B also changes sign under time reversal. This happens because a magnetic field is produced by an electric current, J, which reverses sign under T. Thus, the motion of classical charged particles in electromagnetic fields is also time reversal invariant. (Despite this, it is still useful to consider the time-reversal non-invariance in a *local* sense when the external field is held fixed, as when the magneto-optic effect is analyzed. This allows one to analyze the conditions under which optical phenomena that locally break time-reversal, such as Faraday isolators and directional dichroism, can occur.) The laws of gravity also seem to be time reversal invariant in classical mechanics.

In physics one separates the laws of motion, called kinematics, from the laws of force, called dynamics. Following the classical kinematics of Newton's laws of motion, the kinematics of quantum mechanics is built in such a way that it presupposes nothing about the time reversal symmetry of the dynamics. In other words, if the dynamics are invariant, then the kinematics will allow it to remain invariant; if the dynamics is not, then the kinematics will also show this. The structure of the quantum laws of motion are richer, and we examine these next.

29.6.1   Time reversal in quantum mechanics

This section contains a discussion of the three most important properties of time reversal in quantum mechanics; chiefly,

1. that it must be represented as an anti-unitary operator,

2. that it protects non-degenerate quantum states from having an electric dipole moment,

3. that it has two-dimensional representations with the property $T^2 = -1$.

The strangeness of this result is clear if one compares it with parity. If parity transforms a pair of quantum states into each other, then the sum and difference of these two basis states are states of good parity. Time reversal does not behave like this. It seems to violate the theorem that all abelian groups be represented by one-dimensional irreducible representations. The reason it does this is that it is represented by an anti-unitary operator. It thus opens the way to spinors in quantum mechanics.

29.6.2   Anti-unitary representation of time reversal

Eugene Wigner showed that a symmetry operation S of a Hamiltonian is represented, in quantum mechanics either by a **unitary** operator, $S = U$, or an **antiunitary** one, $S = UK$ where U is unitary, and K denotes complex conjugation. These are the only operations that act on Hilbert space so as to preserve the *length* of the projection of any one state-vector onto another state-vector.

Consider the parity operator. Acting on the position, it reverses the directions of space, so that $P^{-1}xP = -x$. Similarly, it reverses the direction of *momentum*, so that $PpP^{-1} = -p$, where x and p are the position and momentum operators. This preserves the canonical commutator $[x, p] = i\hbar$, where $\hbar$ is the reduced Planck constant, only if P is chosen to be unitary, $PiP^{-1} = i$.

On the other hand, for time reversal, the time-component of the momentum is the energy. If time reversal were implemented as a unitary operator, it would reverse the sign of the energy just as space-reversal reverses the sign of the momentum. This is not possible, because, unlike momentum, energy is always positive. Since energy in quantum mechanics is defined as the phase factor $\exp(-iEt)$ that one gets when one moves forward in time, the way to reverse time while preserving the sign of the energy is to reverse the sense of "i", so that the sense of phases is reversed.

Similarly, any operation that reverses the sense of phase, which changes the sign of i, will turn positive energies into negative energies unless it also changes the direction of time. So every antiunitary symmetry in a theory with positive energy must reverse the direction of time. The only antiunitary symmetry is time reversal, together with a unitary symmetry that does not reverse time.

Given the *time reversal* operator T, it does nothing to the x-operator, $TxT^{-1} = x$, but it reverses the direction of p, so that $TpT^{-1} = -p$. The canonical commutator is invariant only if T is chosen to be anti-unitary, i.e., $TiT^{-1} = -i$. For a particle with spin J, one can use the representation

$$T = e^{-i\pi J_y/\hbar} K,$$

where J_y is the y-component of the spin, and use of $TJT^{-1} = -J$ has been made.

29.6.3 Electric dipole moments

This has an interesting consequence on the electric dipole moment (EDM) of any particle. The EDM is defined through the shift in the energy of a state when it is put in an external electric field: $\Delta e = d \cdot E + E \cdot \delta \cdot E$, where d is called the EDM and δ, the induced dipole moment. One important property of an EDM is that the energy shift due to it changes sign under a parity transformation. However, since **d** is a vector, its expectation value in a state $|\psi\rangle$ must be proportional to $\langle\psi| J |\psi\rangle$. Thus, under time reversal, an invariant state must have vanishing EDM. In other words, a non-vanishing EDM signals both P and T symmetry-breaking.[2]

It is interesting to examine this argument further, since one feels that some molecules, such as water, must have EDM irrespective of whether **T** is a symmetry. This is correct: if a quantum system has degenerate ground states that transform into each other under parity, then time reversal need not be broken to give EDM.

Experimentally observed bounds on the electric dipole moment of the nucleon currently set stringent limits on the violation of time reversal symmetry in the strong interactions, and their modern theory: quantum chromodynamics. Then, using the CPT invariance of a relativistic quantum field theory, this puts strong bounds on strong CP violation.

Experimental bounds on the electron electric dipole moment also place limits on theories of particle physics and their parameters.[3][4]

29.6.4 Kramers' theorem

Main article: Kramers' degeneracy theorem

For T, which is an anti-unitary Z_2 symmetry generator

$$T^2 = UKUK = U U^* = U (U^T)^{-1} = \Phi,$$

where Φ is a diagonal matrix of phases. As a result, $U = \Phi U^T$ and $U^T = U\Phi$, showing that

$$U = \Phi \, U \, \Phi.$$

This means that the entries in Φ are ± 1, as a result of which one may have either $T^2 = \pm 1$. This is specific to the anti-unitarity of T. For a unitary operator, such as the parity, any phase is allowed.

Next, take a Hamiltonian invariant under T. Let $|a\rangle$ and $T|a\rangle$ be two quantum states of the same energy. Now, if $T^2 = -1$, then one finds that the states are orthogonal: a result called **Kramers' theorem**. This implies that if $T^2 = -1$, then there is a twofold degeneracy in the state. This result in non-relativistic quantum mechanics presages the spin statistics theorem of quantum field theory.

Quantum states that give unitary representations of time reversal, i.e., have $\mathbf{T^2 = 1}$, are characterized by a multiplicative quantum number, sometimes called the **T-parity**.

Time reversal transformation for fermions in quantum field theories can be represented by an 8-component spinor in which the above-mentioned **T-parity** can be a complex number with unit radius. The CPT invariance is not a theorem but a **better to have** property in these class of theories.

29.6.5   Time reversal of the known dynamical laws

Particle physics codified the basic laws of dynamics into the standard model. This is formulated as a quantum field theory that has CPT symmetry, i.e., the laws are invariant under simultaneous operation of time reversal, parity and charge conjugation. However, time reversal itself is seen not to be a symmetry (this is usually called CP violation). There are two possible origins of this asymmetry, one through the mixing of different flavours of quarks in their weak decays, the second through a direct CP violation in strong interactions. The first is seen in experiments, the second is strongly constrained by the non-observation of the EDM of a neutron.

It is important to stress that this time reversal violation is unrelated to the second law of thermodynamics, because due to the conservation of the CPT symmetry, the effect of time reversal is to rename particles as antiparticles and *vice versa*. Thus the second law of thermodynamics is thought to originate in the initial conditions in the universe.

29.6.6   Time reversal of noninvasive measurements

Strong measurements (both classical and quantum) are certainly disturbing, causing asymmetry due to second law of thermodynamics. However, noninvasive measurements should not disturb the evolution so they are expected to be time-symmetric. Surprisingly, it is true only in classical physics but not quantum, even in a thermodynamically invariant equilibrium state. [1] This type of asymmetry is independent of CPT symmetry but has not yet been confirmed experimentally due to extreme conditions of the checking proposal.

29.7   See also

- The second law of thermodynamics, Maxwell's demon and the arrow of time (also Loschmidt's paradox).

- Microscopic reversibility

- Detailed balance

- Applications to reversible computing and quantum computing, including limits to computing.

- The standard model of particle physics, CP violation, the CKM matrix and the strong CP problem

- Neutrino masses and CPT invariance.

- Wheeler–Feynman absorber theory

- Teleonomy

29.8   References

[1] Bednorz, Adam; Franke, Kurt; Belzig, Wolfgang (February 2013). "Noninvasiveness and time symmetry of weak measurements". *New Journal of Physics* **15**: 023043. Bibcode:2013NJPh...15b3043B. doi:10.1088/1367-2630/15/2/023043.

[2] Khriplovich, Iosip B.; Lamoreaux, Steve K. (2012). *CP violation without strangeness : electric dipole moments of particles, atoms, and molecules.* [S.l.]: Springer. ISBN 978-3-642-64577-8.

[3] Ibrahim, Tarik; Itani, Ahmad; Nath, Pran (12 Aug 2014). "Electron EDM as a Sensitive Probe of PeV Scale Physics". *arXiv*: 1406.0083.

[4] Kim, Jihn E.; Carosi, Gianpaolo (4 March 2010). "Axions and the strong CP problem". *Reviews of Modern Physics* **82**: 557. arXiv:0807.3125. Bibcode:2010RvMP...82..557K. doi:10.1103/RevModPhys.82.557.

- Maxwell's demon: entropy, information, computing, edited by H.S.Leff and A.F. Rex (IOP publishing, 1990) ISBN 0-7503-0057-4

- Maxwell's demon, 2: entropy, classical and quantum information, edited by H.S.Leff and A.F. Rex (IOP publishing, 2003) ISBN 0-7503-0759-5

- The emperor's new mind: concerning computers, minds, and the laws of physics, by Roger Penrose (Oxford university press, 2002) ISBN 0-19-286198-0

- Sozzi, M.S. (2008). *Discrete symmetries and CP violation*. Oxford University Press. ISBN 978-0-19-929666-8.

- Birss, R. R. (1964). *Symmetry and Magnetism*. John Wiley & Sons, Inc., New York.

- Multiferroic materials with time-reversal breaking optical properties

- CP violation, by I.I. Bigi and A.I. Sanda (Cambridge University Press, 2000) ISBN 0-521-44349-0

- Particle Data Group on CP violation

    - the Babar experiment in SLAC
    - the BELLE experiment in KEK
    - the KTeV experiment in Fermilab
    - the CPLEAR experiment in CERN

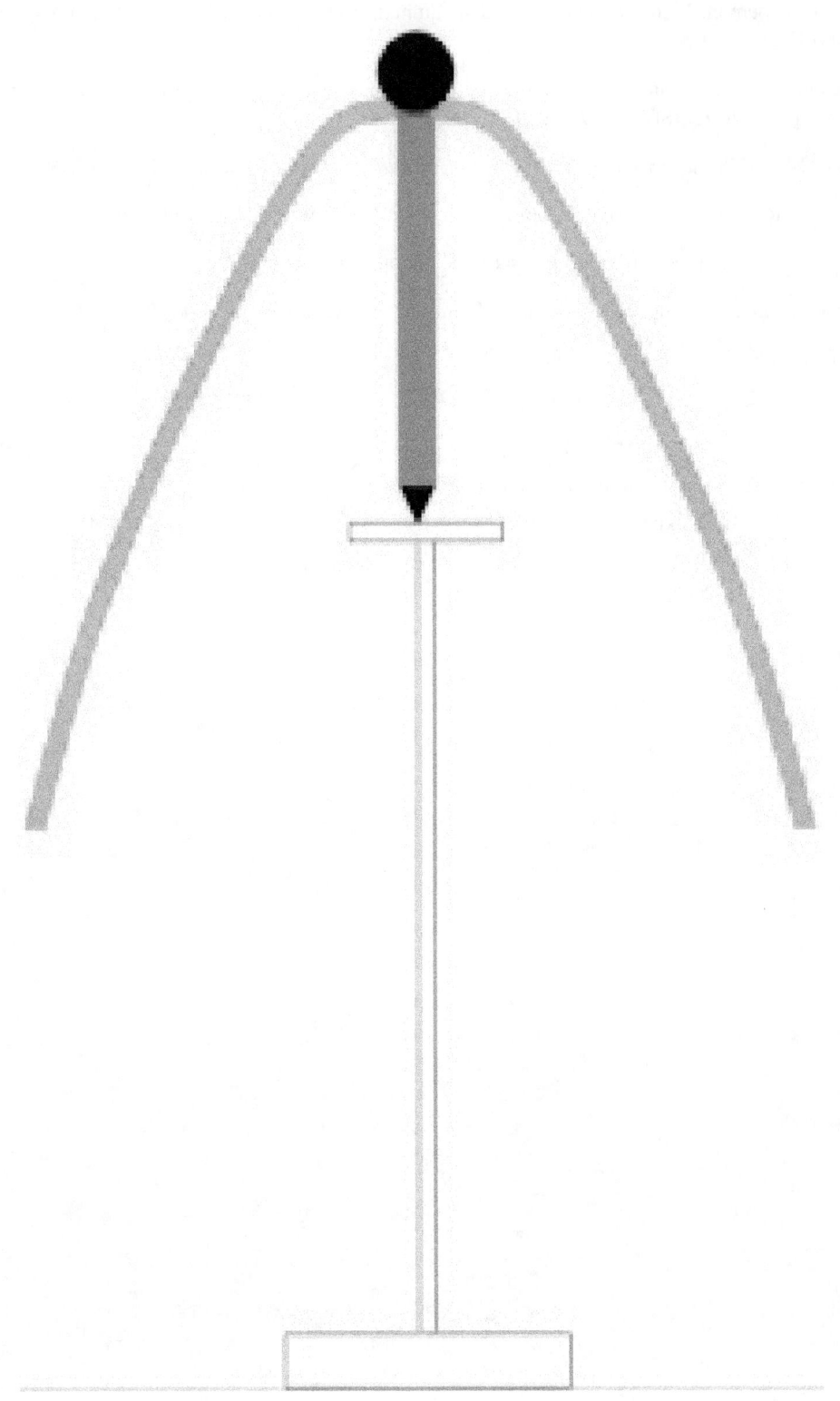

A toy called the teeter-totter illustrates the two aspects of time reversal invariance. When set into motion atop a pedestal, the figure oscillates for a very long time. The toy is engineered to minimize friction and illustrate the reversibility of Newton's laws of motion. However, the mechanically stable state of the toy is when the figure falls down from the pedestal into one of arbitrarily many positions. This is an illustration of the law of increase of entropy through Boltzmann's identification of the logarithm of the number of states with the entropy.

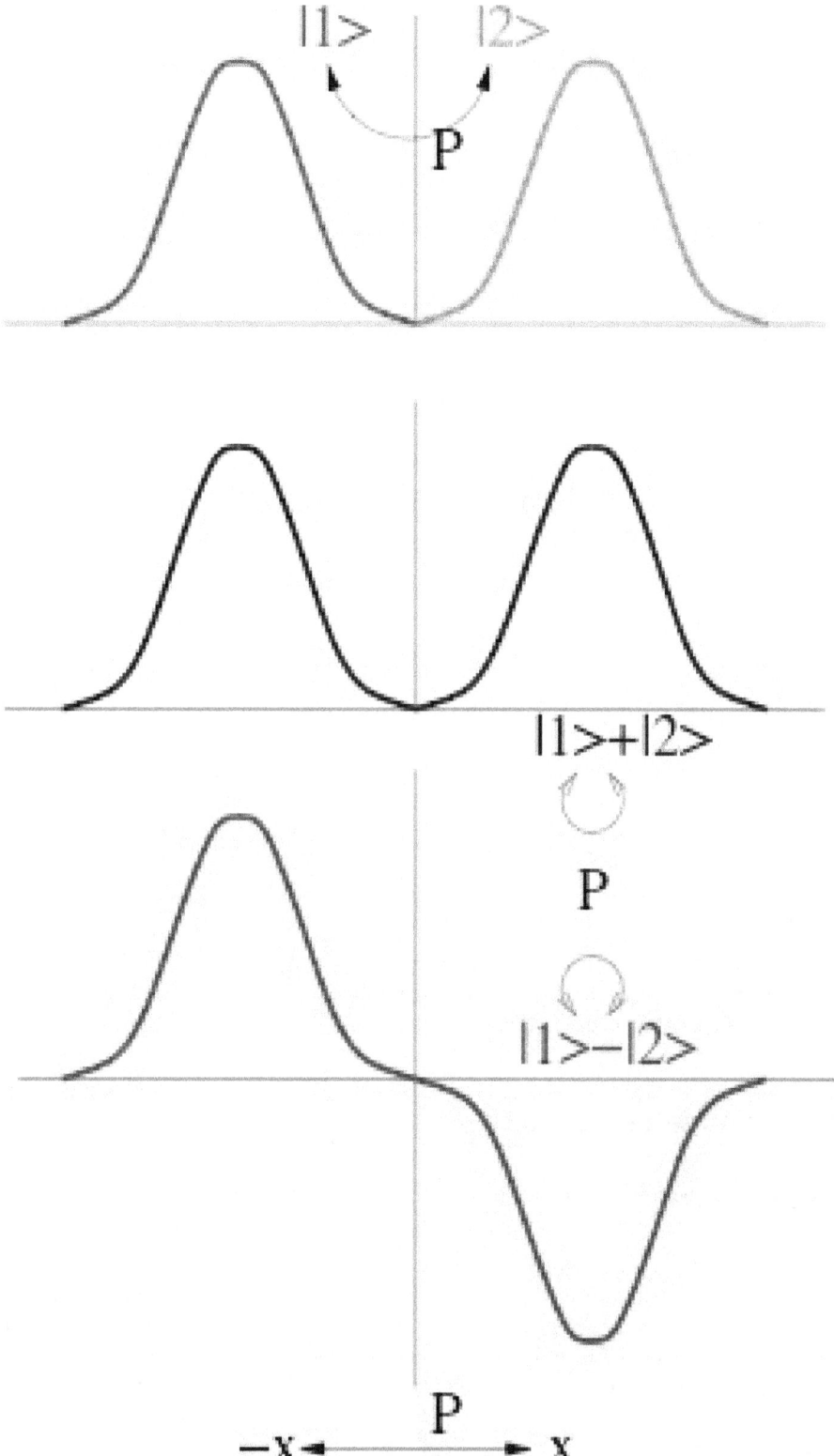

Two-dimensional representations of parity are given by a pair of quantum states that go into each other under parity. However, this representation can always be reduced to linear combinations of states, each of which is either even or odd under parity. One says that all irreducible representations of parity are one-dimensional. Kramers' theorem states that time reversal need not have this property because it is represented by an anti-unitary operator.

Chapter 30

Parity (physics)

In quantum mechanics, a **parity transformation** (also called **parity inversion**) is the flip in the sign of *one* spatial coordinate. In three dimensions, it is also often described by the simultaneous flip in the sign of all three spatial coordinates (a point reflection):

$$\mathbf{P} : \begin{pmatrix} x \\ y \\ z \end{pmatrix} \mapsto \begin{pmatrix} -x \\ -y \\ -z \end{pmatrix}.$$

It can also be thought of as a test for chirality of a physical phenomenon, in that a parity inversion transforms a phenomenon into its mirror image. A parity transformation on something achiral, on the other hand, can be viewed as an identity transformation. All fundamental interactions of elementary particles, with the exception of the weak interaction, are symmetric under parity. The weak interaction is chiral and thus provides a means for probing chirality in physics. In interactions that are symmetric under parity, such as electromagnetism in atomic and molecular physics, parity serves as a powerful controlling principle underlying quantum transitions.

A matrix representation of **P** (in any number of dimensions) has determinant equal to −1, and hence is distinct from a rotation, which has a determinant equal to 1. In a two-dimensional plane, a simultaneous flip of all coordinates in sign is *not* a parity transformation; it is the same as a 180°-rotation.

30.1   Simple symmetry relations

Under rotations, classical geometrical objects can be classified into scalars, vectors, and tensors of higher rank. In classical physics, physical configurations need to transform under representations of every symmetry group.

Quantum theory predicts that states in a Hilbert space do not need to transform under representations of the group of rotations, but only under projective representations. The word *projective* refers to the fact that if one projects out the phase of each state, where we recall that the overall phase of a quantum state is not an observable, then a projective representation reduces to an ordinary representation. All representations are also projective representations, but the converse is not true, therefore the projective representation condition on quantum states is weaker than the representation condition on classical states.

The projective representations of any group are isomorphic to the ordinary representations of a central extension of the group. For example, projective representations of the 3-dimensional rotation group, which is the special orthogonal group SO(3), are ordinary representations of the special unitary group SU(2) (see Representation theory of SU(2)). Projective representations of the rotation group that are not representations are called spinors, and so quantum states may transform not only as tensors but also as spinors.

If one adds to this a classification by parity, these can be extended, for example, into notions of

216

- *scalars* ($P = 1$) and *pseudoscalars* ($P = -1$) which are rotationally invariant.

- *vectors* ($P = -1$) and *axial vectors* (also called *pseudovectors*) ($P = 1$) which both transform as vectors under rotation.

One can define **reflections** such as

$$V_x : \begin{pmatrix} x \\ y \\ z \end{pmatrix} \mapsto \begin{pmatrix} -x \\ y \\ z \end{pmatrix},$$

which also have negative determinant and form a valid parity transformation. Then, combining them with rotations (or successively performing *x*-, *y*-, and *z*-reflections) one can recover the particular parity transformation defined earlier. The first parity transformation given does not work in an even number of dimensions, though, because it results in a positive determinant. In odd number of dimensions only the latter example of a parity transformation (or any reflection of an odd number of coordinates) can be used.

Parity forms the abelian group Z_2 due to the relation $\mathbf{P}^2 = 1$. All Abelian groups have only one-dimensional irreducible representations. For Z_2, there are two irreducible representations: one is even under parity ($\mathbf{P}\varphi = \varphi$), the other is odd ($\mathbf{P}\varphi = -\varphi$). These are useful in quantum mechanics. However, as is elaborated below, in quantum mechanics states need not transform under actual representations of parity but only under projective representations and so in principle a parity transformation may rotate a state by any phase.

30.2 Classical mechanics

Newton's equation of motion $\mathbf{F} = m\mathbf{a}$ (if the mass is constant) equates two vectors, and hence is invariant under parity. The law of gravity also involves only vectors and is also, therefore, invariant under parity.

However, angular momentum $\mathbf{L}$ is an axial vector,

$$\mathbf{L} = \mathbf{r} \times \mathbf{p},$$
$$\mathbf{P}(\mathbf{L}) = (-\mathbf{r}) \times (-\mathbf{p}) = \mathbf{L}.$$

In classical electrodynamics, the charge density ϱ is a scalar, the electric field, $\mathbf{E}$, and current $\mathbf{j}$ are vectors, but the magnetic field, $\mathbf{H}$ is an axial vector. However, Maxwell's equations are invariant under parity because the curl of an axial vector is a vector.

30.3 Effect of spatial inversion on some variables of classical physics

30.3.1 Even

Classical variables, predominantly scalar quantities, which do not change upon spatial inversion include:

t , the time when an event occurs

m , the mass of a particle

E , the energy of the particle

P , power (rate of work done)

ρ , the electric charge density

V , the electric potential (voltage)

ρ , energy density of the electromagnetic field

L , the angular momentum of a particle (both orbital and spin) (axial vector)

B , the magnetic field (axial vector)

H , the auxiliary magnetic field

M , the magnetization

T_{ij} Maxwell stress tensor.

All masses, charges, coupling constants, and other physical constants, except those associated with the weak force

30.3.2   Odd

Classical variables, predominantly vector quantities, which have their sign flipped by spatial inversion include:

h , the helicity

Φ , the magnetic flux

x , the position of a particle in three-space

v , the velocity of a particle

a , the acceleration of the particle

p , the linear momentum of a particle

F , the force exerted on a particle

J , the electric current density

E , the electric field

D , the electric displacement field

P , the electric polarization

A , the electromagnetic vector potential

S , Poynting vector.

30.4   Quantum mechanics

30.4.1   Possible eigenvalues

In quantum mechanics, spacetime transformations act on quantum states. The parity transformation, **P**, is a unitary operator, in general acting on a state ψ as follows: $\mathbf{P}\psi(r) = e^{i\varphi/2}\psi(-r)$.

One must then have $\mathbf{P}^2\psi(r) = e^{i\varphi}\psi(r)$, since an overall phase is unobservable. The operator $\mathbf{P}^2$, which reverses the parity of a state twice, leaves the spacetime invariant, and so is an internal symmetry which rotates its eigenstates by phases $e^{i\varphi}$. If $\mathbf{P}^2$ is an element e^{iQ} of a continuous U(1) symmetry group of phase rotations, then $e^{-iQ/2}$ is part of this U(1) and so is also a symmetry. In particular, we can define $\mathbf{P}' = \mathbf{P}e^{-iQ/2}$, which is also a symmetry, and so we can choose to call $\mathbf{P}'$ our parity operator, instead of **P**. Note that $\mathbf{P}'^2 = 1$ and so $\mathbf{P}'$ has eigenvalues ± 1. However, when no such symmetry group exists, it may be that all parity transformations have some eigenvalues which are phases other than ± 1.

For electronic wavefunctions, even states are usually indicated by a subscript g for *gerade* (German: even) and odd states by a subscript u for *ungerade* (German: odd). For example, the lowest energy level of the hydrogen molecule ion (H_2^+) is labelled $1\sigma_g$ and the next-lowest $1\sigma_u$.[1]

30.4.2 Consequences of parity symmetry

When parity generates the Abelian group $\mathbb{Z}_2$, one can always take linear combinations of quantum states such that they are either even or odd under parity (see the figure). Thus the parity of such states is ± 1. The parity of a multiparticle state is the product of the parities of each state; in other words parity is a multiplicative quantum number

In quantum mechanics, Hamiltonians are invariant (symmetric) under a parity transformation if $\mathbf{P}$ commutes with the Hamiltonian. In non-relativistic quantum mechanics, this happens for any potential which is scalar, i.e., $V = V(r)$, hence the potential is spherically symmetric. The following facts can be easily proven:

- If $|A\rangle$ and $|B\rangle$ have the same parity, then $\langle A| \mathbf{X} |B\rangle = 0$ where $\mathbf{X}$ is the position operator.

- For a state $|L, L_z\rangle$ of orbital angular momentum $\mathbf{L}$ with z-axis projection L_z, $\mathbf{P}|L, L_z\rangle = (-1)^L |L, L_z\rangle$.

- If $[\mathbf{H}, \mathbf{P}] = 0$, then atomic dipole transitions only occur between states of opposite parity.[2]

- If $[\mathbf{H}, \mathbf{P}] = 0$, then a non-degenerate eigenstate of $\mathbf{H}$ is also an eigenstate of the parity operator; i.e., a non-degenerate eigenfunction of $\mathbf{H}$ is either invariant to $\mathbf{P}$ or is changed in sign by $\mathbf{P}$.

Some of the non-degenerate eigenfunctions of $\mathbf{H}$ are unaffected (invariant) by parity $\mathbf{P}$ and the others will be merely reversed in sign when the Hamiltonian operator and the parity operator commute:

$$\mathbf{P}\,\Psi = c\,\Psi,$$

where c is a constant, the eigenvalue of $\mathbf{P}$,

$$\mathbf{P}^2\Psi = c\mathbf{P}\,\Psi.$$

30.5 Quantum field theory

The intrinsic parity assignments in this section are true for relativistic quantum mechanics as well as quantum field theory.

If we can show that the vacuum state is invariant under parity ($\mathbf{P}|0\rangle = |0\rangle$), the Hamiltonian is parity invariant ($[\mathbf{H}, \mathbf{P}] = 0$) and the quantization conditions remain unchanged under parity, then it follows that every state has good parity, and this parity is conserved in any reaction.

To show that quantum electrodynamics is invariant under parity, we have to prove that the action is invariant and the quantization is also invariant. For simplicity we will assume that canonical quantization is used; the vacuum state is then invariant under parity by construction. The invariance of the action follows from the classical invariance of Maxwell's equations. The invariance of the canonical quantization procedure can be worked out, and turns out to depend on the transformation of the annihilation operator:

$$\mathbf{P}a(\mathbf{p}, \pm)\mathbf{P}^+ = -a(-\mathbf{p}, \pm)$$

where $\mathbf{p}$ denotes the momentum of a photon and $\pm$ refers to its polarization state. This is equivalent to the statement that the photon has odd intrinsic parity. Similarly all vector bosons can be shown to have odd intrinsic parity, and all axial-vectors to have even intrinsic parity.

There is a straightforward extension of these arguments to scalar field theories which shows that scalars have even parity, since

$$\mathbf{P}a(\mathbf{p})\mathbf{P}^+ = a(-\mathbf{p}).$$

This is true even for a complex scalar field. (*Details of spinors are dealt with in the article on the* Dirac equation, *where it is shown that fermions and antifermions have opposite intrinsic parity.*)

With fermions, there is a slight complication because there is more than one spin group.

30.6   Parity in the standard model

30.6.1   Fixing the global symmetries

See also: $(-1)^F$

In the Standard Model of fundamental interactions there are precisely three global internal U(1) symmetry groups available, with charges equal to the baryon number B, the lepton number L and the electric charge Q. The product of the parity operator with any combination of these rotations is another parity operator. It is conventional to choose one specific combination of these rotations to define a standard parity operator, and other parity operators are related to the standard one by internal rotations. One way to fix a standard parity operator is to assign the parities of three particles with linearly independent charges B, L and Q. In general one assigns the parity of the most common massive particles, the proton, the neutron and the electron, to be +1.

Steven Weinberg has shown that if $\mathbf{P}^2 = (-1)^F$, where F is the fermion number operator, then, since the fermion number is the sum of the lepton number plus the baryon number, $F = B + L$, for all particles in the Standard Model and since lepton number and baryon number are charges Q of continuous symmetries e^{iQ}, it is possible to redefine the parity operator so that $\mathbf{P}^2 = 1$. However, if there exist Majorana neutrinos, which experimentalists today believe is possible, their fermion number is equal to one because they are neutrinos while their baryon and lepton numbers are zero because they are Majorana, and so $(-1)^F$ would not be embedded in a continuous symmetry group. Thus Majorana neutrinos would have parity $\pm i$.

30.6.2   Parity of the pion

In 1954, a paper by William Chinowsky and Jack Steinberger demonstrated that the pion has negative parity.[3] They studied the decay of an "atom" made from a deuteron (2
1H+) and a negatively charged pion ($\pi-$) in a state with zero orbital angular momentum $L = 0$ into two neutrons (n).

Neutrons are fermions and so obey Fermi–Dirac statistics, which implies that the final state is antisymmetric. Using the fact that the deuteron has spin one and the pion spin zero together with the antisymmetry of the final state they concluded that the two neutrons must have orbital angular momentum $L = 1$. The total parity is the product of the intrinsic parities of the particles and the extrinsic parity of the spherical harmonic function $(-1)^L$. Since the orbital momentum changes from zero to one in this process, if the process is to conserve the total parity then the products of the intrinsic parities of the initial and final particles must have opposite sign. A deuteron nucleus is made from a proton and a neutron, and so using the aforementioned convention that protons and neutrons have intrinsic parities equal to +1 they argued that the parity of the pion is equal to minus the product of the parities of the two neutrons divided by that of the proton and neutron in the deuteron, $(-1)(1)^2/(1)^2$, which is equal to minus one. Thus they concluded that the pion is a pseudoscalar particle.

30.6.3   Parity violation

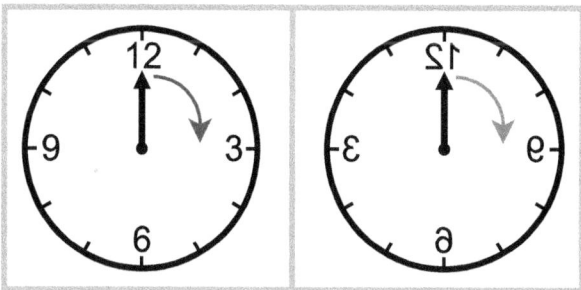

Top: P-symmetry: A clock built like its mirrored image will behave like the mirrored image of the original clock.
Bottom: P-asymmetry: A clock built like its mirrored image will *not* behave like the mirrored image of the original clock.

Although parity is conserved in electromagnetism, strong interactions and gravity, it turns out to be violated in weak interactions. The Standard Model incorporates **parity violation** by expressing the weak interaction as a chiral gauge interaction. Only the left-handed components of particles and right-handed components of antiparticles participate in weak interactions in the Standard Model. This implies that parity is not a symmetry of our universe, unless a hidden mirror sector exists in which parity is violated in the opposite way.

By the mid-20th Century, it had been suggested by several scientists that parity might not be conserved (in different contexts), but without solid evidence these suggestions were not considered important. Then, in 1956, a careful review and analysis by theoretical physicists Tsung Dao Lee and Chen Ning Yang[4] went further, showing that while parity conservation had been verified in decays by the strong or electromagnetic interactions, it was untested in the weak interaction. They proposed several possible direct experimental tests. They were mostly ignored, but Lee was able to convince his Columbia colleague Chien-Shiung Wu to try it. She needed special cryogenic facilities and expertise, so the experiment was done at the National Bureau of Standards.

In 1957 C. S. Wu, E. Ambler, R. W. Hayward, D. D. Hoppes, and R. P. Hudson found a clear violation of parity conservation in the beta decay of cobalt-60.[5] As the experiment was winding down, with double-checking in progress, Wu informed Lee and Yang of their positive results, and saying the results need further examination, she asked them not to publicize the results first. However, Lee revealed the results to his Columbia colleagues on 4 January 1957 at a "Friday Lunch" gathering of the Physics Department of Columbia. Three of them, R. L. Garwin, Leon Lederman, and R. Weinrich modified an existing cyclotron experiment, and they immediately verified the parity violation.[6] They delayed publication of their results until after Wu's group was ready, and the two papers appeared back to back in the same physics journal.

After the fact, it was noted that an obscure 1928 experiment had in effect reported parity violation in weak decays, but since the appropriate concepts had not yet been developed, those results had no impact.[7] The discovery of parity violation immediately explained the outstanding τ–θ puzzle in the physics of kaons.

In 2010, it was reported that physicists working with the Relativistic Heavy Ion Collider (RHIC) had created a short-lived parity symmetry-breaking bubble in quark-gluon plasmas. An experiment conducted by several physicists including Yale's Jack Sandweiss as part of the STAR collaboration, suggested that parity may also be violated in the strong interaction.[8]

30.6.4 Intrinsic parity of hadrons

To every particle one can assign an **intrinsic parity** as long as nature preserves parity. Although weak interactions do not, one can still assign a parity to any hadron by examining the strong interaction reaction that produces it, or through decays not involving the weak interaction, such as rho meson decay to pions.

30.7 See also

- Electroweak theory

- Standard Model

- Mirror matter

30.8   References

General

- Perkins, Donald H. (2000). *Introduction to High Energy Physics*. ISBN 9780521621960.

- Sozzi, M. S. (2008). *Discrete symmetries and CP violation*. Oxford University Press. ISBN 978-0-19-929666-8.

- Bigi, I. I.; Sanda, A. I. (2000). *CP Violation*. Cambridge Monographs on Particle Physics, Nuclear Physics and Cosmology. Cambridge University Press. ISBN 0-521-44349-0.

- Weinberg, S. (1995). *The Quantum Theory of Fields*. Cambridge University Press. ISBN 0-521-67053-5.

Specific

[1] Levine, I.N. *Quantum Chemistry* (Prentice-Hall, 4th edn. 1991), p.355

[2] Bransden, B. H.; Joachain, C. J. (2003). *Physics of Atoms and Molecules* (2nd ed.). Prentice Hall. p. 204. ISBN 978-0-582-35692-4.

[3] Chinowsky, W.; Steinberger, J. (1954). "Absorption of Negative Pions in Deuterium: Parity of the Pion". *Physical Review* **95** (6): 1561–1564. Bibcode:1954PhRv...95.1561C. doi:10.1103/PhysRev.95.1561.

[4] Lee, T. D.; Yang, C. N. (1956). "Question of Parity Conservation in Weak Interactions". *Physical Review* **104** (1): 254–258. Bibcode:1956PhRv..104..254L. doi:10.1103/PhysRev.104.254.

[5] Wu, C. S.; Ambler, E; Hayward, R. W.; Hoppes, D. D.; Hudson, R. P. (1957). "Experimental Test of Parity Conservation in Beta Decay". *Physical Review* **105** (4): 1413–1415. Bibcode:1957PhRv..105.1413W. doi:10.1103/PhysRev.105.1413.

[6] Garwin, R. L.; Lederman, L. M.; Weinrich, M. (1957). "Observations of the Failure of Conservation of Parity and Charge Conjugation in Meson Decays: The Magnetic Moment of the Free Muon". *Physical Review* **105** (4): 1415–1417. Bibcode:1957PhRv..105.1415G.doi:10.1103/PhysRev.105.1415.

[7] Roy, A. (2005). "Discovery of parity violation". *Resonance* **10** (12): 164–175. doi:10.1007/BF02835140.

[8] Muzzin, S. T. (19 March 2010). "For One Tiny Instant, Physicists May Have Broken a Law of Nature". *PhysOrg*. Retrieved 2011-08-05.

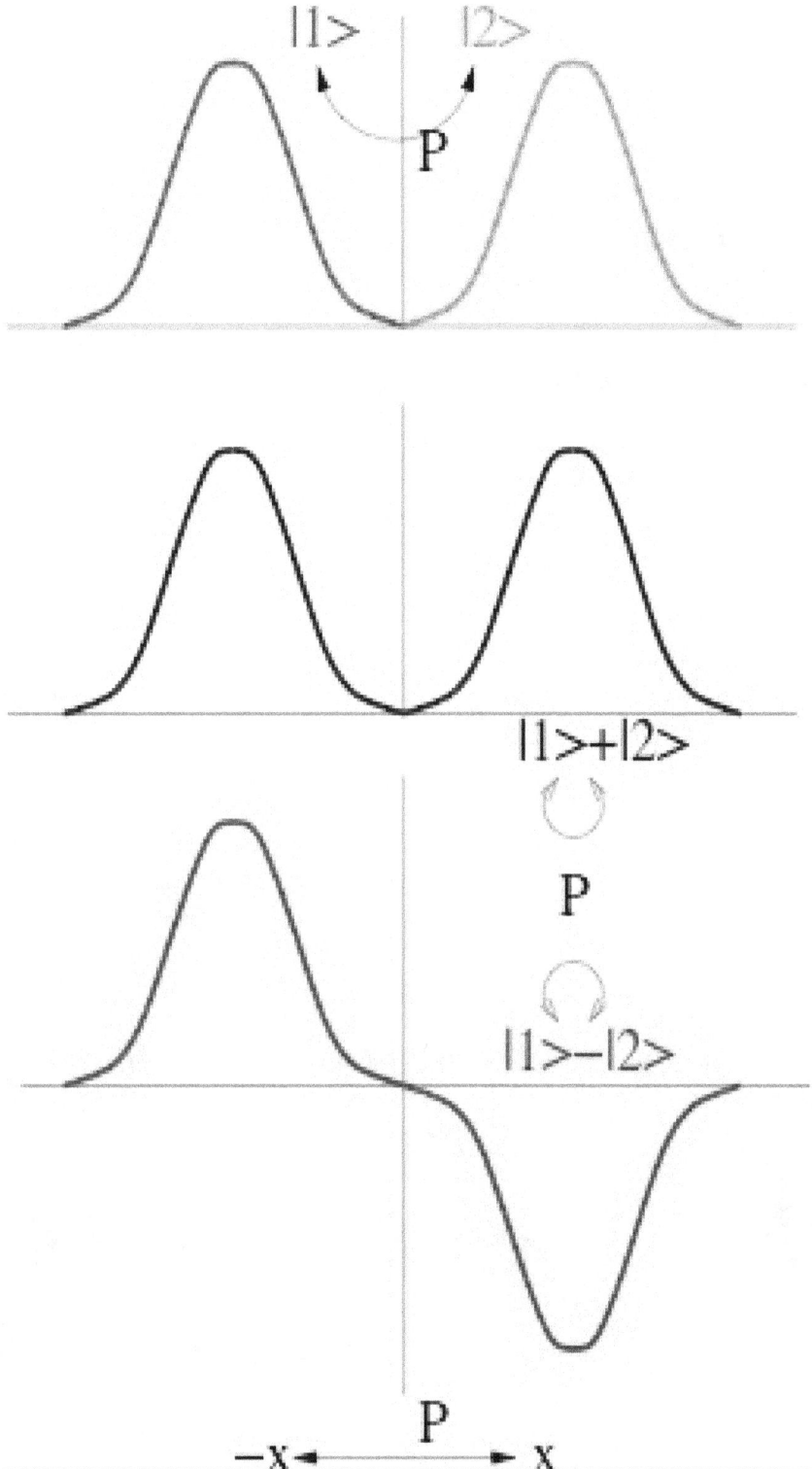

Two-dimensional representations of parity are given by a pair of quantum states that go into
 each other under parity. However, this representation can always be reduced to linear
combinations of states, each of which is either even or odd under parity. One says that all
irreducible representations of parity are one-dimensional. Kramers' theorem states that time
reversal need not have this property because it is represented by an anti-unitary operator.

Chapter 31

Glide reflection

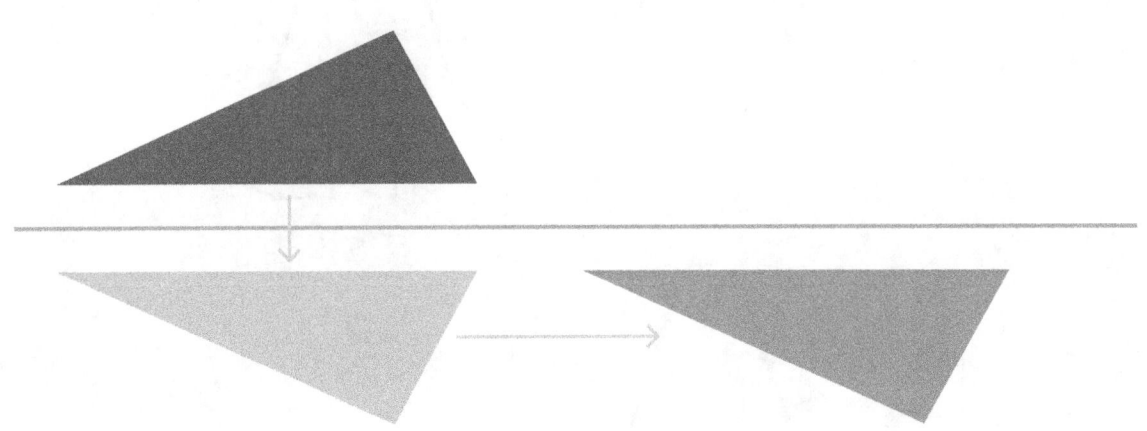

Example of a glide reflection: A composite of a reflection across a line and a translation parallel to the line of reflection

A glide reflection will map a set of left and right footprints into each other

In 2-dimensional geometry, a **glide reflection** (or **transflection**) is a type of opposite isometry of the Euclidean plane: the combination of a reflection in a line and a translation along that line.

A single glide is represented as frieze group p11g. A glide reflection can be seen as a limiting rotoreflection, where the rotation becomes a translation. It can also be given a Schoenflies notation as $S_2\infty$, Coxeter notation as $[\infty^+,2^+]$, and orbifold notation as $\infty\times$.

31.1 Description

The combination of a reflection in a line and a translation in a perpendicular direction is a reflection in a parallel line. However, a glide reflection cannot be reduced like that. Thus the effect of a reflection combined with *any* translation is a glide reflection, with as special case just a reflection. These are the two kinds of indirect isometries in 2D.

For example, there is an isometry consisting of the reflection on the *x*-axis, followed by translation of one unit parallel to it. In coordinates, it takes

$$(x, y) \rightarrow (x + 1, -y).$$

It fixes a system of parallel lines.

The isometry group generated by just a glide reflection is an infinite cyclic group.[1]

Combining two equal glide reflections gives a pure translation with a translation vector that is twice that of the glide reflection, so the even powers of the glide reflection form a translation group.

In the case of **glide reflection symmetry**, the symmetry group of an object contains a glide reflection, and hence the group generated by it. If that is all it contains, this type is frieze group p11g.

Example pattern with this symmetry group:

Frieze group nr. 6 (glide-reflections, translations and rotations) is generated by a glide reflection and a rotation about a point on the line of reflection. It is isomorphic to a semi-direct product of $\mathbf{Z}$ and C_2.

Example pattern with this symmetry group:

A typical example of glide reflection in everyday life would be the track of footprints left in the sand by a person walking on a beach.

For any symmetry group containing some glide reflection symmetry, the translation vector of any glide reflection is one half of an element of the translation group. If the translation vector of a glide reflection is itself an element of the translation group, then the corresponding glide reflection symmetry reduces to a combination of reflection symmetry and translational symmetry.

Glide reflection symmetry with respect to two parallel lines with the same translation implies that there is also translational symmetry in the direction perpendicular to these lines, with a translation distance which is twice the distance between glide reflection lines. This corresponds to wallpaper group pg; with additional symmetry it occurs also in pmg, pgg and p4g.

If there are also true reflection lines in the same direction then they are evenly spaced between the glide reflection lines. A glide reflection line parallel to a true reflection line already implies this situation. This corresponds to wallpaper group cm. The translational symmetry is given by oblique translation vectors from one point on a true reflection line to two points on the next, supporting a rhombus with the true reflection line as one of the diagonals. With additional symmetry it occurs also in cmm, p3m1, p31m, p4m and p6m.

In 3D the glide reflection is called a **glide plane**. It is a reflection in a plane combined with a translation parallel to the plane.

31.2   Wallpaper groups

In the Euclidean plane 3 of 17 wallpaper groups require glide reflection generators. p2gg has orthogonal glide reflections and 2-fold rotations. cm has parallel mirrors and glides, and pg has parallel glides. (Glide reflections are shown below as dashed lines)

31.3   Glide reflection in nature and games

Glide symmetry can be observed in nature among certain fossils of the Ediacara biota; the machaeridians; and certain palaeoscolecid worms.[2]

Glide reflection is common in Conway's Game of Life.

31.4   See also

- Screw axis, glide plane for the corresponding 3D symmetry operations

31.5   References

[1] Martin, George E. (1982), *Transformation Geometry: An Introduction to Symmetry*, Undergraduate Texts in Mathematics, Springer, p. 64, ISBN 9780387906362.

[2] Waggoner, B. M. (1996). "Phylogenetic Hypotheses of the Relationships of Arthropods to Precambrian and Cambrian Problematic Fossil Taxa". *Systematic Biology* **45** (2): 190–222. doi:10.2307/2413615. JSTOR 2413615.

31.6   External links

- Glide Reflection at cut-the-knot

Chapter 32

C-symmetry

In physics, **C-symmetry** means the symmetry of physical laws under a charge-conjugation transformation. Electromagnetism, gravity and the strong interaction all obey C-symmetry, but weak interactions violate C-symmetry.

32.1   Charge reversal in electromagnetism

The laws of electromagnetism (both classical and quantum) are invariant under this transformation: if each charge q were to be replaced with a charge $-q$, and thus the directions of the electric and magnetic fields were reversed, the dynamics would preserve the same form. In the language of quantum field theory, charge conjugation transforms:[1]

1. $\psi \rightarrow -i(\bar{\psi}\gamma^0\gamma^2)^T$

2. $\bar{\psi} \rightarrow -i(\gamma^0\gamma^2\psi)^T$

3. $A^\mu \rightarrow -A^\mu$

Notice that these transformations do not alter the chirality of particles. A left-handed neutrino would be taken by charge conjugation into a left-handed antineutrino, which does not interact in the Standard Model. This property is what is meant by the "maximal violation" of C-symmetry in the weak interaction.

(Some postulated extensions of the Standard Model, like left-right models, restore this C-symmetry.)

32.2   Combination of charge and parity reversal

It was believed for some time that C-symmetry could be combined with the parity-inversion transformation (see P-symmetry) to preserve a combined CP-symmetry. However, violations of this symmetry have been identified in the weak interactions (particularly in the kaons and B mesons). In the Standard Model, this CP violation is due to a single phase in the CKM matrix. If CP is combined with time reversal (T-symmetry), the resulting CPT-symmetry can be shown using only the Wightman axioms to be universally obeyed.

32.3   Charge definition

To give an example, take two real scalar fields, φ and χ. Suppose both fields have even C-parity (even C-parity refers to even symmetry under charge conjugation ex. $C\psi(q) = C\psi(-q)$, as opposed to odd C-parity which refers to antisymmetry under charge conjugation ex. $C\psi(q) = -C\psi(-q)$). Now reformulate things so that $\psi \stackrel{\text{def}}{=} \frac{\phi+i\chi}{\sqrt{2}}$. Now, φ and χ have even C-parities because the imaginary number i has an odd C-parity (C is antiunitary).

In other models, it is possible for both φ and χ to have odd C-parities.

32.4   See also

- C parity

- anti-particle

- antimatter

32.5   References

[1] Peskin, M.E. and Schroeder, D.V. (1997). *An Introduction to Quantum Field Theory*. Addison Wesley. ISBN 0-201-50397-2.

- Sozzi, M.S. (2008). *Discrete symmetries and CP violation*. Oxford University Press. ISBN 978-0-19-929666-8.

Chapter 33

Symmetry in quantum mechanics

Symmetries in quantum mechanics describe features of spacetime and particles which are unchanged under some transformation, in the context of quantum mechanics, relativistic quantum mechanics and quantum field theory, with applications in the mathematical formulation of the standard model and condensed matter physics. In general, symmetry in physics, invariance, and conservation laws, are fundamentally important constraints for formulating physical theories and models. In practice; they are powerful methods for solving problems and predicting what could happen. While conservation laws do not always give the answer to the problem directly and alone, they form the correct constraints and the first steps to solving the problem.

This article outlines the connection between the classical form of continuous symmetries as well as their quantum operators, and relates them to the Lie groups, and relativistic transformations in the Lorentz group, and Poincaré group.

33.1 Notation

The notational conventions used in this article are as follows. Boldface indicates vectors, four vectors, matrices, and vectorial operators, while quantum states use bra–ket notation. Wide hats are for operators, narrow hats are for unit vectors (including their components in tensor index notation). The summation convention on the repeated tensor indices is used, unless stated otherwise. The Minkowski metric signature is (+−−−).

33.2 Symmetry transformations on the wavefunction in non-relativistic quantum mechanics

33.2.1 Continuous symmetries

Generally, the correspondence between continuous symmetries and conservation laws is given by Noether's theorem.

The form of the fundamental quantum operators, for example energy as a partial time derivative and momentum as a spatial gradient, becomes clear when one considers the initial state, then changes one parameter of it slightly. This can be done for displacements (lengths), durations (time), and angles (rotations). Additionally, the invariance of certain quantities can be seen by making such changes in lengths and angles, which illustrates conservation of these quantities.

In what follows, transformations on only one-particle wavefunctions in the form:

$$\widehat{\Omega}\psi(\mathbf{r}, t) = \psi(\mathbf{r}', t')$$

are considered, where $\widehat{\Omega}$ denotes a unitary operator. Unitarity is generally required for operators representing transformations of space, time, and spin, since the norm of a state (representing the total probability of finding the particle somewhere with some spin) must be invariant under these transformations. The inverse is the Hermitian conjugate $\widehat{\Omega}^{-1} = \widehat{\Omega}^{\dagger}$. The results can be extended to many-particle wavefunctions. Written in Dirac notation as standard, the transformations on quantum state vectors are:

$$\widehat{\Omega} \left| \mathbf{r}(t) \right\rangle = \left| \mathbf{r}'(t') \right\rangle$$

Now, the action of $\widehat{\Omega}$ changes $\psi(\mathbf{r}, t)$ to $\psi(\mathbf{r}', t')$, so the inverse $\widehat{\Omega} = \widehat{\Omega}^{\dagger}$ changes $\psi(\mathbf{r}', t')$ back to $\psi(\mathbf{r}, t)$, so an operator $\widehat{A}$ invariant under $\widehat{\Omega}$ satisfies:

$$\widehat{A}\psi = \widehat{\Omega}^{\dagger} \widehat{A} \widehat{\Omega} \psi \quad \Rightarrow \quad \widehat{\Omega} \widehat{A} \psi = \widehat{A} \widehat{\Omega} \psi$$

and thus:

$$[\widehat{\Omega}, \widehat{A}]\psi = 0$$

for any state ψ. Quantum operators representing observables are also required to be Hermitian so that their eigenvalues are real numbers, i.e. the operator equals its Hermitian conjugate, $\widehat{A} = \widehat{A}^{\dagger}$.

33.2.2   Overview of Lie group theory

For a full exposition and details, see Lie group and Generator (mathematics).

Following are the key points of group theory relevant to quantum theory, examples are given throughout the article. For an alternative approach using matrix groups, see the books of Hall[1][2]

Let G be a *Lie group*, which is a group parameterized by a finite number N of real continuously varying parameters ξ_1, ξ_2, ... ξN.

- the *dimension of the group*, N, is the number of parameters it has.

- the *group elements*, g, in G are functions of the parameters:

$$g = G(\xi_1, \xi_2, \cdots)$$

and all parameters set to zero returns the *identity element* of the group:

$$I = G(0, 0 \cdots)$$

Group elements are often matrices which act on vectors, or transformations acting on functions.

- The *generators of the group* are the partial derivatives of the group elements with respect to the group parameters with the result evaluated when the parameter is set to zero:

$$X_j = \left. \frac{\partial g}{\partial \xi_j} \right|_{\xi_j = 0}$$

One aspect of generators in theoretical physics is they can be construed themselves as operators corresponding to symmetries, which may be written as matrices, or as differential operators. In quantum theory, for unitary representations of the group, the generators require a factor of i:

$$X_j = i \left. \frac{\partial g}{\partial \xi_j} \right|_{\xi_j = 0}$$

The generators of the group form a vector space, which means linear combinations of generators also form a generator.

- The generators (whether matrices or differential operators) satisfy the *commutator*:

$$[X_a, X_b] = i f_{abc} X_c$$

where *fabc* are the (basis dependent) *structure constants* of the group. This makes, together with the vector space property, the set of all generators of a group a Lie algebra. Due to the antisymmetry of the bracket, the structure constants of the group are antisymmetric in the first two indices.

- The *representations of the group* are denoted using a capital D and defined by:

$$D[g(\xi_j)] \equiv D(\xi_j) = e^{i \xi_j D(X_j)}$$

without summation on the repeated index j. Representations are linear operators that take in group elements and preserve the composition rule:

$$D(\xi_a) D(\xi_b) = D(\xi_a \xi_b).$$

A representation which cannot be decomposed into a direct sum of other representations, is called *irreducible*. It is conventional to label irreducible representations by a superscripted number n in brackets, as in $D^{(n)}$, or if there is more than one number, we write $D^{(n, m, \dots)}$.

Representations also exist for the generators and the same notation of a capital D is used in this context: $D(X)$. The D in the representation of a generator $D(X)$ is not the same mapping as the D in a representation of a group element, nevertheless this notational abuse of using the same letter to denote two different mappings is used in the literature. An example of this abuse is to be found in the defining equation above.

33.2.3 Momentum and energy as generators of translation and time evolution, and rotation

The space translation operator $\widehat{T}(\Delta \mathbf{r})$ acts on a wavefunction to shift the space coordinates by an infinitesimal displacement $\Delta \mathbf{r}$. The explicit expression $\widehat{T}$ can be quickly determined by a Taylor expansion of $\psi(\mathbf{r} + \Delta \mathbf{r}, t)$ about $\mathbf{r}$, then (keeping the first order term and neglecting second and higher order terms), replace the space derivatives by the momentum operator $\widehat{\mathbf{p}}$. Similarly for the time translation operator acting on the time parameter, the Taylor expansion of $\psi(\mathbf{r}, t + \Delta t)$ is about t, and the time derivative replaced by the energy operator $\widehat{E}$.

The exponential functions arise by definition as those limits, due to Euler, and can be understood physically and mathematically as follows. A net translation can be composed of many small translations, so to obtain the translation operator for a finite increment, replace $\Delta \mathbf{r}$ by $\Delta \mathbf{r}/N$ and Δt by $\Delta t/N$, where N is a positive non-zero integer. Then as N increases, the magnitude of $\Delta \mathbf{r}$ and Δt become even smaller, while leaving the directions unchanged. Acting the infinitesimal operators on the wavefunction N times and taking the limit as N tends to infinity gives the finite operators.

Space and time translations commute, which means the operators and generators commute.

For a time-independent Hamiltonian, energy is conserved in time and quantum states are stationary states: the eigenstates of the Hamiltonian are the energy eigenvalues E:

$$\widehat{U}(t) = \exp\left(-\frac{i\Delta t E}{\hbar}\right)$$

and all stationary states have the form

$$\psi(\mathbf{r}, t + t_0) = \widehat{U}(t - t_0)\psi(\mathbf{r}, t_0)$$

where t_0 is the initial time, usually set to zero since there is no loss of continuity when the initial time is set. An alternative notation is $\widehat{U}(t - t_0) \equiv U(t, t_0)$.

33.2.4 Angular momentum as the generator of rotations

Orbital angular momentum

The rotation operator acts on a wavefunction to rotate the spatial coordinates of a particle by a constant angle $\Delta\theta$:

$$R(\Delta\theta, \hat{\mathbf{a}})\psi(\mathbf{r}, t) = \psi(\mathbf{r}', t)$$

where $\mathbf{r}'$ are the rotated coordinates about an axis defined by a unit vector $\hat{\mathbf{a}} = (a_1, a_2, a_3)$ through an angular increment $\Delta\theta$, given by:

$$\mathbf{r}' = \widehat{R}(\Delta\theta, \hat{\mathbf{a}})\mathbf{r}.$$

where $\widehat{R}(\Delta\theta, \hat{\mathbf{a}})$ is a rotation matrix dependent on the axis and angle. In group theoretic language, the rotation matrices are group elements, and the angles and axis $\Delta\theta\hat{\mathbf{a}} = \Delta\theta(a_1, a_2, a_3)$ are the parameters, of the three-dimensional special orthogonal group, SO(3). The rotation matrices about the standard Cartesian basis vector $\hat{\mathbf{e}}_x, \hat{\mathbf{e}}_y, \hat{\mathbf{e}}_z$ through angle $\Delta\theta$, and the corresponding generators of rotations $\mathbf{J} = (Jx, Jy, Jz)$, are:

More generally for rotations about an axis defined by $\hat{\mathbf{a}}$, the rotation matrix elements are:[3]

$$[\widehat{R}(\theta, \hat{\mathbf{a}})]_{ij} = (\delta_{ij} - a_i a_j)\cos\theta - \varepsilon_{ijk}a_k\sin\theta + a_i a_j$$

where δij is the Kronecker delta, and εijk is the Levi-Civita symbol.

It is not as obvious how to determine the rotational operator compared to space and time translations. We may consider a special case (rotations about the x, y, or z-axis) then infer the general result, or use the general rotation matrix directly and tensor index notation with δij and εijk. To derive the infinitesimal rotation operator, which corresponds to small $\Delta\theta$, we use the small angle approximations $\sin(\Delta\theta) \approx \Delta\theta$ and $\cos(\Delta\theta) \approx 1$, then Taylor expand about $\mathbf{r}$ or ri, keep the first order term, and substitute the angular momentum operator components.

The z-component of angular momentum can be replaced by the component along the axis defined by $\hat{\mathbf{a}}$, using the dot product $\hat{\mathbf{a}} \cdot \widehat{\mathbf{L}}$.

Again, a finite rotation can be made from lots of small rotations, replacing $\Delta\theta$ by $\Delta\theta/N$ and taking the limit as N tends to infinity gives the rotation operator for a finite rotation.

Rotations about the *same* axis do commute, for example a rotation through angles θ_1 and θ_2 about axis i can be written

$$R(\theta_1 + \theta_2, \mathbf{e}_i) = R(\theta_1 \mathbf{e}_i)R(\theta_2 \mathbf{e}_i), \quad [R(\theta_1 \mathbf{e}_i), R(\theta_2 \mathbf{e}_i)] = 0.$$

However, rotations about *different* axes do not commute. The general commutation rules are summarized by

$$[L_i, L_j] = i\hbar\varepsilon_{ijk}L_k.$$

In this sense, orbital angular momentum has the common sense properties of rotations. Each of the above commutators can be easily demonstrated by holding an everyday object and rotating it through the same angle about any two different axes in both possible orderings; the final configurations are different.

In quantum mechanics, there is another form of rotation which mathematically appears similar to the orbital case, but has different properties, described next.

Spin angular momentum

All previous quantities have classical definitions. Spin is a quantity possessed by particles in quantum mechanics without any classical analogue, having the units of angular momentum. The spin vector operator is denoted $\widehat{\mathbf{S}} = (\widehat{S_x}, \widehat{S_y}, \widehat{S_z})$. The eigenvalues of its components are the possible outcomes (in units of $\hbar$) of a measurement of the spin projected onto one of the basis directions.

Rotations (of ordinary space) about an axis $\hat{\mathbf{a}}$ through angle θ about the unit vector $\hat{a}$ in space acting on a multicomponent wave function (spinor) at a point in space is represented by:

However, unlike orbital angular momentum in which the z-projection quantum number ℓ can only take positive or negative integer values (including zero), the z-projection spin quantum number s can take all positive and negative half-integer values. There are rotational matrices for each spin quantum number.

Evaluating the exponential for a given z-projection spin quantum number s gives a $(2s + 1)$-dimensional spin matrix. This can be used to define a spinor as a column vector of $2s + 1$ components which transforms to a rotated coordinate system according to the spin matrix at a fixed point in space.

For the simplest non-trivial case of $s = 1/2$, the spin operator is given by

$$\widehat{\mathbf{S}} = \frac{\hbar}{2}\sigma$$

where the Pauli matrices in the standard representation are:

$$\sigma_1 = \sigma_x = \begin{pmatrix} 0 & 1 \\ 1 & 0 \end{pmatrix}, \quad \sigma_2 = \sigma_y = \begin{pmatrix} 0 & -i \\ i & 0 \end{pmatrix}, \quad \sigma_3 = \sigma_z = \begin{pmatrix} 1 & 0 \\ 0 & -1 \end{pmatrix}$$

Total angular momentum

The total angular momentum operator is the sum of the orbital and spin

$$\widehat{\mathbf{J}} = \widehat{\mathbf{L}} + \widehat{\mathbf{S}}$$

and is an important quantity for multi-particle systems, especially in nuclear physics and the quantum chemistry of multi-electron atoms and molecules.

We have a similar rotation matrix:

$$\widehat{J}(\theta, \hat{\mathbf{a}}) = \exp\left(-\frac{i}{\hbar}\theta \hat{\mathbf{a}} \cdot \widehat{\mathbf{J}}\right)$$

33.3 Lorentz group in relativistic quantum mechanics

Following is an overview of the Lorentz group; a treatment of boosts and rotations in spacetime. Throughout this section, see (for example) T. Ohlsson (2011)[4] and E. Abers (2004).[5]

Lorentz transformations can be parametrized by rapidity φ for a boost in the direction of a three-dimensional unit vector $\hat{\mathbf{n}} = (n_1, n_2, n_3)$, and a rotation angle θ about a three-dimensional unit vector $\hat{\mathbf{a}} = (a_1, a_2, a_3)$ defining an axis, so $\varphi\hat{\mathbf{n}} = \varphi(n_1, n_2, n_3)$ and $\theta\hat{\mathbf{a}} = \theta(a_1, a_2, a_3)$ are together six parameters of the Lorentz group (three for rotations and three for boosts). The Lorentz group is 6-dimensional.

33.3.1 Pure rotations in spacetime

The rotation matrices and rotation generators considered above form the spacelike part of a four-dimensional matrix, representing pure-rotation Lorentz transformations. Three of the Lorentz group elements $\widehat{R}_x, \widehat{R}_y, \widehat{R}_z$ and generators $\mathbf{J} = (J_1, J_2, J_3)$ for pure rotations are:

The rotation matrices act on any four vector $\mathbf{A} = (A_0, A_1, A_2, A_3)$ and rotate the space-like components according to

$$\mathbf{A}' = \widehat{R}(\Delta\theta, \hat{\mathbf{n}})\mathbf{A}$$

leaving the time-like coordinate unchanged. In matrix expressions, $\mathbf{A}$ is treated as a column vector.

33.3.2 Pure boosts in spacetime

A boost with velocity $c\tanh\varphi$ in the x, y, or z directions given by the standard Cartesian basis vector $\hat{\mathbf{e}}_x, \hat{\mathbf{e}}_y, \hat{\mathbf{e}}_z$, are the boost transformation matrices. These matrices $\widehat{B}_x, \widehat{B}_y, \widehat{B}_z$ and the corresponding generators $\mathbf{K} = (K_1, K_2, K_3)$ are the remaining three group elements and generators of the Lorentz group:

The boost matrices act on any four vector $\mathbf{A} = (A_0, A_1, A_2, A_3)$ and mix the time-like and the space-like components, according to:

$$\mathbf{A}' = \widehat{B}(\varphi, \hat{\mathbf{n}})\mathbf{A}$$

The term "boost" refers to the relative velocity between two frames, and is not to be conflated with momentum as the *generator of translations*, as explained below.

33.3.3 Combining boosts and rotations

Products of rotations give another rotation (a frequent exemplification of a subgroup), while products of boosts and boosts or of rotations and boosts cannot be expressed as pure boosts or pure rotations. In general, any Lorentz transformation can be expressed as a product of a pure rotation and a pure boost. For more background see (for example) B.R. Durney (2011)[6] and H.L. Berk et al.[7] and references therein.

The boost and rotation generators have representations denoted $D(\mathbf{K})$ and $D(\mathbf{J})$ respectively, the capital D in this context indicates a group representation.

For the Lorentz group, the representations $D(\mathbf{K})$ and $D(\mathbf{J})$ of the generators $\mathbf{K}$ and $\mathbf{J}$ fulfill the following commutation rules.

In all commutators, the boost entities mixed with those for rotations, although rotations alone simply give another rotation. Exponentiating the generators gives the boost and rotation operators which combine into the general Lorentz transformation, under which the spacetime coordinates transform from one rest frame to another boosted and/or rotating frame. Likewise, exponentiating the representations of the generators gives the representations of the boost and rotation operators, under which a particle's spinor field transforms.

In the literature, the boost generators $\mathbf{K}$ and rotation generators $\mathbf{J}$ are sometimes combined into one generator for Lorentz transformations $\mathbf{M}$, an antisymmetric four-dimensional matrix with entries:

$$M^{0a} = -M^{a0} = K_a, \quad M^{ab} = \varepsilon_{abc}J_c.$$

and correspondingly, the boost and rotation parameters are collected into another antisymmetric four-dimensional matrix $\boldsymbol{\omega}$, with entries:

$$\omega_{0a} = -\omega_{a0} = \varphi n_a, \quad \omega_{ab} = \theta \varepsilon_{abc} a_c,$$

The general Lorentz transformation is then:

$$\Lambda(\varphi, \hat{\mathbf{n}}, \theta, \hat{\mathbf{a}}) = \exp\left(-\frac{i}{2}\omega_{\alpha\beta}M^{\alpha\beta}\right) = \exp\left[-\frac{i}{2}\left(\varphi \hat{\mathbf{n}} \cdot \mathbf{K} + \theta \hat{\mathbf{a}} \cdot \mathbf{J}\right)\right]$$

with summation over repeated matrix indices α and β. The Λ matrices act on any four vector $\mathbf{A} = (A_0, A_1, A_2, A_3)$ and mix the time-like and the space-like components, according to:

$$\mathbf{A}' = \Lambda(\varphi, \hat{\mathbf{n}}, \theta, \hat{\mathbf{a}})\mathbf{A}$$

33.3.4   Transformations of spinor wavefunctions in relativistic quantum mechanics

In relativistic quantum mechanics, wavefunctions are no longer single-component scalar fields, but now $2(2s + 1)$ component spinor fields, where s is the spin of the particle. The transformations of these functions in spacetime are given below.

Under a proper orthochronous Lorentz transformation $(\mathbf{r}, t) \to \Lambda(\mathbf{r}, t)$ in Minkowski space, all one-particle quantum states $\psi\sigma$ locally transform under some representation D of the Lorentz group:[8] [9]

$$\psi_\sigma(\mathbf{r}, t) \to D(\Lambda)\psi_\sigma(\Lambda^{-1}(\mathbf{r}, t))$$

where $D(\Lambda)$ is a finite-dimensional representation, in other words a $(2s + 1) \times (2s + 1)$ dimensional square matrix, and ψ is thought of as a column vector containing components with the $(2s + 1)$ allowed values of σ:

$$\psi(\mathbf{r}, t) = \begin{bmatrix} \psi_{\sigma=s}(\mathbf{r}, t) \\ \psi_{\sigma=s-1}(\mathbf{r}, t) \\ \vdots \\ \psi_{\sigma=-s+1}(\mathbf{r}, t) \\ \psi_{\sigma=-s}(\mathbf{r}, t) \end{bmatrix} \rightleftharpoons \psi(\mathbf{r}, t)^\dagger = \begin{bmatrix} \psi_{\sigma=s}(\mathbf{r}, t)^\star & \psi_{\sigma=s-1}(\mathbf{r}, t)^\star & \cdots & \psi_{\sigma=-s+1}(\mathbf{r}, t)^\star & \psi_{\sigma=-s}(\mathbf{r}, t)^\star \end{bmatrix}$$

33.3.5   Real irreducible representations and spin

The irreducible representations of $D(\mathbf{K})$ and $D(\mathbf{J})$, in short "irreps", can be used to build to spin representations of the Lorentz group. Defining new operators:

$$\mathbf{A} = \frac{\mathbf{J} + i\mathbf{K}}{2}, \quad \mathbf{B} = \frac{\mathbf{J} - i\mathbf{K}}{2},$$

so $\mathbf{A}$ and $\mathbf{B}$ are simply complex conjugates of each other, it follows they satisfy the symmetrically formed commutators:

$$[A_i, A_j] = \varepsilon_{ijk}A_k, \quad [B_i, B_j] = \varepsilon_{ijk}B_k, \quad [A_i, B_j] = 0,$$

and these are essentially the commutators the orbital and spin angular momentum operators satisfy. Therefore $\mathbf{A}$ and $\mathbf{B}$ form operator algebras analogous to angular momentum; same ladder operators, z-projections, etc., independently of each other as each of their components mutually commute. By the analogy to the spin quantum number, we can introduce positive integers or half integers, a, b, with corresponding sets of values $m = a, a - 1, \ldots -a + 1, -a$ and $n = b, b - 1, \ldots -b + 1, -b$. The matrices satisfying the above commutation relations are the same as for spins a and b have components given by multiplying Kronecker delta values with angular momentum matrix elements:

$$(A_x)_{m'n',mn} = \delta_{n'n}\left(J_x^{(m)}\right)_{m'm} \qquad (B_x)_{m'n',mn} = \delta_{m'm}\left(J_x^{(n)}\right)_{n'n}$$

$$(A_y)_{m'n',mn} = \delta_{n'n}\left(J_y^{(m)}\right)_{m'm} \qquad (B_y)_{m'n',mn} = \delta_{m'm}\left(J_y^{(n)}\right)_{n'n}$$

$$(A_z)_{m'n',mn} = \delta_{n'n}\left(J_z^{(m)}\right)_{m'm} \qquad (B_z)_{m'n',mn} = \delta_{m'm}\left(J_z^{(n)}\right)_{n'n}$$

where in each case the row number $m'n'$ and column number mn are separated by a comma, and in turn:

$$\left(J_z^{(m)}\right)_{m'm} = m\delta_{m'm} \qquad \left(J_x^{(m)} \pm iJ_y^{(m)}\right)_{m'm} = m\delta_{a',a\pm1}\sqrt{(a \mp m)(a \pm m + 1)}$$

and similarly for $\mathbf{J}^{(n)}$.[note 1] The three $\mathbf{J}^{(m)}$ matrices are each $(2m+1)\times(2m+1)$ square matrices, and the three $\mathbf{J}^{(n)}$ are each $(2n+1)\times(2n+1)$ square matrices. The integers or half-integers m and n numerate all the irreducible representations by, in equivalent notations used by authors: $D^{(m,n)} \equiv (m,n) \equiv D^{(m)} \otimes D^{(n)}$, which are each $[(2m+1)(2n+1)]\times[(2m+1)(2n+1)]$ square matrices.

Applying this to particles with spin s;

- left-handed $(2s+1)$-component spinors transform under the real irreps $D^{(s,0)}$,

- right-handed $(2s+1)$-component spinors transform under the real irreps $D^{(0,s)}$,

- taking direct sums symbolized by $\oplus$ (see direct sum of matrices for the simpler matrix concept), one obtains the representations under which $2(2s+1)$-component spinors transform: $D^{(m,n)} \oplus D^{(n,m)}$ where $m+n=s$. These are also real irreps, but as shown above, they split into complex conjugates.

In these cases the D refers to any of $D(\mathbf{J})$, $D(\mathbf{K})$, or a full Lorentz transformation $D(\Lambda)$.

33.3.6 Relativistic wave equations

In the context of the Dirac equation and Weyl equation, the Weyl spinors satisfying the Weyl equation transform under the simplest irreducible spin representations of the Lorentz group, since the spin quantum number in this case is the smallest non-zero number allowed: 1/2. The 2-component left-handed Weyl spinor transforms under $D^{(1/2,0)}$ and the 2-component right-handed Weyl spinor transforms under $D^{(0,1/2)}$. Dirac spinors satisfying the Dirac equation transform under the representation $D^{(1/2,0)} \oplus D^{(0,1/2)}$, the direct sum of the irreps for the Weyl spinors.

33.4 The Poincaré group in relativistic quantum mechanics and field theory

Space translations, time translations, rotations, and boosts, all taken together, constitute the Poincaré group. The group elements are the three rotation matrices and three boost matrices (as in the Lorentz group), and one for time translations and three for space translations in spacetime. There is a generator for each. Therefore the Poincaré group is 10-dimensional.

In special relativity, space and time can be collected into a four-position vector $\mathbf{X} = (ct, -\mathbf{r})$, and in parallel so can energy and momentum which combine into a four-momentum vector $\mathbf{P} = (E/c, -\mathbf{p})$. With relativistic quantum mechanics in mind, the time duration and spatial displacement parameters (four in total, one for time and three for space) combine into a spacetime displacement $\Delta\mathbf{X} = (c\Delta t, -\Delta\mathbf{r})$, and the energy and momentum operators are inserted in the four-momentum to obtain a four-momentum operator,

$$\widehat{\mathbf{P}} = \left(\frac{\widehat{E}}{c}, -\widehat{\mathbf{p}}\right) = i\hbar\left(\frac{1}{c}\frac{\partial}{\partial t}, \nabla\right),$$

which are the generators of spacetime translations (four in total, one time and three space):

$$\widehat{X}(\Delta\mathbf{X}) = \exp\left(-\frac{i}{\hbar}\Delta\mathbf{X}\cdot\widehat{\mathbf{P}}\right) = \exp\left[-\frac{i}{\hbar}\left(\Delta t\widehat{E} + \Delta\mathbf{r}\cdot\widehat{\mathbf{p}}\right)\right].$$

There are commutation relations between the components four-momentum $\mathbf{P}$ (generators of spacetime translations), and angular momentum $\mathbf{M}$ (generators of Lorentz transformations), that define the Poincaré algebra:[10][11]

- $[P_\mu, P_\nu] = 0$
- $\frac{1}{i}[M_{\mu\nu}, P_\rho] = \eta_{\mu\rho}P_\nu - \eta_{\nu\rho}P_\mu$

- $\frac{1}{i}[M_{\mu\nu}, M_{\rho\sigma}] = \eta_{\mu\rho}M_{\nu\sigma} - \eta_{\mu\sigma}M_{\nu\rho} - \eta_{\nu\rho}M_{\mu\sigma} + \eta_{\nu\sigma}M_{\mu\rho}$

where η is the Minkowski metric tensor. (It is common to drop any hats for the four-momentum operators in the commutation relations). These equations are an expression of the fundamental properties of space and time as far as they are known today. They have a classical counterpart where the commutators are replaced by Poisson brackets.

To describe spin in relativistic quantum mechanics, the Pauli–Lubanski pseudovector

$$W_\mu = \frac{1}{2}\varepsilon_{\mu\nu\rho\sigma}J^{\nu\rho}P^\sigma ,$$

a Casimir operator, is the constant spin contribution to the total angular momentum, and there are commutation relations between **P** and **W** and between **M** and **W**:

$$[P^\mu, W^\nu] = 0 ,$$

$$[J^{\mu\nu}, W^\rho] = i\left(\eta^{\rho\nu}W^\mu - \eta^{\rho\mu}W^\nu\right) ,$$

$$[W_\mu, W_\nu] = -i\epsilon_{\mu\nu\rho\sigma}W^\rho P^\sigma .$$

Invariants constructed from **W**, instances of Casimir invariants can be used to classify irreducible representations of the Lorentz group.

33.5   Symmetries in quantum field theory and particle physics

33.5.1   Unitary groups in quantum field theory

Group theory is an abstract way of mathematically analyzing symmetries. Unitary operators are paramount to quantum theory, so unitary groups are important in particle physics. The group of N dimensional unitary square matrices is denoted U(N). Unitary operators preserve inner products which means probabilities are also preserved, so the quantum mechanics of the system is invariant under unitary transformations. Let $\widehat{U}$ be a unitary operator, so the inverse is the Hermitian adjoint $\widehat{U} = \widehat{U}^\dagger$, which commutes with the Hamiltonian:

$$\left[\widehat{U}, \widehat{H}\right] = 0$$

then the observable corresponding to the operator $\widehat{U}$ is conserved, and the Hamiltonian is invariant under the transformation $\widehat{U}$.

Since the predictions of quantum mechanics should be invariant under the action of a group, physicists look for unitary transformations to represent the group.

Important subgroups of each U(N) are those unitary matrices which have unit determinant (or are "unimodular"): these are called the special unitary groups and are denoted SU(N).

U(1) and SU(1)

The simplest unitary group is U(1), which is just a complex number of modulus 1. This one-dimensional matrix entry is of the form:

$$U = e^{-i\theta}$$

in which θ is the parameter of the group, and the group is Abelian since one-dimensional matrices always commute under matrix multiplication. Lagrangians in quantum field theory for complex scalar fields are often invariant under U(1) transformations. If there is a quantum number a associated with the U(1) symmetry, for example baryon and the three lepton numbers in electromagnetic interactions, we have:

$$U = e^{-ia\theta}$$

U(2) and SU(2)

The general form of an element of a U(2) element is parametrized by two complex numbers a and b:

$$U = \begin{pmatrix} a & b \\ -b^\star & a^\star \end{pmatrix}$$

and for SU(2), the determinant is restricted to 1:

$$\det(U) = aa^\star + bb^\star = |a|^2 + |b|^2 = 1$$

In group theoretic language, the Pauli matrices are the generators of the special unitary group in two dimensions, denoted SU(2). Their commutation relation is the same as for orbital angular momentum, aside from a factor of 2:

$$[\sigma_a, \sigma_b] = 2i\hbar\varepsilon_{abc}\sigma_c$$

A group element of SU(2) can be written:

$$U(\theta, \hat{\mathbf{e}}_j) = e^{i\theta\sigma_j/2}$$

where σj is a Pauli matrix, and the group parameters are the angles turned through about an axis.

U(3) and SU(3)

The eight Gell-Mann matrices λn (see article for them and the structure constants) are important for quantum chromodynamics. They originally arose in the theory SU(3) of flavor which is still of practical importance in nuclear physics. They are the generators for the SU(3) group, so an element of SU(3) can be written analogously to an element of SU(2):

$$U(\theta, \hat{\mathbf{e}}_j) = \exp\left(-\frac{i}{2}\sum_{n=1}^{8}\theta_n\lambda_n\right)$$

where θn are eight independent parameters. The λn matrices satisfy the commutator:

$$[\lambda_a, \lambda_b] = 2if_{abc}\lambda_c$$

where the indices a, b, c take the values 1, 2, 3... 8. The structure constants $fabc$ are totally antisymmetric in all indices analogous to those of SU(2). In the standard colour charge basis (r for red, g for green, b for blue):

$$|r\rangle = \begin{pmatrix} 1 \\ 0 \\ 0 \end{pmatrix}, \quad |g\rangle = \begin{pmatrix} 0 \\ 1 \\ 0 \end{pmatrix}, \quad |b\rangle = \begin{pmatrix} 0 \\ 0 \\ 1 \end{pmatrix}$$

the colour states are eigenstates of the λ_3 and λ_8 matrices, while the other matrices mix colour states together.

The eight gluons states (8-dimensional column vectors) are simultaneous eigenstates of the adjoint representation of SU(3), the 8-dimensional representation acting on its own Lie algebra su(3), for the λ_3 and λ_8 matrices. By forming tensor products of representations (the standard representation and its dual) and taking appropriate quotients, protons and neutrons, and other hadrons are eigenstates of various representations of SU(3) of color. The representations of SU(3) can be described by a "theorem of the highest weight".[12]

33.5.2   Matter and antimatter

In relativistic quantum mechanics, relativistic wave equations predict a remarkable symmetry of nature: that every particle has a corresponding antiparticle. This is mathematically contained in the spinor fields which are the solutions of the relativistic wave equations.

Charge conjugation switches particles and antiparticles. Physical laws and interactions unchanged by this operation have C symmetry.

33.5.3   Discrete spacetime symmetries

- Parity mirrors the orientation of the spatial coordinates from left-handed to right-handed. Informally, space is "reflected" into its mirror image. Physical laws and interactions unchanged by this operation have P symmetry.

- Time reversal negates the time coordinate, which amounts to time running from future to past. A curious property of time, which space does not have, is that it is unidirectional: particles traveling forwards in time are equivalent to antiparticles traveling back in time. Physical laws and interactions unchanged by this operation have T symmetry.

33.5.4   *C, P, T* symmetries

- CPT theorem

- CP violation

- Lorentz violation

33.5.5   Gauge theory

Main article: Gauge theory

In quantum electrodynamics, the symmetry group is U(1) and is abelian. In quantum chromodynamics, the symmetry group is SU(3) and is non-abelian.

The electromagnetic interaction is mediated by photons, which have no electric charge. The electromagnetic tensor has an electromagnetic four-potential field possessing gauge symmetry.

The strong (color) interaction is mediated by gluons, which can have eight color charges. There are eight gluon field strength tensors with corresponding gluon four potentials field, each possessing gauge symmetry.

33.5.6 The strong (color) interaction

Color charge

Analogous to the spin operator, there are color charge operators in terms of the Gell-Mann matrices λj:

$$\hat{F}_j = \frac{1}{2}\lambda_j$$

and since color charge is a conserved charge, all color charge operators must commute with the Hamiltonian:

$$\left[\hat{F}_j, \hat{H}\right] = 0$$

Isospin

Isospin is conserved in strong interactions.

33.5.7 The weak and electromagnetic interactions

Duality transformation

Magnetic monopoles can be theoretically realized, although current observations and theory are consistent with them existing or not existing. Electric and magnetic charges can effectively be "rotated into one another" by a duality transformation.

Electroweak symmetry

- Electroweak symmetry
- Electroweak symmetry breaking

33.5.8 Supersymmetry

Main article: Supersymmetry

A Lie superalgebra is an algebra in which (suitable) basis elements either have a commutation relation or have an anti-commutation relation. Symmetries have been proposed to the effect that all fermionic particles have bosonic analogues, and vice versa. These symmetry have theoretical appeal in that no extra assumptions (such as existence of strings) barring symmetries are made. In addition, by assuming supersymmetry, a number puzzling issues can be resolved. These symmetries, which are represented by Lie superalgebras, have not been confirmed experimentally. It is now believed that they are broken symmetries, if they exist. But it has been speculated that dark matter is constitutes gravitinos, a spin 3/2 particle with mass, its supersymmetric partner being the graviton.

33.6 Exchange symmetry

See also: Exchange interaction, Identical particles and Holstein–Herring method

The concept of **exchange symmetry** is derived from a fundamental postulate of quantum statistics, which states that no observable physical quantity should change after exchanging two identical particles. It states that because all observables are proportional to $|\psi|^2$ for a system of identical particles, the wave function ψ must either remain the same or change sign upon such an exchange.

Because the exchange of two identical particles is mathematically equivalent to the rotation of each particle by 180 degrees (and so to the rotation of one particle's frame by 360 degrees),[13] the symmetric nature of the wave function depends on the particle's spin after the rotation operator is applied to it. Integer spin particles do not change the sign of their wave function upon a 360 degree rotation—therefore the sign of the wave function of the entire system does not change. Semi-integer spin particles change the sign of their wave function upon a 360 degree rotation (see more in spin–statistics theorem).

Particles for which the wave function does not change sign upon exchange are called bosons, or particles with a symmetric wave function. The particles for which the wave function of the system changes sign are called fermions, or particles with an antisymmetric wave function.

Fermions therefore obey different statistics (called Fermi–Dirac statistics) than bosons (which obey Bose–Einstein statistics). One of the consequences of Fermi–Dirac statistics is the exclusion principle for fermions—no two identical fermions can share the same quantum state (in other words, the wave function of two identical fermions in the same state is zero). This in turn results in degeneracy pressure for fermions—the strong resistance of fermions to compression into smaller volume. This resistance gives rise to the "stiffness" or "rigidity" of ordinary atomic matter (as atoms contain electrons which are fermions).

33.7   See also

- Casimir operator

- Pauli–Lubanski pseudovector

- Symmetries in general relativity

- Renormalization group

- Center of mass (relativistic)

- Representation of a Lie group

- Representation theory of the Poincaré group

- Representation theory of the Lorentz group

33.8   Footnotes

[1] Sometimes the tuple abbreviations:

$$(\mathbf{A})_{m'n',mn} \equiv \left[(A_x)_{m'n',mn}, (A_y)_{m'n',mn}, (A_z)_{m'n',mn} \right]$$

$$(\mathbf{B})_{m'n',mn} \equiv \left[(B_x)_{m'n',mn}, (B_y)_{m'n',mn}, (B_z)_{m'n',mn} \right]$$

$$\left(\mathbf{J}^{(m)} \right)_{m'm} \equiv \left[\left(J_x^{(m)} \right)_{m'm}, \left(J_y^{(m)} \right)_{m'm}, \left(J_z^{(m)} \right)_{m'm} \right]$$

are used.

33.9   References

[1] Hall, Brian C. (2015). *Lie Groups, Lie Algebras, and Representations: An Elementary Introduction.* Graduate Texts in Mathematics **222** (2nd ed.). Springer.

[2] Hall, Brian C. (2013). *Quantum Theory for Mathematicians.* Graduate Texts in Mathematics. Springer.

[3] C.B. Parker (1994). *McGraw Hill Encyclopaedia of Physics* (2nd ed.). McGraw Hill. p. 1333. ISBN 0-07-051400-3.

[4] T. Ohlsson (2011). *Relativistic Quantum Physics: From Advanced Quantum Mechanics to Introductory Quantum Field Theory.* Cambridge University Press. pp. 7–10. ISBN 1-13950-4320.

[5] E. Abers (2004). *Quantum Mechanics.* Addison Wesley. pp. 11, 104, 105, 410–411. ISBN 978-0-13-146100-0.

[6] B.R. Durney. "Lorentz Transformations". arXiv:1103.0156.

[7] H.L. Berk, K. Chaicherdsakul, T. Udagawa. "The Proper Homogeneous Lorentz Transformation Operator $e^L = e^{-\omega \cdot S - \xi \cdot K}$, Where's It Going, What's the Twist" (PDF). Texas, Austin.

[8] Weinberg, S. (1964). "Feynman Rules *for Any* spin" (PDF). *Phys. Rev.* **133** (5B): B1318–B1332. Bibcode:1964PhRv..133.1318W. doi:10.1103/PhysRev.133.B1318.; Weinberg, S. (1964). "Feynman Rules *for Any* spin. II. Massless Particles" (PDF). *Phys. Rev.* **134** (4B): B882–B896. Bibcode:1964PhRv..134..882W. doi:10.1103/PhysRev.134.B882.; Weinberg, S. (1969). "Feynman Rules *for Any* spin. III" (PDF). *Phys. Rev.* **181** (5): 1893–1899. Bibcode:1969PhRv..181.1893W. doi:10.1103/PhysRev.181.1893.

[9] K. Masakatsu (2012). "Superradiance Problem of Bosons and Fermions for Rotating Black Holes in Bargmann–Wigner Formulation". Nara, Japan. arXiv:1208.0644.

[10] N.N. Bogolubov (1989). *General Principles of Quantum Field Theory* (2nd ed.). Springer. p. 272. ISBN 0-7923-0540-X.

[11] T. Ohlsson (2011). *Relativistic Quantum Physics: From Advanced Quantum Mechanics to Introductory Quantum Field Theory.* Cambridge University Press. p. 10. ISBN 1-13950-4320.

[12] Hall, Brian C. (2015). *Lie Groups, Lie Algebras, and Representations: An Elementary Introduction.* Graduate Texts in Mathematics **222** (2nd ed.). Springer. Chapter 6

[13] Feynman, Richard. *The 1986 Dirac Memorial Lectures.* Cambridge University Press. p. 57. ISBN 978-0-521-65862-1.

- D. McMahon (2008). *Quantum Field Theory.* Mc Graw Hill. ISBN 978-0-07-154382-8.

- B. R. Martin, G.Shaw. *Particle Physics* (3rd ed.). Manchester Physics Series, John Wiley & Sons. p. 3. ISBN 978-0-470-03294-7.

- M. Chaichian, R. Hagedorn (1998). *Symmetry in quantum mechanics: From angular momentum to supersymmetry.* Graduate student series in physics. Institute of physic s (Bristol and Philadelphia). ISBN 0-7503-0408-1.

- W. Ludwig, C. Falter (1996). *Symmetries in physics.* Solid state science (2nd ed.). Springer. ISBN 3-540-60284-4.

- M. F. C. Ladd (1989). *Symmetry in molecules and crystals.* Solid state science. Ellis Horwood series in physical chemistry. ISBN 0-85312-255-5.

- K. J. Barnes (2010). *Group theory for the standard model and beyond.* Series in high energy physics, cosmology, and gravitation. Taylor & Francis. ISBN 142-007-874-7.

- S. Haywood (2011). *Symmetries and Conservation Laws in Particle Physics: An Introduction to Group Theory for Particle Physicists.* World Scientific. ISBN 184-816-703-2.

33.10   External links

- (2010) *Irreducible Tensor Operators and the Wigner-Eckart Theorem*

- R.D. Reece (2006) *A Derivation of the Quantum Mechanical Momentum Operator in the Position Representation*

- D. E. Soper (2011) *Position and momentum in quantum mechanics*

- *Lie groups*

- F. Porter (2009) *Lie Groups and Lie Algebras*

- *Continuous Groups, Lie Groups, and Lie Algebras*

- P.J. Mulders (2011) *Quantum field theory*

- arXiv:math-ph/0005032v1 B.C. Hall (2000) *An Elementary Introduction to Groups and Representations*

Chapter 34

Noether's theorem

This article is about Emmy Noether's first theorem, which derives conserved quantities from symmetries. For other uses, see Noether's theorem (disambiguation).

Noether's (first)[1] **theorem** states that every differentiable symmetry of the action of a physical system has a corresponding conservation law. The theorem was proven by German mathematician Emmy Noether in 1915 and published in 1918.[2] The action of a physical system is the integral over time of a Lagrangian function (which may or may not be an integral over space of a Lagrangian density function), from which the system's behavior can be determined by the principle of least action.

Noether's theorem is used in theoretical physics and the calculus of variations. A generalization of the formulations on constants of motion in Lagrangian and Hamiltonian mechanics (developed in 1788 and 1833, respectively), it does not apply to systems that cannot be modeled with a Lagrangian alone (e.g. systems with a Rayleigh dissipation function). In particular, dissipative systems with continuous symmetries need not have a corresponding conservation law.

34.1 Basic illustrations and background

As an illustration, if a physical system behaves the same regardless of how it is oriented in space, its Lagrangian is rotationally symmetric: from this symmetry, Noether's theorem dictates that the angular momentum of the system be conserved, as a consequence of its laws of motion. The physical system itself need not be symmetric; a jagged asteroid tumbling in space conserves angular momentum despite its asymmetry. It is the laws of its motion that are symmetric.

As another example, if a physical process exhibits the same outcomes regardless of place or time, then its Lagrangian is symmetric under continuous translations in space and time: by Noether's theorem, these symmetries account for the conservation laws of linear momentum and energy within this system, respectively.

Noether's theorem is important, both because of the insight it gives into conservation laws, and also as a practical calculational tool. It allows investigators to determine the conserved quantities (invariants) from the observed symmetries of a physical system. Conversely, it allows researchers to consider whole classes of hypothetical Lagrangians with given invariants, to describe a physical system. As an illustration, suppose that a physical theory is proposed which conserves a quantity X. A researcher can calculate the types of Lagrangians that conserve X through a continuous symmetry. Due to Noether's theorem, the properties of these Lagrangians provide further criteria to understand the implications and judge the fitness of the new theory.

There are numerous versions of Noether's theorem, with varying degrees of generality. The original version only applied to ordinary differential equations (particles) and not partial differential equations (fields). The original versions also assume that the Lagrangian only depends upon the first derivative, while later versions generalize the theorem to Lagrangians depending on the nth derivative. There are natural quantum counterparts of this theorem, expressed in the Ward–Takahashi identities. Generalizations of Noether's theorem to superspaces are also available.

34.2   Informal statement of the theorem

All fine technical points aside, Noether's theorem can be stated informally

> If a system has a continuous symmetry property, then there are corresponding quantities whose values are conserved in time.[3]

A more sophisticated version of the theorem involving fields states that:

> To every differentiable symmetry generated by local actions, there corresponds a conserved current.

The word "symmetry" in the above statement refers more precisely to the covariance of the form that a physical law takes with respect to a one-dimensional Lie group of transformations satisfying certain technical criteria. The conservation law of a physical quantity is usually expressed as a continuity equation.

The formal proof of the theorem utilizes the condition of invariance to derive an expression for a current associated with a conserved physical quantity. In modern (since ca. 1980[4]) terminology, the conserved quantity is called the *Noether charge*, while the flow carrying that charge is called the *Noether current*. The Noether current is defined up to a solenoidal (divergenceless) vector field.

In the context of gravitation, Felix Klein's statement of Noether's theorem for action I stipulates for the invariants:[5]

> If an integral I is invariant under a continuous group G_ρ with ρ parameters, then ρ linearly independent combinations of the Lagrangian expressions are divergences.

34.3   Historical context

Main articles: Constant of motion, conservation law and conserved current

A conservation law states that some quantity X in the mathematical description of a system's evolution remains constant throughout its motion — it is an invariant. Mathematically, the rate of change of X (its derivative with respect to time) vanishes,

$$\frac{dX}{dt} = 0 \,.$$

Such quantities are said to be conserved; they are often called constants of motion (although motion *per se* need not be involved, just evolution in time). For example, if the energy of a system is conserved, its energy is invariant at all times, which imposes a constraint on the system's motion and may help in solving for it. Aside from insights that such constants of motion give into the nature of a system, they are a useful calculational tool; for example, an approximate solution can be corrected by finding the nearest state that satisfies the suitable conservation laws.

The earliest constants of motion discovered were momentum and energy, which were proposed in the 17th century by René Descartes and Gottfried Leibniz on the basis of collision experiments, and refined by subsequent researchers. Isaac Newton was the first to enunciate the conservation of momentum in its modern form, and showed that it was a consequence of Newton's third law. According to general relativity, the conservation laws of linear momentum, energy and angular momentum are only exactly true globally when expressed in terms of the sum of the stress–energy tensor (non-gravitational stress–energy) and the Landau–Lifshitz stress–energy–momentum pseudotensor (gravitational stress–energy). The local conservation of non-gravitational linear momentum and energy in a free-falling reference frame is expressed by the vanishing of the covariant divergence of the stress–energy tensor. Another important conserved quantity, discovered in studies of the celestial mechanics of astronomical bodies, is the Laplace–Runge–Lenz vector.

In the late 18th and early 19th centuries, physicists developed more systematic methods for discovering invariants. A major advance came in 1788 with the development of Lagrangian mechanics, which is related to the principle of least action. In this approach, the state of the system can be described by any type of generalized coordinates **q**; the laws of motion need not be expressed in a Cartesian coordinate system, as was customary in Newtonian mechanics. The action is defined as the time integral I of a function known as the Lagrangian L

$$I = \int L(\mathbf{q}, \dot{\mathbf{q}}, t)\, dt \,,$$

where the dot over **q** signifies the rate of change of the coordinates **q**,

$$\dot{\mathbf{q}} = \frac{d\mathbf{q}}{dt} \,.$$

Hamilton's principle states that the physical path $\mathbf{q}(t)$—the one actually taken by the system—is a path for which infinitesimal variations in that path cause no change in I, at least up to first order. This principle results in the Euler–Lagrange equations,

$$\frac{d}{dt}\left(\frac{\partial L}{\partial \dot{\mathbf{q}}}\right) = \frac{\partial L}{\partial \mathbf{q}} \,.$$

Thus, if one of the coordinates, say qk, does not appear in the Lagrangian, the right-hand side of the equation is zero, and the left-hand side requires that

$$\frac{d}{dt}\left(\frac{\partial L}{\partial \dot{q}_k}\right) = \frac{dp_k}{dt} = 0 \,,$$

where the momentum

$$p_k = \frac{\partial L}{\partial \dot{q}_k}$$

is conserved throughout the motion (on the physical path).

Thus, the absence of the **ignorable** coordinate qk from the Lagrangian implies that the Lagrangian is unaffected by changes or transformations of qk; the Lagrangian is invariant, and is said to exhibit a symmetry under such transformations. This is the seed idea generalized in Noether's theorem.

Several alternative methods for finding conserved quantities were developed in the 19th century, especially by William Rowan Hamilton. For example, he developed a theory of canonical transformations which allowed changing coordinates so that some coordinates disappeared from the Lagrangian, as above, resulting in conserved canonical momenta. Another approach, and perhaps the most efficient for finding conserved quantities, is the Hamilton–Jacobi equation.

34.4 Mathematical expression

See also: Perturbation theory

34.4.1   Simple form using perturbations

The essence of Noether's theorem is generalizing the ignorable coordinates outlined.

Imagine that the action I defined above is invariant under small perturbations (warpings) of the time variable t and the generalized coordinates $\mathbf{q}$; in a notation commonly used in physics,

$$t \to t' = t + \delta t$$

$$\mathbf{q} \to \mathbf{q}' = \mathbf{q} + \delta \mathbf{q} \, ,$$

where the perturbations δt and $\delta \mathbf{q}$ are both small, but variable. For generality, assume there are (say) N such symmetry transformations of the action, i.e. transformations leaving the action unchanged; labelled by an index $r = 1, 2, 3, \ldots, N$.

Then the resultant perturbation can be written as a linear sum of the individual types of perturbations,

$$\delta t = \sum_r \varepsilon_r T_r$$

$$\delta \mathbf{q} = \sum_r \varepsilon_r \mathbf{Q}_r \, ,$$

where εr are infinitesimal parameter coefficients corresponding to each:

- generator $T r$ of time evolution, and

- generator $\mathbf{Q}r$ of the generalized coordinates.

For translations, $\mathbf{Q}r$ is a constant with units of length; for rotations, it is an expression linear in the components of $\mathbf{q}$, and the parameters make up an angle.

Using these definitions, Noether showed that the N quantities

$$\left(\frac{\partial L}{\partial \dot{\mathbf{q}}} \cdot \dot{\mathbf{q}} - L \right) T_r - \frac{\partial L}{\partial \dot{\mathbf{q}}} \cdot \mathbf{Q}_r$$

(which have the dimensions of [energy]·[time] + [momentum]·[length] = [action]) are conserved (constants of motion).

Examples

Time invariance

For illustration, consider a Lagrangian that does not depend on time, i.e., that is invariant (symmetric) under changes $t \to t + \delta t$, without any change in the coordinates $\mathbf{q}$. In this case, $N = 1$, $T = 1$ and $\mathbf{Q} = 0$; the corresponding conserved quantity is the total energy H[6]

$$H = \frac{\partial L}{\partial \dot{\mathbf{q}}} \cdot \dot{\mathbf{q}} - L.$$

Translational invariance

Consider a Lagrangian which does not depend on an ("ignorable", as above) coordinate qk; so it is invariant (symmetric) under changes $qk \rightarrow qk + \delta qk$. In that case, $N = 1$, $T = 0$, and $Qk = 1$; the conserved quantity is the corresponding momentum pk[7]

$$p_k = \frac{\partial L}{\partial \dot{q}_k}.$$

In special and general relativity, these apparently separate conservation laws are aspects of a single conservation law, that of the stress–energy tensor,[8] that is derived in the next section.

Rotational invariance

The conservation of the angular momentum $\mathbf{L} = \mathbf{r} \times \mathbf{p}$ is analogous to its linear momentum counterpart.[9] It is assumed that the symmetry of the Lagrangian is rotational, i.e., that the Lagrangian does not depend on the absolute orientation of the physical system in space. For concreteness, assume that the Lagrangian does not change under small rotations of an angle $\delta\theta$ about an axis $\mathbf{n}$; such a rotation transforms the Cartesian coordinates by the equation

$$\mathbf{r} \rightarrow \mathbf{r} + \delta\theta\mathbf{n} \times \mathbf{r}.$$

Since time is not being transformed, $T=0$. Taking $\delta\theta$ as the ε parameter and the Cartesian coordinates $\mathbf{r}$ as the generalized coordinates $\mathbf{q}$, the corresponding $\mathbf{Q}$ variables are given by

$$\mathbf{Q} = \mathbf{n} \times \mathbf{r}.$$

Then Noether's theorem states that the following quantity is conserved,

$$\frac{\partial L}{\partial \dot{\mathbf{q}}} \cdot \mathbf{Q}_r = \mathbf{p} \cdot (\mathbf{n} \times \mathbf{r}) = \mathbf{n} \cdot (\mathbf{r} \times \mathbf{p}) = \mathbf{n} \cdot \mathbf{L}.$$

In other words, the component of the angular momentum $\mathbf{L}$ along the $\mathbf{n}$ axis is conserved.

If $\mathbf{n}$ is arbitrary, i.e., if the system is insensitive to any rotation, then every component of $\mathbf{L}$ is conserved; in short, angular momentum is conserved.

34.4.2 Field theory version

Although useful in its own right, the version of Noether's theorem just given is a special case of the general version derived in 1915. To give the flavor of the general theorem, a version of the Noether theorem for continuous fields in four-dimensional space–time is now given. Since field theory problems are more common in modern physics than mechanics problems, this field theory version is the most commonly used version (or most often implemented) of Noether's theorem.

Let there be a set of differentiable fields φ defined over all space and time; for example, the temperature $T(\mathbf{x}, t)$ would be representative of such a field, being a number defined at every place and time. The principle of least action can be applied to such fields, but the action is now an integral over space and time

$$I = \int L\left(\phi, \partial_\mu \phi, x^\mu\right) d^4 x$$

(the theorem can actually be further generalized to the case where the Lagrangian depends on up to the nth derivative using jet bundles)

Let the action be invariant under certain transformations of the space–time coordinates x^μ and the fields φ

$$x^\mu \to x^\mu + \delta x^\mu$$

$$\phi \to \phi + \delta\phi$$

where the transformations can be indexed by $r = 1, 2, 3, \ldots, N$

$$\delta x^\mu = \varepsilon_r X_r^\mu$$

$$\delta\phi = \varepsilon_r \Psi_r \, .$$

For such systems, Noether's theorem states that there are N conserved current densities

$$j_r^\nu = -\left(\frac{\partial L}{\partial \phi_{,\nu}}\right) \cdot \Psi_r + \left[\left(\frac{\partial L}{\partial \phi_{,\nu}}\right) \cdot \phi_{,\sigma} - L\delta_\sigma^\nu\right] X_r^\sigma$$

In such cases, the conservation law is expressed in a four-dimensional way

$$\partial_\mu j^\mu = 0$$

which expresses the idea that the amount of a conserved quantity within a sphere cannot change unless some of it flows out of the sphere. For example, electric charge is conserved; the amount of charge within a sphere cannot change unless some of the charge leaves the sphere.

For illustration, consider a physical system of fields that behaves the same under translations in time and space, as considered above; in other words, $L\left(\phi, \partial_\mu\phi, x^\mu\right)$ is constant in its third argument. In that case, $N = 4$, one for each dimension of space and time. Since only the positions in space–time are being warped, not the fields, the Ψ are all zero and the $X\mu^\nu$ equal the Kronecker delta $\delta\mu^\nu$, where we have used μ instead of r for the index. In that case, Noether's theorem corresponds to the conservation law for the stress–energy tensor $T\mu^{\nu}$[8]

$$T_\mu{}^\nu = \left[\left(\frac{\partial L}{\partial \phi_{,\nu}}\right) \cdot \phi_{,\sigma} - L\,\delta_\sigma^\nu\right]\delta_\mu^\sigma = \left(\frac{\partial L}{\partial \phi_{,\nu}}\right) \cdot \phi_{,\mu} - L\,\delta_\mu^\nu$$

The conservation of electric charge, by contrast, can be derived by considering zero $X\mu^\nu = 0$ and Ψ linear in the fields φ themselves.[10] In quantum mechanics, the probability amplitude $\psi(\mathbf{x})$ of finding a particle at a point $\mathbf{x}$ is a complex field φ, because it ascribes a complex number to every point in space and time. The probability amplitude itself is physically unmeasurable; only the probability $p = |\psi|^2$ can be inferred from a set of measurements. Therefore, the system is invariant under transformations of the ψ field and its complex conjugate field ψ^* that leave $|\psi|^2$ unchanged, such as

$$\psi \to e^{i\theta}\psi \, , \ \psi^* \to e^{-i\theta}\psi^* \, ,$$

a complex rotation. In the limit when the phase θ becomes infinitesimally small, $\delta\theta$, it may be taken as the parameter ε, while the Ψ are equal to $i\psi$ and $-i\psi^*$, respectively. A specific example is the Klein–Gordon equation, the relativistically correct version of the Schrödinger equation for spinless particles, which has the Lagrangian density

$$L = \psi_{,\nu}\psi_{,\mu}^*\eta^{\nu\mu} + m^2\psi\psi^* \, .$$

In this case, Noether's theorem states that the conserved $(\partial \cdot j = 0)$ current equals

$$j^\nu = i \left(\frac{\partial \psi}{\partial x^\mu} \psi^* - \frac{\partial \psi^*}{\partial x^\mu} \psi \right) \eta^{\nu\mu} ,$$

which, when multiplied by the charge on that species of particle, equals the electric current density due to that type of particle. This "gauge invariance" was first noted by Hermann Weyl, and is one of the prototype gauge symmetries of physics.

34.5 Derivations

34.5.1 One independent variable

Consider the simplest case, a system with one independent variable, time. Suppose the dependent variables $\mathbf{q}$ are such that the action integral

$$I = \int_{t_1}^{t_2} L[\mathbf{q}[t], \dot{\mathbf{q}}[t], t] \, dt$$

is invariant under brief infinitesimal variations in the dependent variables. In other words, they satisfy the Euler–Lagrange equations

$$\frac{d}{dt} \frac{\partial L}{\partial \dot{\mathbf{q}}}[t] = \frac{\partial L}{\partial \mathbf{q}}[t].$$

And suppose that the integral is invariant under a continuous symmetry. Mathematically such a symmetry is represented as a flow, $\boldsymbol{\varphi}$, which acts on the variables as follows

$$t \to t' = t + \varepsilon T$$

$$\mathbf{q}[t] \to \mathbf{q}'[t'] = \phi[\mathbf{q}[t], \varepsilon] = \phi[\mathbf{q}[t' - \varepsilon T], \varepsilon]$$

where ε is a real variable indicating the amount of flow, and T is a real constant (which could be zero) indicating how much the flow shifts time.

$$\dot{\mathbf{q}}[t] \to \dot{\mathbf{q}}'[t'] = \frac{d}{dt} \phi[\mathbf{q}[t], \varepsilon] = \frac{\partial \phi}{\partial \mathbf{q}}[\mathbf{q}[t' - \varepsilon T], \varepsilon] \dot{\mathbf{q}}[t' - \varepsilon T].$$

The action integral flows to

$$I'[\varepsilon] = \int_{t_1 + \varepsilon T}^{t_2 + \varepsilon T} L[\mathbf{q}'[t'], \dot{\mathbf{q}}'[t'], t'] \, dt'$$

$$= \int_{t_1 + \varepsilon T}^{t_2 + \varepsilon T} L[\phi[\mathbf{q}[t' - \varepsilon T], \varepsilon], \frac{\partial \phi}{\partial \mathbf{q}}[\mathbf{q}[t' - \varepsilon T], \varepsilon] \dot{\mathbf{q}}[t' - \varepsilon T], t'] \, dt'$$

which may be regarded as a function of ε. Calculating the derivative at $\varepsilon = 0$ and using the symmetry, we get

$$0 = \frac{dI'}{d\varepsilon}[0] = L[\mathbf{q}[t_2], \dot{\mathbf{q}}[t_2], t_2]T - L[\mathbf{q}[t_1], \dot{\mathbf{q}}[t_1], t_1]T$$

$$+ \int_{t_1}^{t_2} \frac{\partial L}{\partial \mathbf{q}} \left(-\frac{\partial \phi}{\partial \mathbf{q}} \dot{\mathbf{q}} T + \frac{\partial \phi}{\partial \varepsilon} \right) + \frac{\partial L}{\partial \dot{\mathbf{q}}} \left(-\frac{\partial^2 \phi}{(\partial \mathbf{q})^2} \dot{\mathbf{q}}^2 T + \frac{\partial^2 \phi}{\partial \varepsilon \partial \mathbf{q}} \dot{\mathbf{q}} - \frac{\partial \phi}{\partial \mathbf{q}} \ddot{\mathbf{q}} T \right) \, dt.$$

Notice that the Euler–Lagrange equations imply

$$\frac{d}{dt}\left(\frac{\partial L}{\partial \dot{\mathbf{q}}}\frac{\partial \phi}{\partial \mathbf{q}}\dot{\mathbf{q}}T\right) = \left(\frac{d}{dt}\frac{\partial L}{\partial \dot{\mathbf{q}}}\right)\frac{\partial \phi}{\partial \mathbf{q}}\dot{\mathbf{q}}T + \frac{\partial L}{\partial \dot{\mathbf{q}}}\left(\frac{d}{dt}\frac{\partial \phi}{\partial \mathbf{q}}\right)\dot{\mathbf{q}}T + \frac{\partial L}{\partial \dot{\mathbf{q}}}\frac{\partial \phi}{\partial \mathbf{q}}\ddot{\mathbf{q}}T$$

$$= \frac{\partial L}{\partial \mathbf{q}}\frac{\partial \phi}{\partial \mathbf{q}}\dot{\mathbf{q}}T + \frac{\partial L}{\partial \dot{\mathbf{q}}}\left(\frac{\partial^2 \phi}{(\partial \mathbf{q})^2}\dot{\mathbf{q}}\right)\dot{\mathbf{q}}T + \frac{\partial L}{\partial \dot{\mathbf{q}}}\frac{\partial \phi}{\partial \mathbf{q}}\ddot{\mathbf{q}}T.$$

Substituting this into the previous equation, one gets

$$0 = \frac{dI'}{d\varepsilon}[0] = L[\mathbf{q}[t_2], \dot{\mathbf{q}}[t_2], t_2]T - L[\mathbf{q}[t_1], \dot{\mathbf{q}}[t_1], t_1]T - \frac{\partial L}{\partial \dot{\mathbf{q}}}\frac{\partial \phi}{\partial \mathbf{q}}\dot{\mathbf{q}}[t_2]T + \frac{\partial L}{\partial \dot{\mathbf{q}}}\frac{\partial \phi}{\partial \mathbf{q}}\dot{\mathbf{q}}[t_1]T$$

$$+ \int_{t_1}^{t_2}\frac{\partial L}{\partial \mathbf{q}}\frac{\partial \phi}{\partial \varepsilon} + \frac{\partial L}{\partial \dot{\mathbf{q}}}\frac{\partial^2 \phi}{\partial \varepsilon \partial \mathbf{q}}\dot{\mathbf{q}}\, dt.$$

Again using the Euler–Lagrange equations we get

$$\frac{d}{dt}\left(\frac{\partial L}{\partial \dot{\mathbf{q}}}\frac{\partial \phi}{\partial \varepsilon}\right) = \left(\frac{d}{dt}\frac{\partial L}{\partial \dot{\mathbf{q}}}\right)\frac{\partial \phi}{\partial \varepsilon} + \frac{\partial L}{\partial \dot{\mathbf{q}}}\frac{\partial^2 \phi}{\partial \varepsilon \partial \mathbf{q}}\dot{\mathbf{q}} = \frac{\partial L}{\partial \mathbf{q}}\frac{\partial \phi}{\partial \varepsilon} + \frac{\partial L}{\partial \dot{\mathbf{q}}}\frac{\partial^2 \phi}{\partial \varepsilon \partial \mathbf{q}}\dot{\mathbf{q}}.$$

Substituting this into the previous equation, one gets

$$0 = L[\mathbf{q}[t_2], \dot{\mathbf{q}}[t_2], t_2]T - L[\mathbf{q}[t_1], \dot{\mathbf{q}}[t_1], t_1]T - \frac{\partial L}{\partial \dot{\mathbf{q}}}\frac{\partial \phi}{\partial \mathbf{q}}\dot{\mathbf{q}}[t_2]T + \frac{\partial L}{\partial \dot{\mathbf{q}}}\frac{\partial \phi}{\partial \mathbf{q}}\dot{\mathbf{q}}[t_1]T$$

$$+ \frac{\partial L}{\partial \dot{\mathbf{q}}}\frac{\partial \phi}{\partial \varepsilon}[t_2] - \frac{\partial L}{\partial \dot{\mathbf{q}}}\frac{\partial \phi}{\partial \varepsilon}[t_1].$$

From which one can see that

$$\left(\frac{\partial L}{\partial \dot{\mathbf{q}}}\frac{\partial \phi}{\partial \mathbf{q}}\dot{\mathbf{q}} - L\right)T - \frac{\partial L}{\partial \dot{\mathbf{q}}}\frac{\partial \phi}{\partial \varepsilon}$$

is a constant of the motion, i.e., it is a conserved quantity. Since $\varphi[\mathbf{q}, 0] = \mathbf{q}$, we get $\frac{\partial \phi}{\partial \mathbf{q}} = 1$ and so the conserved quantity simplifies to

$$\left(\frac{\partial L}{\partial \dot{\mathbf{q}}}\dot{\mathbf{q}} - L\right)T - \frac{\partial L}{\partial \dot{\mathbf{q}}}\frac{\partial \phi}{\partial \varepsilon}.$$

To avoid excessive complication of the formulas, this derivation assumed that the flow does not change as time passes. The same result can be obtained in the more general case.

34.5.2   Field-theoretic derivation

Noether's theorem may also be derived for tensor fields φ^A where the index A ranges over the various components of the various tensor fields. These field quantities are functions defined over a four-dimensional space whose points are labeled by coordinates x^μ where the index μ ranges over time ($\mu = 0$) and three spatial dimensions ($\mu = 1, 2, 3$). These four coordinates are the independent variables; and the values of the fields at each event are the dependent variables. Under an infinitesimal transformation, the variation in the coordinates is written

$$x^\mu \to \xi^\mu = x^\mu + \delta x^\mu$$

whereas the transformation of the field variables is expressed as

$$\phi^A \to \alpha^A(\xi^\mu) = \phi^A(x^\mu) + \delta\phi^A(x^\mu).$$

By this definition, the field variations $\delta\varphi^A$ result from two factors: intrinsic changes in the field themselves and changes in coordinates, since the transformed field α^A depends on the transformed coordinates ξ^μ. To isolate the intrinsic changes, the field variation at a single point x^μ may be defined

$$\alpha^A(x^\mu) = \phi^A(x^\mu) + \bar\delta\phi^A(x^\mu).$$

If the coordinates are changed, the boundary of the region of space–time over which the Lagrangian is being integrated also changes; the original boundary and its transformed version are denoted as Ω and Ω', respectively.

Noether's theorem begins with the assumption that a specific transformation of the coordinates and field variables does not change the action, which is defined as the integral of the Lagrangian density over the given region of spacetime. Expressed mathematically, this assumption may be written as

$$\int_{\Omega'} L\left(\alpha^A, \alpha^A{}_{,\nu}, \xi^\mu\right) d^4\xi - \int_\Omega L\left(\phi^A, \phi^A{}_{,\nu}, x^\mu\right) d^4x = 0$$

where the comma subscript indicates a partial derivative with respect to the coordinate(s) that follows the comma, e.g.

$$\phi^A{}_{,\sigma} = \frac{\partial\phi^A}{\partial x^\sigma}.$$

Since ξ is a dummy variable of integration, and since the change in the boundary Ω is infinitesimal by assumption, the two integrals may be combined using the four-dimensional version of the divergence theorem into the following form

$$\int_\Omega \left\{ \left[L\left(\alpha^A, \alpha^A{}_{,\nu}, x^\mu\right) - L\left(\phi^A, \phi^A{}_{,\nu}, x^\mu\right)\right] + \frac{\partial}{\partial x^\sigma}\left[L\left(\phi^A, \phi^A{}_{,\nu}, x^\mu\right)\delta x^\sigma\right]\right\} d^4x = 0.$$

The difference in Lagrangians can be written to first-order in the infinitesimal variations as

$$\left[L\left(\alpha^A, \alpha^A{}_{,\nu}, x^\mu\right) - L\left(\phi^A, \phi^A{}_{,\nu}, x^\mu\right)\right] = \frac{\partial L}{\partial\phi^A}\bar\delta\phi^A + \frac{\partial L}{\partial\phi^A{}_{,\sigma}}\bar\delta\phi^A{}_{,\sigma}.$$

However, because the variations are defined at the same point as described above, the variation and the derivative can be done in reverse order; they commute

$$\bar\delta\phi^A{}_{,\sigma} = \bar\delta\frac{\partial\phi^A}{\partial x^\sigma} = \frac{\partial}{\partial x^\sigma}(\bar\delta\phi^A).$$

Using the Euler–Lagrange field equations

$$\frac{\partial}{\partial x^\sigma}\left(\frac{\partial L}{\partial\phi^A{}_{,\sigma}}\right) = \frac{\partial L}{\partial\phi^A}$$

the difference in Lagrangians can be written neatly as

$$\left[L\left(\alpha^A, \alpha^A{}_{,\nu}, x^\mu\right) - L\left(\phi^A, \phi^A{}_{,\nu}, x^\mu\right)\right] = \frac{\partial}{\partial x^\sigma}\left(\frac{\partial L}{\partial \phi^A{}_{,\sigma}}\right)\bar{\delta}\phi^A + \frac{\partial L}{\partial \phi^A{}_{,\sigma}}\bar{\delta}\phi^A{}_{,\sigma} = \frac{\partial}{\partial x^\sigma}\left(\frac{\partial L}{\partial \phi^A{}_{,\sigma}}\bar{\delta}\phi^A\right).$$

Thus, the change in the action can be written as

$$\int_\Omega \frac{\partial}{\partial x^\sigma}\left\{\frac{\partial L}{\partial \phi^A{}_{,\sigma}}\bar{\delta}\phi^A + L\left(\phi^A, \phi^A{}_{,\nu}, x^\mu\right)\delta x^\sigma\right\}d^4x = 0.$$

Since this holds for any region Ω, the integrand must be zero

$$\frac{\partial}{\partial x^\sigma}\left\{\frac{\partial L}{\partial \phi^A{}_{,\sigma}}\bar{\delta}\phi^A + L\left(\phi^A, \phi^A{}_{,\nu}, x^\mu\right)\delta x^\sigma\right\} = 0.$$

For any combination of the various symmetry transformations, the perturbation can be written

$$\delta x^\mu = \varepsilon X^\mu$$

$$\delta \phi^A = \varepsilon \Psi^A = \bar{\delta}\phi^A + \varepsilon \mathcal{L}_X \phi^A$$

where $\mathcal{L}_X \phi^A$ is the Lie derivative of φ^A in the X^μ direction. When φ^A is a scalar or $X^\mu{}_{,\nu} = 0$,

$$\mathcal{L}_X \phi^A = \frac{\partial \phi^A}{\partial x^\mu}X^\mu.$$

These equations imply that the field variation taken at one point equals

$$\bar{\delta}\phi^A = \varepsilon \Psi^A - \varepsilon \mathcal{L}_X \phi^A.$$

Differentiating the above divergence with respect to ε at $\varepsilon = 0$ and changing the sign yields the conservation law

$$\frac{\partial}{\partial x^\sigma}j^\sigma = 0$$

where the conserved current equals

$$j^\sigma = \left[\frac{\partial L}{\partial \phi^A{}_{,\sigma}}\mathcal{L}_X \phi^A - L X^\sigma\right] - \left(\frac{\partial L}{\partial \phi^A{}_{,\sigma}}\right)\Psi^A.$$

34.5.3 Manifold/fiber bundle derivation

Suppose we have an n-dimensional oriented Riemannian manifold, M and a target manifold T. Let $\mathcal{C}$ be the configuration space of smooth functions from M to T. (More generally, we can have smooth sections of a fiber bundle over M.)

Examples of this M in physics include:

- In classical mechanics, in the Hamiltonian formulation, M is the one-dimensional manifold **R**, representing time and the target space is the cotangent bundle of space of generalized positions.

- In field theory, M is the spacetime manifold and the target space is the set of values the fields can take at any given point. For example, if there are m real-valued scalar fields, $\phi_1, \ldots, \phi_m$, then the target manifold is $\mathbf{R}^m$. If the field is a real vector field, then the target manifold is isomorphic to $\mathbf{R}^3$.

Now suppose there is a functional

$$\mathcal{S} : \mathcal{C} \to \mathbf{R},$$

called the action. (Note that it takes values into $\mathbf{R}$, rather than $\mathbf{C}$; this is for physical reasons, and doesn't really matter for this proof.)

To get to the usual version of Noether's theorem, we need additional restrictions on the action. We assume $\mathcal{S}[\phi]$ is the integral over M of a function

$$\mathcal{L}(\phi, \partial_\mu \phi, x)$$

called the Lagrangian density, depending on φ, its derivative and the position. In other words, for φ in $\mathcal{C}$

$$\mathcal{S}[\phi] = \int_M \mathcal{L}[\phi(x), \partial_\mu \phi(x), x] \mathrm{d}^n x.$$

Suppose we are given boundary conditions, i.e., a specification of the value of φ at the boundary if M is compact, or some limit on φ as x approaches ∞. Then the subspace of $\mathcal{C}$ consisting of functions φ such that all functional derivatives of $\mathcal{S}$ at φ are zero, that is:

$$\frac{\delta \mathcal{S}[\phi]}{\delta \phi(x)} \approx 0$$

and that φ satisfies the given boundary conditions, is the subspace of on shell solutions. (See principle of stationary action)

Now, suppose we have an infinitesimal transformation on $\mathcal{C}$, generated by a functional derivation, Q such that

$$Q \left[\int_N \mathcal{L} \, \mathrm{d}^n x \right] \approx \int_{\partial N} f^\mu [\phi(x), \partial \phi, \partial \partial \phi, \ldots] \, \mathrm{d}s_\mu$$

for all compact submanifolds N or in other words,

$$Q[\mathcal{L}(x)] \approx \partial_\mu f^\mu(x)$$

for all x, where we set

$$\mathcal{L}(x) = \mathcal{L}[\phi(x), \partial_\mu \phi(x), x].$$

If this holds on shell and off shell, we say Q generates an off-shell symmetry. If this only holds on shell, we say Q generates an on-shell symmetry. Then, we say Q is a generator of a one parameter symmetry Lie group.

Now, for any N, because of the Euler–Lagrange theorem, on shell (and only on-shell), we have

Since this is true for any N, we have

$$\partial_\mu \left[\frac{\partial \mathcal{L}}{\partial(\partial_\mu \phi)} Q[\phi] - f^\mu \right] \approx 0.$$

But this is the continuity equation for the current J^μ defined by:[11]

$$J^\mu = \frac{\partial \mathcal{L}}{\partial(\partial_\mu \phi)} Q[\phi] - f^\mu,$$

which is called the **Noether current** associated with the symmetry. The continuity equation tells us that if we integrate this current over a space-like slice, we get a conserved quantity called the Noether charge (provided, of course, if M is noncompact, the currents fall off sufficiently fast at infinity).

34.5.4 Comments

Noether's theorem is an on shell theorem: it relies on use of the equations of motion—the classical path. It reflects the relation between the boundary conditions and the variational principle. Assuming no boundary terms in the action, Noether's theorem implies that

$$\int_{\partial N} J^\mu \mathrm{d}s_\mu \approx 0 \, .$$

The quantum analogs of Noether's theorem involving expectation values, e.g. $\langle \int d^4 x \, \partial \cdot J \rangle = 0$, probing off shell quantities as well are the Ward–Takahashi identities.

34.5.5 Generalization to Lie algebras

Suppose say we have two symmetry derivations Q_1 and Q_2. Then, $[Q_1, Q_2]$ is also a symmetry derivation. Let's see this explicitly. Let's say

$$Q_1[\mathcal{L}] \approx \partial_\mu f_1^\mu$$

and

$$Q_2[\mathcal{L}] \approx \partial_\mu f_2^\mu$$

Then,

$$[Q_1, Q_2][\mathcal{L}] = Q_1[Q_2[\mathcal{L}]] - Q_2[Q_1[\mathcal{L}]] \approx \partial_\mu f_{12}^\mu$$

where $f_{12} = Q_1[f_2{}^\mu] - Q_2[f_1{}^\mu]$. So,

$$j_{12}^\mu = \left(\frac{\partial}{\partial(\partial_\mu \phi)} \mathcal{L} \right) (Q_1[Q_2[\phi]] - Q_2[Q_1[\phi]]) - f_{12}^\mu.$$

This shows we can extend Noether's theorem to larger Lie algebras in a natural way.

34.5.6 Generalization of the proof

This applies to *any* local symmetry derivation Q satisfying $QS \approx 0$, and also to more general local functional differentiable actions, including ones where the Lagrangian depends on higher derivatives of the fields. Let ε be any arbitrary smooth function of the spacetime (or time) manifold such that the closure of its support is disjoint from the boundary. ε is a test function. Then, because of the variational principle (which does *not* apply to the boundary, by the way), the derivation distribution q generated by $q[\varepsilon][\Phi(x)] = \varepsilon(x)Q[\Phi(x)]$ satisfies $q[\varepsilon][S] \approx 0$ for every ε, or more compactly, $q(x)[S] \approx 0$ for all x not on the boundary (but remember that $q(x)$ is a shorthand for a derivation *distribution*, not a derivation parametrized by x in general). This is the generalization of Noether's theorem.

To see how the generalization is related to the version given above, assume that the action is the spacetime integral of a Lagrangian that only depends on φ and its first derivatives. Also, assume

$$Q[\mathcal{L}] \approx \partial_\mu f^\mu$$

Then,

$$q[\varepsilon][\mathcal{S}] = \int q[\varepsilon][\mathcal{L}] \, \mathrm{d}^n x$$
$$= \int \left\{ \left(\frac{\partial}{\partial \phi} \mathcal{L} \right) \varepsilon Q[\phi] + \left[\frac{\partial}{\partial (\partial_\mu \phi)} \mathcal{L} \right] \partial_\mu (\varepsilon Q[\phi]) \right\} \mathrm{d}^n x$$
$$= \int \left\{ \varepsilon Q[\mathcal{L}] + \partial_\mu \varepsilon \left[\frac{\partial}{\partial (\partial_\mu \phi)} \mathcal{L} \right] Q[\phi] \right\} \mathrm{d}^n x$$
$$\approx \int \varepsilon \partial_\mu \left\{ f^\mu - \left[\frac{\partial}{\partial (\partial_\mu \phi)} \mathcal{L} \right] Q[\phi] \right\} \mathrm{d}^n x$$

for all ε.

More generally, if the Lagrangian depends on higher derivatives, then

$$\partial_\mu \left[f^\mu - \left[\frac{\partial}{\partial (\partial_\mu \phi)} \mathcal{L} \right] Q[\phi] - 2 \left[\frac{\partial}{\partial (\partial_\mu \partial_\nu \phi)} \mathcal{L} \right] \partial_\nu Q[\phi] + \partial_\nu \left[\left[\frac{\partial}{\partial (\partial_\mu \partial_\nu \phi)} \mathcal{L} \right] Q[\phi] \right] - \cdots \right] \approx 0.$$

34.6 Examples

34.6.1 Example 1: Conservation of energy

Looking at the specific case of a Newtonian particle of mass m, coordinate x, moving under the influence of a potential V, coordinatized by time t. The action, S, is:

$$\mathcal{S}[x] = \int L[x(t), \dot{x}(t)] \, dt$$
$$= \int \left(\frac{m}{2} \sum_{i=1}^{3} \dot{x}_i^2 - V(x(t)) \right) dt.$$

The first term in the brackets is the kinetic energy of the particle, whilst the second is its potential energy. Consider the generator of time translations $Q = \partial/\partial t$. In other words, $Q[x(t)] = \dot{x}(t)$. Note that x has an explicit dependence on time, whilst V does not; consequently:

$$Q[L] = m \sum_i \dot{x}_i \ddot{x}_i - \sum_i \frac{\partial V(x)}{\partial x_i} \dot{x}_i = \frac{d}{dt}\left[\frac{m}{2}\sum_i \dot{x}_i^2 - V(x)\right]$$

so we can set

$$f = \frac{m}{2}\sum_i \dot{x}_i^2 - V(x).$$

Then,

$$j = \sum_{i=1}^{3} \frac{\partial L}{\partial \dot{x}_i} Q[x_i] - f$$

$$= m \sum_i \dot{x}_i^2 - \left[\frac{m}{2}\sum_i \dot{x}_i^2 - V(x)\right]$$

$$= \frac{m}{2}\sum_i \dot{x}_i^2 + V(x).$$

The right hand side is the energy, and Noether's theorem states that $\dot{j} = 0$ (i.e. the principle of conservation of energy is a consequence of invariance under time translations).

More generally, if the Lagrangian does not depend explicitly on time, the quantity

$$\sum_{i=1}^{3} \frac{\partial L}{\partial \dot{x}_i} \dot{x}_i - L$$

(called the Hamiltonian) is conserved.

34.6.2   Example 2: Conservation of center of momentum

Still considering 1-dimensional time, let

$$\mathcal{S}[\vec{x}] = \int \mathcal{L}[\vec{x}(t), \dot{\vec{x}}(t)]\, dt$$

$$= \int \left[\sum_{\alpha=1}^{N} \frac{m_\alpha}{2}(\dot{\vec{x}}_\alpha)^2 - \sum_{\alpha<\beta} V_{\alpha\beta}(\vec{x}_\beta - \vec{x}_\alpha)\right] dt$$

i.e. N Newtonian particles where the potential only depends pairwise upon the relative displacement.

For $\vec{Q}$, let's consider the generator of Galilean transformations (i.e. a change in the frame of reference). In other words,

$$Q_i[x_\alpha^j(t)] = t\delta_i^j.$$

Note that

$$Q_i[\mathcal{L}] = \sum_\alpha m_\alpha \dot{x}^i_\alpha - \sum_{\alpha<\beta} \partial_i V_{\alpha\beta}(\vec{x}_\beta - \vec{x}_\alpha)(t - t)$$

$$= \sum_\alpha m_\alpha \dot{x}^i_\alpha.$$

This has the form of $\frac{d}{dt} \sum_\alpha m_\alpha x^i_\alpha$ so we can set

$$\vec{f} = \sum_\alpha m_\alpha \vec{x}_\alpha.$$

Then,

$$\vec{j} = \sum_\alpha \left(\frac{\partial}{\partial \dot{\vec{x}}_\alpha} \mathcal{L} \right) \cdot \vec{Q}[\vec{x}_\alpha] - \vec{f}$$

$$= \sum_\alpha (m_\alpha \dot{\vec{x}}_\alpha t - m_\alpha \vec{x}_\alpha)$$

$$= \vec{P}t - M\vec{x}_{CM}$$

where $\vec{P}$ is the total momentum, M is the total mass and $\vec{x}_{CM}$ is the center of mass. Noether's theorem states:

$$\dot{\vec{j}} = 0 \Rightarrow \vec{P} - M\dot{\vec{x}}_{CM} = 0.$$

34.6.3 Example 3: Conformal transformation

Both examples 1 and 2 are over a 1-dimensional manifold (time). An example involving spacetime is a conformal transformation of a massless real scalar field with a quartic potential in (3 + 1)-Minkowski spacetime.

For Q, consider the generator of a spacetime rescaling. In other words,

$$Q[\phi(x)] = x^\mu \partial_\mu \phi(x) + \phi(x).$$

The second term on the right hand side is due to the "conformal weight" of φ. Note that

$$Q[\mathcal{L}] = \partial^\mu \phi \left(\partial_\mu \phi + x^\nu \partial_\mu \partial_\nu \phi + \partial_\mu \phi \right) - 4\lambda \phi^3 \left(x^\mu \partial_\mu \phi + \phi \right).$$

This has the form of

$$\partial_\mu \left[\frac{1}{2} x^\mu \partial^\nu \phi \partial_\nu \phi - \lambda x^\mu \phi^4 \right] = \partial_\mu \left(x^\mu \mathcal{L} \right)$$

(where we have performed a change of dummy indices) so set

$f^\mu = x^\mu \mathcal{L}.$

Then,

$$j^\mu = \left[\frac{\partial}{\partial(\partial_\mu \phi)} \mathcal{L} \right] Q[\phi] - f^\mu$$

$$= \partial^\mu \phi \left(x^\nu \partial_\nu \phi + \phi \right) - x^\mu \left(\frac{1}{2} \partial^\nu \phi \partial_\nu \phi - \lambda \phi^4 \right).$$

Noether's theorem states that $\partial_\mu j^\mu = 0$ (as one may explicitly check by substituting the Euler–Lagrange equations into the left hand side).

Note that if one tries to find the Ward–Takahashi analog of this equation, one runs into a problem because of anomalies.

34.7   Applications

Application of Noether's theorem allows physicists to gain powerful insights into any general theory in physics, by just analyzing the various transformations that would make the form of the laws involved invariant. For example:

- the invariance of physical systems with respect to spatial translation (in other words, that the laws of physics do not vary with locations in space) gives the law of conservation of linear momentum;

- invariance with respect to rotation gives the law of conservation of angular momentum;

- invariance with respect to time translation gives the well-known law of conservation of energy

In quantum field theory, the analog to Noether's theorem, the Ward–Takahashi identity, yields further conservation laws, such as the conservation of electric charge from the invariance with respect to a change in the phase factor of the complex field of the charged particle and the associated gauge of the electric potential and vector potential.

The Noether charge is also used in calculating the entropy of stationary black holes.[12]

34.8   See also

- Charge (physics)

- Gauge symmetry

- Gauge symmetry (mathematics)

- Invariant (physics)

- Goldstone boson

- Symmetry in physics

34.9 Notes

[1] See also Noether's second theorem.

[2] Noether E (1918). "Invariante Variationsprobleme". *Nachr. D. König. Gesellsch. D. Wiss. Zu Göttingen, Math-phys. Klasse* **1918**: 235–257.

[3] Thompson, W.J. (1994). *Angular Momentum: an illustrated guide to rotational symmetries for physical systems* **1**. Wiley. p. 5. ISBN 0-471-55264-X.

[4] The term "Noether charge" occurs in Seligman, *Group theory and its applications in physics, 1980: Latin American School of Physics, Mexico City*, American Institute of Physics, 1981. It comes enters wider use during the 1980s, e.g. by G. Takeda in: Errol Gotsman, Gerald Tauber (eds.) *From SU(3) to Gravity: Festschrift in Honor of Yuval Ne'eman*, 1985, p. 196.

[5] Nina Byers (1998) "E. Noether's Discovery of the Deep Connection Between Symmetries and Conservation Laws." in Proceedings of a Symposium on the Heritage of Emmy Noether, held on 2–4 December 1996, at the Bar-Ilan University, Israel, Appendix B.

[6] Lanczos 1970, pp. 401–3

[7] Lanczos 1970, pp. 403–4

[8] Goldstein 1980, pp. 592–3

[9] Lanczos 1970, pp. 404–5

[10] Goldstein 1980, pp. 593–4

[11] Michael E. Peskin, Daniel V. Schroeder (1995). *An Introduction to Quantum Field Theory*. Basic Books. p. 18. ISBN 0-201-50397-2.

[12] Vivek Iyer; Wald (1995). "A comparison of Noether charge and Euclidean methods for Computing the Entropy of Stationary Black Holes". *Physical Review D* **52** (8): 4430–9. arXiv:gr-qc/9503052. Bibcode:1995PhRvD..52.4430I. doi:10.1103/Phy.

34.10 References

- Goldstein, Herbert (1980). *Classical Mechanics* (2nd ed.). Reading, MA: Addison-Wesley. pp. 588–596. ISBN 0-201-02918-9.

- Kosmann-Schwarzbach, Yvette (2010). *The Noether theorems: Invariance and conservation laws in the twentieth century*. Sources and Studies in the History of Mathematics and Physical Sciences. Springer-Verlag. ISBN 978-0-387-87867-6

- Lanczos, C. (1970). *The Variational Principles of Mechanics* (4th ed.). New York: Dover Publications. pp. 401–5. ISBN 0-486-65067-7.

- Olver, Peter (1993). *Applications of Lie groups to differential equations*. Graduate Texts in Mathematics **107** (2nd ed.). Springer-Verlag. ISBN 0-387-95000-1

34.11 External links

- Emmy Noether; Mort Tavel (translator) (1971). "Invariant Variation Problems". *Transport Theory and StatisticalPhysics* **1** (3): 186–207. arXiv:physics/0503066. Bibcode:1971TTSP....1..186N. doi:10.1080/00411457108231446.(Original in *Gott. Nachr.* 1918:235–257)

- Emmy Noether (1918). "Invariante Variationenprobleme" (in German).

- *Emmy Noether and The Fabric of Reality* (video) on YouTube

- Byers, Nina (1998). "E. Noether's Discovery of the Deep Connection Between Symmetries and Conservation Laws". arXiv:physics/9807044 [physics.hist-ph].

- John Baez (2002) "Noether's Theorem in a Nutshell."

- Hanca, J.; Tulejab, S.; Hancova, M. (2004). "Symmetries and conservation laws: Consequences of Noether's theorem". *American Journal of Physics* **72** (4): 428–35. Bibcode:2004AmJPh..72..428H. doi:10.1119/1.1591764.

- Merced Montesinos; Ernesto Flores (2006). "Symmetric energy–momentum tensor in Maxwell, Yang–Mills, and Proca theories obtained using only Noether's theorem" (PDF). *Revista Mexicana de Física* **52**: 29–36. arXiv:hep-th/0602190. Bibcode:2006RMxF...52...29M.

- Vladimir Cuesta; Merced Montesinos; José David Vergara (2007). "Gauge invariance of the action principle for gauge systems with noncanonical symplectic structures". *Physical Review D* **76**: 025025. Bibcode:2007PhRvD..76b5025C doi:10.1103/PhysRevD.76.025025.

- Sardanashvily (2009). "Gauge conservation laws in a general setting. Superpotential". *International Journal of Geometric Methods in Modern Physics* **6** (06): 1047. arXiv:0906.1732. Bibcode:2009arXiv0906.1732S. doi:10.

- Neuenschwander, Dwight E. (2010). *Emmy Noether's Wonderful Theorem*. Johns Hopkins University Press. ISBN 978-0-8018-9694-1.

- Noether's Theorem at MathPages.

Emmy Noether was an influential German mathematician known for her groundbreaking contributions to abstract algebra and theoretical physics.

Chapter 35

Lorentz covariance

In physics, **Lorentz symmetry**, named for Hendrik Lorentz, is "the feature of nature that says experimental results are independent of the orientation or the boost velocity of the laboratory through space".[1] **Lorentz covariance**, a related concept, is a key property of spacetime following from the special theory of relativity. Lorentz covariance has two distinct, but closely related meanings:

1. A physical quantity is said to be Lorentz covariant if it transforms under a given representation of the Lorentz group. According to the representation theory of the Lorentz group, these quantities are built out of scalars, four-vectors, four-tensors, and spinors. In particular, a scalar (e.g., the space-time interval) remains the same under Lorentz transformations and is said to be a "Lorentz invariant" (i.e., they transform under the trivial representation).

2. An equation is said to be Lorentz covariant if it can be written in terms of Lorentz covariant quantities (confusingly, some use the term "invariant" here). The key property of such equations is that if they hold in one inertial frame, then they hold in any inertial frame; this follows from the result that if all the components of a tensor vanish in one frame, they vanish in every frame. This condition is a requirement according to the principle of relativity, i.e., all non-gravitational laws must make the same predictions for identical experiments taking place at the same spacetime event in two different inertial frames of reference.

This usage of the term *covariant* should not be confused with the related concept of a *covariant vector*. On manifolds, the words *covariant* and *contravariant* refer to how objects transform under general coordinate transformations. Confusingly, both covariant and contravariant four-vectors can be Lorentz covariant quantities.

Local Lorentz covariance, which follows from general relativity, refers to Lorentz covariance applying only *locally* in an infinitesimal region of spacetime at every point. There is a generalization of this concept to cover Poincaré covariance and Poincaré invariance.

35.1 Examples

In general, the nature of a Lorentz tensor can be identified by its tensor order, which is the number of free indices it has. No indices implies it is a scalar, one implies that it is a vector, etc. Furthermore, any number of new scalars, vectors etc. can be made by contracting or creating an outer product of any kinds of tensors together, but many of these may not have any real physical meaning. Some of those tensors that do have a physical interpretation are listed (by no means exhaustively) below.

Please note, the metric sign convention such that $\eta = \mathrm{diag}\,(1, -1, -1, -1)$ is used throughout the article.

35.1.1 Scalars

Spacetime interval:

$$\Delta s^2 = \Delta x^a \Delta x^b \eta_{ab} = c^2 \Delta t^2 - \Delta x^2 - \Delta y^2 - \Delta z^2$$

Proper time (for timelike intervals):

$$\Delta \tau = \sqrt{\frac{\Delta s^2}{c^2}}, \ \Delta s^2 > 0$$

Proper distance (for spacelike intervals):

$$L = \sqrt{-\Delta s^2}, \ \Delta s^2 < 0$$

Rest mass:

$$m_0^2 c^2 = P^a P^b \eta_{ab} = \frac{E^2}{c^2} - p_x^2 - p_y^2 - p_z^2$$

Electromagnetism invariants:

$$F_{ab} F^{ab} = 2 \left(B^2 - \frac{E^2}{c^2} \right)$$

$$G_{cd} F^{cd} = \frac{1}{2} \epsilon_{abcd} F^{ab} F^{cd} = -\frac{4}{c} \left(\vec{B} \cdot \vec{E} \right)$$

D'Alembertian/wave operator:

$$\Box = \eta^{\mu\nu} \partial_\mu \partial_\nu = \frac{1}{c^2} \frac{\partial^2}{\partial t^2} - \frac{\partial^2}{\partial x^2} - \frac{\partial^2}{\partial y^2} - \frac{\partial^2}{\partial z^2}$$

35.1.2 Four-vectors

4-Displacement:

$$\Delta X^a = (c\Delta t, \vec{\Delta x}) = (c\Delta t, \Delta x, \Delta y, \Delta z)$$

4-Position:

$$X^a = (ct, \vec{x}) = (ct, x, y, z)$$

4-Gradient: with is the 4D Partial derivative:

$$\partial^a = \left(\frac{\partial_t}{c}, -\vec{\nabla} \right) = \left(\frac{1}{c} \frac{\partial}{\partial t}, -\frac{\partial}{\partial x}, -\frac{\partial}{\partial y}, -\frac{\partial}{\partial z} \right)$$

4-Velocity:

$$U^a = \gamma(c, \vec{u}) = \gamma\left(c, \frac{dx}{dt}, \frac{dy}{dt}, \frac{dz}{dt}\right)$$

where $U^a = \frac{dX^a}{d\tau}$

4-Momentum:

$$P^a = (mc, \vec{p}) = \left(\frac{E}{c}, \vec{p}\right) = \left(\frac{E}{c}, p_x, p_y, p_z\right)$$

where $P^a = m_o U^a$

4-Current:

$$J^a = (c\rho, \vec{j}) = (c\rho, j_x, j_y, j_z)$$

where $J^a = \rho_o U^a$

35.1.3   Four-tensors

The Kronecker delta:

$$\delta_b^a = \begin{cases} 1 & \text{if } a = b, \\ 0 & \text{if } a \neq b. \end{cases}$$

The Minkowski metric (the metric of flat space according to general relativity):

$$\eta_{ab} = \eta^{ab} = \begin{cases} 1 & \text{if } a = b = 0, \\ -1 & \text{if } a = b = 1, 2, 3, \\ 0 & \text{if } a \neq b. \end{cases}$$

The Levi-Civita symbol:

$$\epsilon_{abcd} = -\epsilon^{abcd} = \begin{cases} +1 & \text{if } \{abcd\} \text{ is an even permutation of } \{0123\}, \\ -1 & \text{if } \{abcd\} \text{ is an odd permutation of } \{0123\}, \\ 0 & \text{otherwise.} \end{cases}$$

Electromagnetic field tensor (using a metric signature of $+---$):

$$F_{ab} = \begin{bmatrix} 0 & E_x/c & E_y/c & E_z/c \\ -E_x/c & 0 & -B_z & B_y \\ -E_y/c & B_z & 0 & -B_x \\ -E_z/c & -B_y & B_x & 0 \end{bmatrix}$$

Dual electromagnetic field tensor:

$$G_{cd} = \frac{1}{2}\epsilon_{abcd}F^{ab} = \begin{bmatrix} 0 & B_x & B_y & B_z \\ -B_x & 0 & E_z/c & -E_y/c \\ -B_y & -E_z/c & 0 & E_x/c \\ -B_z & E_y/c & -E_x/c & 0 \end{bmatrix}$$

35.2 Lorentz violating models

See also: Modern searches for Lorentz violation

In standard field theory, there are very strict and severe constraints on marginal and relevant Lorentz violating operators within both QED and the Standard Model. Irrelevant Lorentz violating operators may be suppressed by a high cutoff scale, but they typically induce marginal and relevant Lorentz violating operators via radiative corrections. So, we also have very strict and severe constraints on irrelevant Lorentz violating operators.

Since some approaches to quantum gravity lead to violations of Lorentz invariance,[2] these studies are part of Phenomenological Quantum Gravity.

Lorentz violating models typically fall into four classes:

- The laws of physics are exactly Lorentz covariant but this symmetry is spontaneously broken. In special relativistic theories, this leads to phonons, which are the Goldstone bosons. The phonons travel at *less* than the speed of light.

- Similar to the approximate Lorentz symmetry of phonons in a lattice (where the speed of sound plays the role of the critical speed), the Lorentz symmetry of special relativity (with the speed of light as the critical speed in vacuum) is only a low-energy limit of the laws of physics, which involve new phenomena at some fundamental scale. Bare conventional "elementary" particles are not point-like field-theoretical objects at very small distance scales, and a nonzero fundamental length must be taken into account. Lorentz symmetry violation is governed by an energy-dependent parameter which tends to zero as momentum decreases.[3] Such patterns require the existence of a privileged local inertial frame (the "vacuum rest frame"). They can be tested, at least partially, by ultra-high energy cosmic ray experiments like the Pierre Auger Observatory.[4]

- The laws of physics are symmetric under a deformation of the Lorentz or more generally, the Poincaré group, and this deformed symmetry is exact and unbroken. This deformed symmetry is also typically a quantum group symmetry, which is a generalization of a group symmetry. Deformed special relativity is an example of this class of models. It is not accurate to call such models Lorentz-violating as much as Lorentz deformed any more than special relativity can be called a violation of Galilean symmetry rather than a deformation of it. The deformation is scale dependent, meaning that at length scales much larger than the Planck scale, the symmetry looks pretty much like the Poincaré group. Ultra-high energy cosmic ray experiments cannot test such models.

- This is a class of its own; a subgroup of the Lorentz group is sufficient to give us all the standard predictions if CP is an exact symmetry. However, CP isn't exact. This is called Very Special Relativity.

Models belonging to the first two classes can be consistent with experiment if Lorentz breaking happens at Planck scale or beyond it, or even before it in suitable preonic models,[5] and if Lorentz symmetry violation is governed by a suitable energy-dependent parameter. One then has a class of models which deviate from Poincaré symmetry near the Planck scale but still flows towards an exact Poincaré group at very large length scales. This is also true for the third class, which is furthermore protected from radiative corrections as one still has an exact (quantum) symmetry.

Even though there is no evidence of the violation of Lorentz invariance, several experimental searches for such violations have been performed during recent years. A detailed summary of the results of these searches is given in the Data Tables for Lorentz and CPT Violation.[6]

35.3 See also

- Antimatter tests of Lorentz violation
- General covariance
- Lorentz invariance in loop quantum gravity
- Lorentz-violating neutrino oscillations

- Symmetry in physics

35.4 References

[1] "Framing Lorentz symmetry". CERN Courier. 2004-11-24. Retrieved 2013-05-26.

[2] Mattingly, Davi d (2005). "Modern Tests of Lorentz Invariance". *Living Reviews in Relativity* **8**. arXiv:gr-qc/0502097. doi:10.12942/lrr-2005-5.

[3] Luis Gonzalez-Mestres (1995-05-25). "Properties of a possible class of particles able to travel faster than light".

[4] Luis Gonzalez-Mestres (1997-05-26). "Absence of Greisen-Zatsepin-Kuzmin Cutoff and Stability of Unstable Particles at Very High Energy, as a Consequence of Lorentz Symmetry Violation".

[5] Luis Gonzalez-Mestres (2014). "Ultra-high energy physics and standard basic principles. Do Planck units really make sense?" (PDF). EPJ Web of Conferences (ICNFP 2013 Conference). doi:10.1051/epjconf/20147100062.

[6] Kostelecky, V.A.; Russell, N. (2010). "Data Tables for Lorentz and CPT Violation". arXiv:0801.0287v3.

- Background information on Lorentz and CPT violation: http://www.physics.indiana.edu/~{}kostelec/faq.html

- Mattingly, Davi d (2005). "Modern Tests of Lorentz Invariance". *Living Reviews in Relativity* **8**. arXiv:gr-Bibcode:2005LRR.....8....5M. doi:10.12942/lrr-2005-5.

- Amelino-Camelia G, Ellis J, Mavromatos N E, Nanopoulos D V, and Sarkar S (June 1998). "Tests of quantum gravity from observations of bold gamma-ray bursts". *Nature* **393** (6687): 763–765. arXiv:astro-ph/9712103. Bibcode:1998Natur.393..763A. doi:10.1038/31647. Retrieved 2007-12-22.

- Jacobson T, Liberati S, and Mattingly D (August 2003). "A strong astrophysical constraint on the violation of special relativity by quantum gravity". *Nature* **424** (6952): 1019–1021. arXiv:astro-ph/0212190. Bibcode:2003Natu .424.1019J.doi:10.1038/nature01882. PMID 12944959. Retrieved 2007-12-22.

- Carroll S (August 2003). "Quantum gravity: An astrophysical constraint". *Nature* **424** (6952): 1007–1008. Bibcode:2003Natur.424.1007C. doi:10.1038/4241007a. PMID 12944951. Retrieved 2007-12-22.

- Jacobson, T.; Liberati, S.; Mattingly, D. (2003). "Threshold effects and Planck scale Lorentz violation: Combined constraints from high energy astrophysics". *Physical Review D* **67** (12). arXiv:hep-ph/0209264. Bibcode:2003PhR vD..67l4011J.doi:10.1103/PhysRevD.67.124011.

35.5 External links

Chapter 36

Wheeler–Feynman absorber theory

The **Wheeler–Feynman absorber theory** (also called the **Wheeler–Feynman time-symmetric theory**), named after its originators, the physicists Richard Feynman and John Archibald Wheeler, is an interpretation of electrodynamics derived from the assumption that the solutions of the electromagnetic field equations must be invariant under time-reversal transformation, as are the field equations themselves. Indeed, there is no apparent reason for the time-reversal symmetry breaking which singles out a preferential time direction and thus makes a distinction between past and future. A time-reversal invariant theory is more logical and elegant. Another key principle, resulting from this interpretation and reminiscent of Mach's principle due to Tetrode, is that elementary particles are not self-interacting. This immediately removes the problem of self-energies.

36.1 T-symmetry and causality

The requirement of time reversal symmetry, in general, is difficult to conjugate with the principle of causality. Maxwell's equations and the equations for electromagnetic waves have, in general, two possible solutions: a delayed solution and an advanced one. Accordingly, any charged particle generates waves, say at time $t_0 = 0$ and point $x_0 = 0$, which will arrive at point x_1 at the instant $t_1 = x_1/c$ (here c is the speed of light) after the emission (retarded solution), and other waves which will arrive at the same place at the instant $t_2 = x_1/c$ before the emission (advanced solution). The latter, however, violates the causality principle: advanced waves could be detected before their emission. Thus the advanced solutions are usually discarded in the interpretation of electromagnetic waves. In the absorber theory, instead charged particles are considered as both emitters and absorbers, and the emission process is connected with the absorption process as follows: Both the retarded waves from emitter to absorber and the advanced waves from absorber to emitter are considered. The sum of the two, however, results in *causal waves*, although the anti-causal (advanced) solutions are not discarded *a priori*.

Feynman and Wheeler obtained this result in a very simple and elegant way. They considered all the charged particles (emitters) present in our universe, and assumed all of them to generate time-reversal symmetric waves. The resulting field is

$$E_{\text{tot}}(\mathbf{x}, t) = \sum_n \frac{E_n^{\text{ret}}(\mathbf{x}, t) + E_n^{\text{adv}}(\mathbf{x}, t)}{2}.$$

Then they observed that, if the relation

$$E_{\text{free}}(\mathbf{x}, t) = \sum_n \frac{E_n^{\text{ret}}(\mathbf{x}, t) - E_n^{\text{adv}}(\mathbf{x}, t)}{2} = 0$$

holds, E_{free}, being a solution of the homogeneous Maxwell equation, can be used to obtain the total field

$$E_{\text{tot}}(\mathbf{x}, t) = \sum_n \frac{E_n^{\text{ret}}(\mathbf{x}, t) + E_n^{\text{adv}}(\mathbf{x}, t)}{2} + \sum_n \frac{E_n^{\text{ret}}(\mathbf{x}, t) - E_n^{\text{adv}}(\mathbf{x}, t)}{2} = \sum_n E_n^{\text{ret}}(\mathbf{x}, t).$$

The total field is retarded and causality is not violated.

The assumption that the *free field* is identically zero is the core of the absorber idea. It means that the radiation emitted by each particle is completely absorbed by all other particles present in the universe. To better understand this point, it may be useful to consider how the absorption mechanism works in common materials. At the microscopic scale, it results from the sum of the incoming electromagnetic wave and the waves generated from the electrons of the material, which react to the external perturbation. If the incoming wave is absorbed, the result is a zero outcoming field. In the absorber theory the same concept is used, however in presence of both retarded and advanced waves.

The resulting wave appears to have a preferred time direction, because it respects causality. However, this is only an illusion. Indeed it is always possible to reverse the time direction by simply exchanging the labels *emitter* and *absorber*. Thus, the apparently preferred time direction results from the arbitrary labelling.

36.2   T-symmetry and self-interaction

One of the major results of the absorber theory is the elegant and clear interpretation of the electromagnetic radiation process. A charged particle which experiences acceleration is known to emit electromagnetic waves, i.e., to lose energy. Thus, the Newtonian equation for the particle ($F = ma$) must contain a dissipative force (damping term), which takes into account this energy loss. In the causal interpretation of electromagnetism, Lorentz and Abraham proposed that such a force, later called Abraham–Lorentz force, is due to the retarded self-interaction of the particle with its own field. This first interpretation, however, is not completely satisfactory, as it leads to divergences in the theory and needs some assumptions on the structure of charge distribution of the particle. Dirac generalized the formula to make it relativistically invariant. While doing so, he also suggested a different interpretation. He showed that the damping term can be expressed in terms of a free field acting on the particle at its own position.

$$E^{\text{damping}}(\mathbf{x}_j, t) = \frac{E_j^{\text{ret}}(\mathbf{x}_j, t) - E_j^{\text{adv}}(\mathbf{x}_j, t)}{2}$$

However Dirac did not propose any physical explanation of this interpretation.

A clear and simple explanation can instead be obtained in the framework of absorber theory, starting from the simple idea that each particle does not interact with itself. This is actually the opposite of the first Abraham–Lorentz proposal. The field acting on the particle j at its own position (the point x_j) is then:

$$E^{\text{tot}}(\mathbf{x}_j, t) = \sum_{n \neq j} \frac{E_n^{\text{ret}}(\mathbf{x}_j, t) + E_n^{\text{adv}}(\mathbf{x}_j, t)}{2}.$$

If we sum the *free field term* of this expression we obtain

$$E^{\text{tot}}(\mathbf{x}_j, t) = \sum_{n \neq j} \frac{E_n^{\text{ret}}(\mathbf{x}_j, t) + E_n^{\text{adv}}(\mathbf{x}_j, t)}{2} + \sum_n \frac{E_n^{\text{ret}}(\mathbf{x}_j, t) - E_n^{\text{adv}}(\mathbf{x}_j, t)}{2}$$

and, thanks to Dirac's result,

$$E^{\text{tot}}(\mathbf{x}_j, t) = \sum_{n \neq j} E_n^{\text{ret}}(\mathbf{x}_j, t) + E^{\text{damping}}(\mathbf{x}_j, t).$$

Thus, the damping force is obtained without the need for self-interaction, which is known to lead to divergences, and also giving a physical justification to the expression derived by Dirac.

36.3 Criticism

The Abraham–Lorentz force is, however, not free of problems. Written in the non-relativistic limit, it gives:

$$E^{\text{damping}}(\mathbf{x}_j, t) = \frac{e}{6\pi c^3} \frac{\mathrm{d}^3}{\mathrm{d}t^3} x$$

Since the third derivative with respect to the time (also called the "jerk" or "jolt") enters in the equation of motion, to derive a solution one needs not only the initial position and velocity of the particle, but also its initial acceleration. This apparent problem however can be solved in the absorber theory, by observing that the equation of motion for the particle has to be solved together with the Maxwell equations for the field. In this case, instead of the initial acceleration, one only needs to specify the initial field and the boundary condition. This interpretation restores the coherence of the physical interpretation of the theory.

Other difficulties may arise trying to solve the equation of motion for a charged particle in the presence of this damping force. It is commonly stated that the Maxwell equations are classical and cannot correctly account for microscopic phenomena, such as the behavior of a point-like particle, where quantum mechanical effects should appear. Nevertheless with absorber theory, Wheeler and Feynman were able to create a coherent classical approach to the problem (see also the "paradoxes" section in the Abraham–Lorentz force).

Also, the time-symmetric interpretation of the electromagnetic waves appears to be in contrast with the experimental evidence that time flows in a given direction and, thus, that the T-symmetry is broken in our world. It is commonly believed, however, that this symmetry breaking appears only in the thermodynamical limit (see, for example, the arrow of time). Wheeler himself accepted that the expansion of the universe is not time symmetric in the thermodynamic limit . This however does not imply that the T-symmetry must be broken also at the microscopic level.

Finally, the main drawback of the theory turned out to be the result that particles are not self-interacting. Indeed, as demonstrated by Hans Bethe, the Lamb shift necessitated a self-energy term to be explained. Feynman and Bethe had an intense discussion over that issue and eventually Feynman himself stated that self-interaction is needed to correctly account for this effect. [citation needed]

36.4 Developments since original formulation

36.4.1 Gravity theory

Main article: Hoyle–Narlikar theory of gravity

Inspired by the Machian nature of the Wheeler–Feynman absorber theory for electrodynamics, Fred Hoyle and Jayant Narliker proposed their own theory of gravity[1][2][3] in the context of general relativity. This model still exists in spite of recent astronomical observations that have challenged the theory.

36.4.2 Transactional interpretation of quantum mechanics

Main article: Transactional interpretation

Again inspired by the Wheeler–Feynman absorber theory, the transactional interpretation of quantum mechanics (TIQM) first proposed in 1986 by John G. Cramer[4] describes quantum interactions in terms of a standing wave formed by retarded (forward-in-time) and advanced (backward-in-time) waves. J. Cramer claims it avoids the philosophical problems with the Copenhagen interpretation and the role of the observer, and resolves various quantum paradoxes, such as quantum nonlocality, quantum entanglement and retrocausality.[5]

36.4.3   Attempted resolution of causality

T. C. Scott and R. A. Moore demonstrated that the apparent acausality suggested by the presence of advanced Liénard–Wiechert potentials could be removed by recasting the theory in terms of retarded potentials only, without the complications of the absorber idea.[6][7] The Lagrangian describing a particle (p_1) under the influence of the time-symmetric potential generated by another particle (p_1) is:

$$L_1 = T_1 - \frac{1}{2}\left((V_R)_1^2 + (V_A)_1^2\right)$$

where T_i is the relativistic kinetic energy functional of particle p_i , and, $(V_R)_i^j$ and $(V_A)_i^j$ are respectively the retarded and advanced Liénard–Wiechert potentials acting on particle p_i and generated by particle p_j . The corresponding Lagrangian for particle p_1 is:

$$L_2 = T_2 - \frac{1}{2}\left((V_R)_2^1 + (V_A)_2^1\right).$$

It was originally demonstrated with computer algebra[8] and then proven analytically[9] that:

$$(V_R)_j^i - (V_A)_i^j$$

is a total time derivative, i.e. a *divergence* in the calculus of variations, and thus it gives no contribution to the Euler–Lagrange equations. Thanks to this result the advanced potentials can be eliminated; here the total derivative plays the same role as the *free field*. The Lagrangian for the N-body system is therefore:

$$L = \sum_{i=1}^{N} T_i - \frac{1}{2}\sum_{i\neq j}^{N}(V_R)_j^i$$

The resulting lagrangian is symmetric under the exchange of p_i with p_j . For $N = 2$ this Lagrangian will generate *exactly* the same equations of motion of L_1 and L_2 . Therefore, from the point of view of an *outside* observer, everything is causal. Only if we isolate the forces acting on a particular body do the advanced potentials make their appearance. This recasting of the problem comes at a price: the N-body Lagrangian depends on all the time derivatives of the curves traced by all particles i.e. the Lagrangian is infinite order. However, much progress was made in examining the unresolved issue of quantizing the theory.[10][11] Also, this formulation recovers the Darwin Lagrangian from which the Breit equation was originally derived, but without the dissipative terms.[9] This ensures agreement with theory and experiment, up to but not including the Lamb shift. Numerical solutions for the classical problem were also found.[12] Finally, Moore and Scott[6] showed that the radiation reaction can be alternatively derived using the notion that, on average, the net dipole moment is zero for a collection of charged particles, thereby avoiding the complications of the absorber theory. An important bonus from their approach is the formulation of a total preserved canonical generalized momentum, as presented in a comprehensive review article in the light of quantum nonlocality[13]

This apparent acausality may be viewed as merely apparent, & this entire problem goes away. For the opposing view as held by Einstein see.[14] This result in no way determines this debate in favor of Ritz.

36.4.4   Alternative Lamb shift calculation

As mentioned previously, a serious criticism against the absorber theory is that its Machian assumption that point particles do not act on themselves does not allow (infinite) self-energies and consequently an explanation for the Lamb shift according to Quantum electrodynamics (QED). Ed Jaynes proposed an alternate model where the Lamb-like shift is due instead to the interaction with *other particles* very much along the same notions of the Wheeler–Feynman absorber theory itself. One simple model is to calculate the motion of an oscillator coupled directly with many other oscillators. Jaynes has shown that it is easy to get both spontaneous emission and Lamb shift behavior in classical mechanics.[15] Furthermore, Jayne's alternatives provides a solution to the process of "addition and subtraction of infinities" associated with renormalization.[13][16]

This model leads to essentially the same type of Bethe Logarithm an essential part of the Lamb shift calculation vindicating Jaynes' claim that two different physical models can be mathematically isomorphic to each other and therefore yield the same results, a point also apparently made by Scott and Moore on the issue of causality.

36.5 Conclusions

This universal absorber theory is mentioned in the chapter titled "Monster Minds" in Feynman's autobiographical work *Surely You're Joking, Mr. Feynman!* as well as in Vol. II of the Feynman Lectures on Physics. It led to the formulation of a framework of quantum mechanics using a Lagrangian and action as starting points, rather than a Hamiltonian, namely the formulation using Feynman path integrals which proved useful in Feynman's earliest calculations in quantum electrodynamics and quantum field theory in general. Both retarded and advanced fields appear respectively as retarded and advanced propagators, and also, in the Feynman propagator and the Dyson propagator. In hindsight, the relationship between retarded and advanced potentials shown here is not so surprising in view of the fact that, in field theory, the advanced propagator can be obtained from the retarded propagator by exchanging the roles of field source and test particle (usually within the kernel of a Green's function formalism). In field theory, advanced as well as retarded fields are simply viewed as *mathematical* solutions of Maxwell's equations whose combinations are decided by the boundary conditions.

36.6 See also

- Causality

- Symmetry in physics and T-symmetry

- Transactional interpretation

- Abraham–Lorentz force

- Retrocausality

- Two-state vector formalism

- Paradox of a charge in a gravitational field

- Ni Guangjiong

- Woodward effect

36.7 Notes

[1] F. Hoyle and J. V. Narlikar (1964). "A Ne w Theory of Gravitation". *Proceedings o f the Royal Society A.* Bibcode:1964 doi:10.1098/rspa.1964.0227.

[2] "Cosmology: Math Plus Mach Equals Far-Out Gravity". Time. June 26, 1964. Retrieved 7 August 2010.

[3] Hoyle, F.; Narlikar, J. V. (1995). "Cosmology and action-at-a-distance electrodynamics". *Reviews of Modern Physics* **67** (1): 113–155. Bibcode:1995RvMP...67..113H. doi:10.1103/RevModPhys.67.113.

[4] The Transactional Interpretation of Quantum Mechanics by John Cramer. *Reviews of Modern Physics* 58, 647-688, July (1986)

[5] John G. Cramer, "Quantum Entanglement, Nonlocality, Back-in-Time Messages, *(April 3, 2010)*.

[6] Moore, R. A.; Scott, T. C.; Monagan, M. B. (1987). "Relativistic, many-particle Lagrangean for electromagnetic interactions". *Phys. Rev. Lett.* **59** (5): 525–527. Bibcode:1987PhRvL..59..525M. doi:10.1103/PhysRevLett.59.525.

[7] Moore, R. A.; Scott, T. C.; Monagan, M. B. (1988). "A Model for a Relativistic Many-Particle Lagrangian with Electromagnetic Interactions". *Can. J. Phys.* **66** (3): 206–211. Bibcode:1988CaJPh..66..206M. doi:10.1139/p88-032.

[8] Scott, T. C.; Moore, R. A.; Monagan, M. B. (1989). "Resolution of Many Particle Electrodynamics by Symbolic Manipulation". *Comput. Phys. Commun.* **52** (2): 261–281. Bibcode:1989CoPhC..52..261S. doi:10.1016/0010-4655(89)90009-X.

[9] Scott, T. C. (1986). "Relativistic Classical and Quantum Mechanical Treatment of the Two-body Problem". *MMath thesis* (U. of Waterloo, Canada).

[10] Scott, T. C.; Moore, R. A. (1989). "Quantization of Hamiltonians from High-Order Lagrangians". *Nucl. Phys. B* **6** (Proc. Suppl.): 455–457. Bibcode:1989NuPhS...6..455S. doi:10.1016/0920-5632(89)90498-2.

[11] Moore, R. A.; Scott, T. C. (1991). "Quantization of Second-Order Lagrangians: Model Problem". *Phys. Rev. A* **44** (3): 1477–1484. Bibcode:1991PhRvA..44.1477M. doi:10.1103/PhysRevA.44.1477.

[12] Moore, R. A.; Qi, D.; Scott, T. C. (1992). "Causality of Relativistic Many-Particle Classical Dynamics Theories". *Can. J. Phys.* **70** (9): 772–781. Bibcode:1992CaJPh..70..772M. doi:10.1139/p92-122.

[13] Scott, T. C.; Andrae, D. (2015). "Quantum Nonlocality and Conservation of momentum". *Phys. Essays* **28** (3): 374–385.

[14] http://www.datasync.com/~{}rsf1/rtzein.htm

[15] E.T. Jaynes, ``The Lamb Shift in Classical Mechanics *in* ``*Probability in Quantum Theory*, pp. 13-15, (1996) Jaynes' analysis of Lamb shift.

[16] E.T. Jaynes, ``Classical Subtraction Physics in ``Probability in Quantum Theory, *pp. 15-18, (1996)* Jaynes' analysis of handing the infinities of the Lamb shift calculation.

36.8 Key papers

- Wheeler, J. A.; Feynman, R. P. (1945). "Interaction with the Absorber as the Mechanism of Radiation". *Reviews of Modern Physics* **17** (2–3): 157–161. Bibcode:1945RvMP...17..157W. doi:10.1103/RevModPhys.17.157.

- Wheeler, J. A.; Feynman, R. P. (1949). "Classical Electrodynamics in Terms of Direct Interparticle Action". *Reviews of Modern Physics* **21** (3): 425–433. Bibcode:1949RvMP...21..425W. doi:10.1103/RevModPhys.21.425.

36.9 External links

- J. A. Wheeler and R. P. Feynman, "Interaction with the Absorber as the Mechanism of Radiation" Caltech Library of Authors

Chapter 37

Standard Model

This article is about the Standard Model of particle physics. For other uses, see Standard model (disambiguation).

This article is a non-mathematical general overview of the Standard Model. For a mathematical description, see the article Standard Model (mathematical formulation).

For the Standard Model of Big Bang cosmology, Lambda-CDM model.

The **Standard Model** of particle physics is a theory concerning the electromagnetic, weak, and strong nuclear inter-

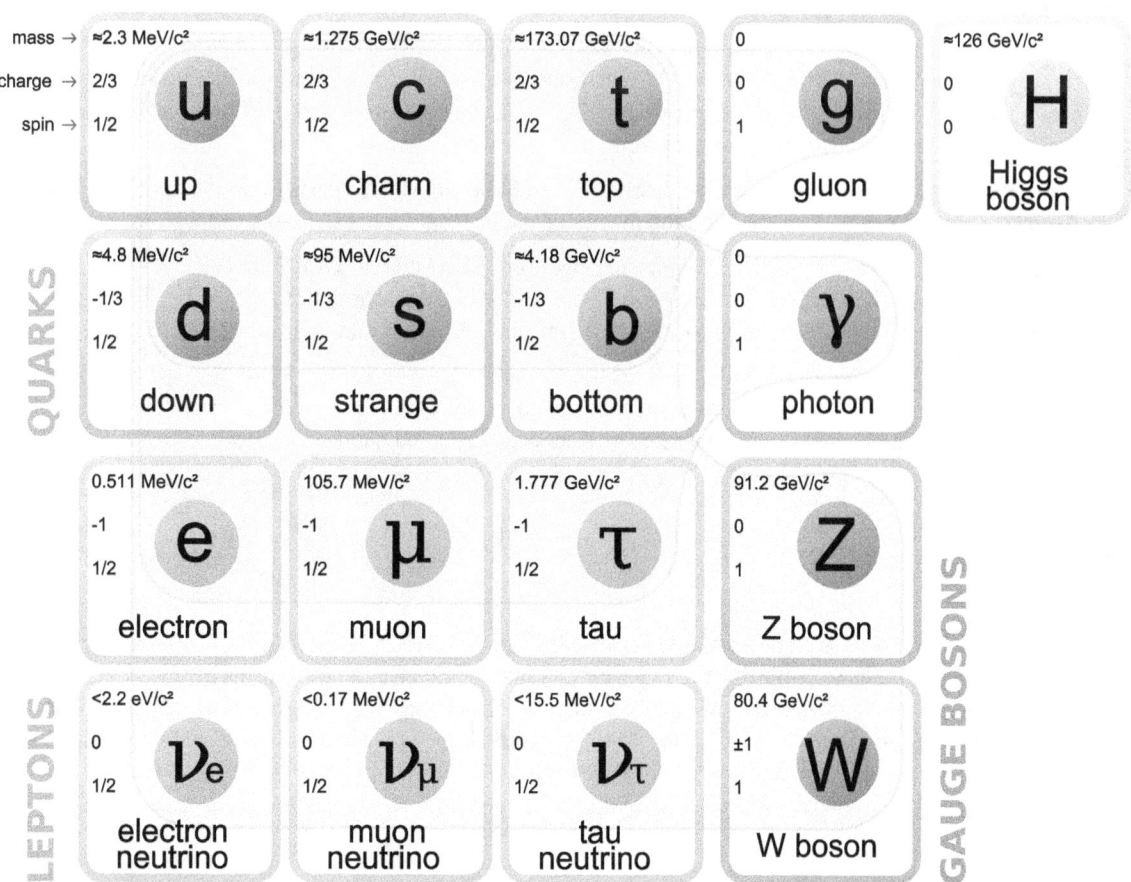

The Standard Model of elementary particles (more schematic depiction), with the three generations of matter, gauge bosons in the fourth column, and the Higgs boson in the fifth.

actions, as well as classifying all the subatomic particles known. It was developed throughout the latter half of the 20th century, as a collaborative effort of scientists around the world.[1] The current formulation was finalized in the mid-1970s upon experimental confirmation of the existence of quarks. Since then, discoveries of the top quark (1995), the tau neutrino (2000), and more recently the Higgs boson (2012), have given further credence to the Standard Model. Because of its success in explaining a wide variety of experimental results, the Standard Model is sometimes regarded as a "theory of almost everything".

Although the Standard Model is believed to be theoretically self-consistent[2] and has demonstrated huge and continued successes in providing experimental predictions, it does leave some phenomena unexplained and it falls short of being a complete theory of fundamental interactions. It does not incorporate the full theory of gravitation[3] as described by general relativity, or account for the accelerating expansion of the universe (as possibly described by dark energy). The model does not contain any viable dark matter particle that possesses all of the required properties deduced from observational cosmology. It also does not incorporate neutrino oscillations (and their non-zero masses).

The development of the Standard Model was driven by theoretical and experimental particle physicists alike. For theorists, the Standard Model is a paradigm of a quantum field theory, which exhibits a wide range of physics including spontaneous symmetry breaking, anomalies, non-perturbative behavior, etc. It is used as a basis for building more exotic models that incorporate hypothetical particles, extra dimensions, and elaborate symmetries (such as supersymmetry) in an attempt to explain experimental results at variance with the Standard Model, such as the existence of dark matter and neutrino oscillations.

37.1   Historical background

The first step towards the Standard Model was Sheldon Glashow's discovery in 1961 of a way to combine the electromagnetic and weak interactions.[4] In 1967 Steven Weinberg[5] and Abdus Salam[6] incorporated the Higgs mechanism[7][8][9] into Glashow's electroweak theory, giving it its modern form.

The Higgs mechanism is believed to give rise to the masses of all the elementary particles in the Standard Model. This includes the masses of the W and Z bosons, and the masses of the fermions, i.e. the quarks and leptons.

After the neutral weak currents caused by Z boson exchange were discovered at CERN in 1973,[10][11][12][13] the electroweak theory became widely accepted and Glashow, Salam, and Weinberg shared the 1979 Nobel Prize in Physics for discovering it. The W and Z bosons were discovered experimentally in 1981, and their masses were found to be as the Standard Model predicted.

The theory of the strong interaction, to which many contributed, acquired its modern form around 1973–74, when experiments confirmed that the hadrons were composed of fractionally charged quarks.

37.2   Overview

At present, matter and energy are best understood in terms of the kinematics and interactions of elementary particles. To date, physics has reduced the laws governing the behavior and interaction of all known forms of matter and energy to a small set of fundamental laws and theories. A major goal of physics is to find the "common ground" that would unite all of these theories into one integrated theory of everything, of which all the other known laws would be special cases, and from which the behavior of all matter and energy could be derived (at least in principle).[14]

37.3   Particle content

The Standard Model includes members of several classes of elementary particles (fermions, gauge bosons, and the Higgs boson), which in turn can be distinguished by other characteristics, such as color charge.

37.3.1 Fermions

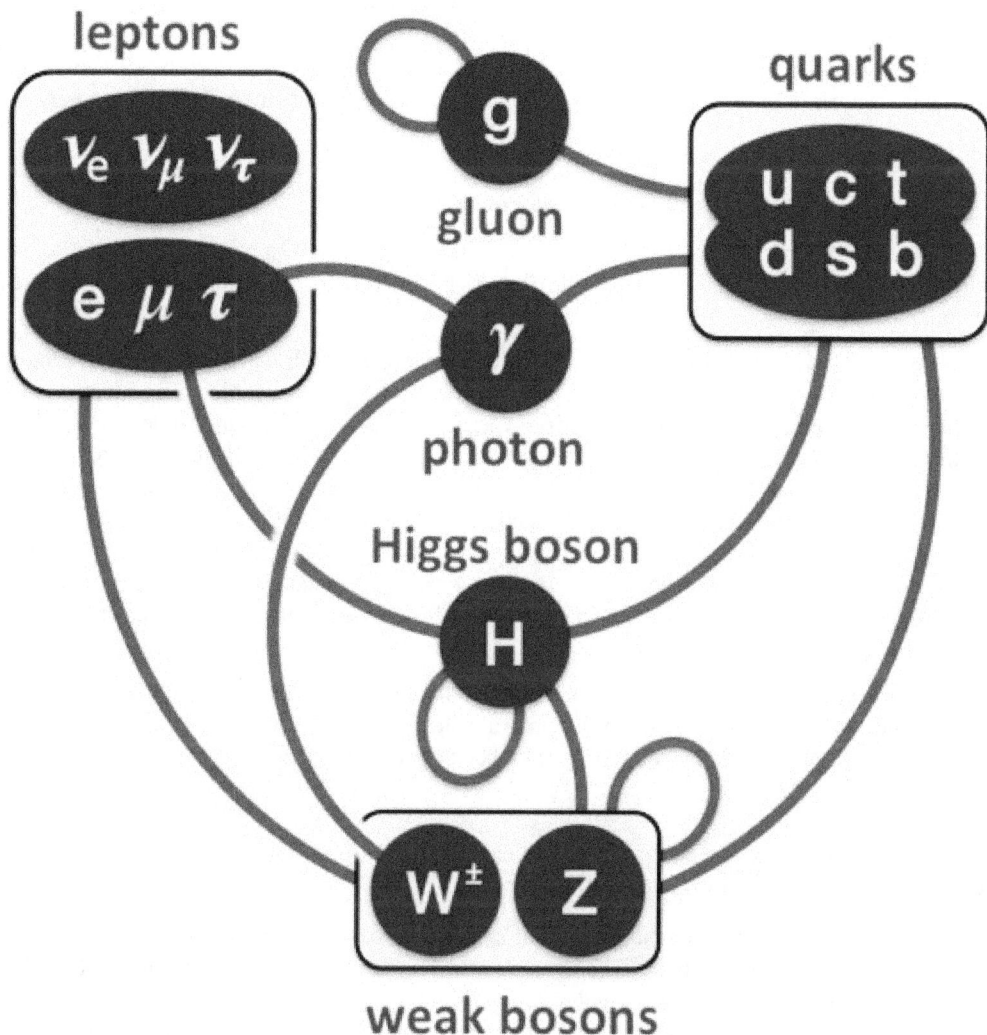

Summary of interactions between particles described by the Standard Model.

The Standard Model includes 12 elementary particles of spin-½ known as fermions. According to the spin-statistics theorem, fermions respect the Pauli exclusion principle. Each fermion has a corresponding antiparticle.

The fermions of the Standard Model are classified according to how they interact (or equivalently, by what charges they carry). There are six quarks (up, down, charm, strange, top, bottom), and six leptons (electron, electron neutrino, muon, muon neutrino, tau, tau neutrino). Pairs from each classification are grouped together to form a generation, with corresponding particles exhibiting similar physical behavior (see table).

The defining property of the quarks is that they carry color charge, and hence, interact via the strong interaction. A phenomenon called color confinement results in quarks being very strongly bound to one another, forming color-neutral composite particles (hadrons) containing either a quark and an antiquark (mesons) or three quarks (baryons). The familiar proton and the neutron are the two baryons having the smallest mass. Quarks also carry electric charge and weak isospin. Hence they interact with other fermions both electromagnetically and via the weak interaction.

The remaining six fermions do not carry colour charge and are called leptons. The three neutrinos do not carry electric charge either, so their motion is directly influenced only by the weak nuclear force, which makes them notoriously difficult to detect. However, by virtue of carrying an electric charge, the electron, muon, and tau all interact electromagnetically.

Each member of a generation has greater mass than the corresponding particles of lower generations. The first generation charged particles do not decay; hence all ordinary (baryonic) matter is made of such particles. Specifically, all atoms consist of electrons orbiting around atomic nuclei, ultimately constituted of up and down quarks. Second and third generation charged particles, on the other hand, decay with very short half lives, and are observed only in very high-energy environments. Neutrinos of all generations also do not decay, and pervade the universe, but rarely interact with baryonic matter.

37.3.2   Gauge bosons

In the Standard Model, gauge bosons are defined as force carriers that mediate the strong, weak, and electromagnetic fundamental interactions.

Interactions in physics are the ways that particles influence other particles. At a macroscopic level, electromagnetism allows particles to interact with one another via electric and magnetic fields, and gravitation allows particles with mass to attract one another in accordance with Einstein's theory of general relativity. The Standard Model explains such forces as resulting from matter particles exchanging other particles, generally referred to as *force mediating particles*. When a force-mediating particle is exchanged, at a macroscopic level the effect is equivalent to a force influencing both of them, and the particle is therefore said to have *mediated* (i.e., been the agent of) that force. The Feynman diagram calculations, which are a graphical representation of the perturbation theory approximation, invoke "force mediating particles", and when applied to analyze high-energy scattering experiments are in reasonable agreement with the data. However, perturbation theory (and with it the concept of a "force-mediating particle") fails in other situations. These include low-energy quantum chromodynamics, bound states, and solitons.

The gauge bosons of the Standard Model all have spin (as do matter particles). The value of the spin is 1, making them bosons. As a result, they do not follow the Pauli exclusion principle that constrains fermions: thus bosons (e.g. photons) do not have a theoretical limit on their spatial density (number per volume). The different types of gauge bosons are described below.

- Photons mediate the electromagnetic force between electrically charged particles. The photon is massless and is well-described by the theory of quantum electrodynamics.

- The W+, W−, and Z gauge bosons mediate the weak interactions between particles of different flavors (all quarks and leptons). They are massive, with the Z being more massive than the W±. The weak interactions involving the W± exclusively act on *left-handed* particles and *right-handed* antiparticles. Furthermore, the W± carries an electric charge of +1 and −1 and couples to the electromagnetic interaction. The electrically neutral Z boson interacts with both left-handed particles and antiparticles. These three gauge bosons along with the photons are grouped together, as collectively mediating the electroweak interaction.

- The eight gluons mediate the strong interactions between color charged particles (the quarks). Gluons are massless. The eightfold multiplicity of gluons is labeled by a combination of color and anticolor charge (e.g. red–antigreen).[nb 1] Because the gluons have an effective color charge, they can also interact among themselves. The gluons and their interactions are described by the theory of quantum chromodynamics.

The interactions between all the particles described by the Standard Model are summarized by the diagrams on the right of this section.

37.3.3   Higgs boson

Main article: Higgs boson

The Higgs particle is a massive scalar elementary particle theorized by Robert Brout, François Englert, Peter Higgs, Gerald Guralnik, C. R. Hagen, and Tom Kibble in 1964 (see 1964 PRL symmetry breaking papers) and is a key building

Standard Model Interactions
(Forces Mediated by Gauge Bosons)

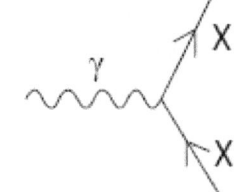

X is any fermion in
the Standard Model.

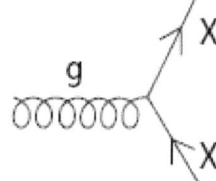

X is electrically charged.

X is any quark.

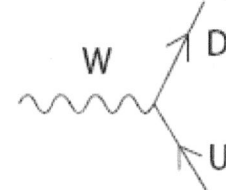

U is a up-type quark;
D is a down-type quark.

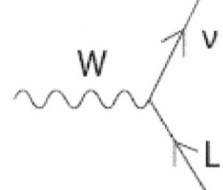

L is a lepton and ν is the
corresponding neutrino.

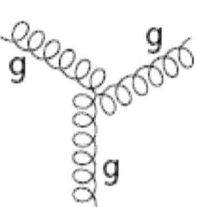

X is a photon or Z-boson.

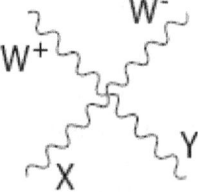

X and Y are any two
electroweak bosons such
that charge is conserved.

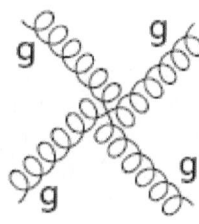

The above interactions form the basis of the standard model. Feynman diagrams in the standard model are built from these vertices. Modifications involving Higgs boson interactions and neutrino oscillations are omitted. The charge of the W bosons is dictated by the fermions they interact with; the conjugate of each listed vertex (i.e. reversing the direction of arrows) is also allowed.

block in the Standard Model.[7][8][9][15] It has no intrinsic spin, and for that reason is classified as a boson (like the gauge bosons, which have integer spin).

The Higgs boson plays a unique role in the Standard Model, by explaining why the other elementary particles, except the photon and gluon, are massive. In particular, the Higgs boson explains why the photon has no mass, while the W and Z bosons are very heavy. Elementary particle masses, and the differences between electromagnetism (mediated by the photon) and the weak force (mediated by the W and Z bosons), are critical to many aspects of the structure of microscopic (and hence macroscopic) matter. In electroweak theory, the Higgs boson generates the masses of the leptons (electron, muon, and tau) and quarks. As the Higgs boson is massive, it must interact with itself.

Because the Higgs boson is a very massive particle and also decays almost immediately when created, only a very high-energy particle accelerator can observe and record it. Experiments to confirm and determine the nature of the Higgs boson using the Large Hadron Collider (LHC) at CERN began in early 2010, and were performed at Fermilab's Tevatron until its closure in late 2011. Mathematical consistency of the Standard Model requires that any mechanism capable of generating the masses of elementary particles become visible at energies above 1.4 TeV;[16] therefore, the LHC (designed to collide two 7 to 8 TeV proton beams) was built to answer the question of whether the Higgs boson actually exists.[17]

On 4 July 2012, the two main experiments at the LHC (ATLAS and CMS) both reported independently that they found a new particle with a mass of about 125 GeV/c^2 (about 133 proton masses, on the order of 10^{-25} kg), which is "consistent with the Higgs boson." Although it has several properties similar to the predicted "simplest" Higgs,[18] they acknowledged that further work would be needed to conclude that it is indeed the Higgs boson, and exactly which version of the Standard Model Higgs is best supported if confirmed.[19][20][21][22][23]

On 14 March 2013 the Higgs Boson was tentatively confirmed to exist.[24]

37.3.4 Total particle count

Counting particles by a rule that distinguishes between particles and their corresponding antiparticles, and among the many color states of quarks and gluons, gives a total of 61 elementary particles.[25] (If neutrinos are their own antiparticles,[26] then by the same counting conventions the total number of elementary particles would be 58.)

37.4 Theoretical aspects

Main article: Standard Model (mathematical formulation)

37.4.1 Construction of the Standard Model Lagrangian

Technically, quantum field theory provides the mathematical framework for the Standard Model, in which a Lagrangian controls the dynamics and kinematics of the theory. Each kind of particle is described in terms of a dynamical field that pervades space-time. The construction of the Standard Model proceeds following the modern method of constructing most field theories: by first postulating a set of symmetries of the system, and then by writing down the most general renormalizable Lagrangian from its particle (field) content that observes these symmetries.

The global Poincaré symmetry is postulated for all relativistic quantum field theories. It consists of the familiar translational symmetry, rotational symmetry and the inertial reference frame invariance central to the theory of special relativity. The local SU(3)×SU(2)×U(1) gauge symmetry is an internal symmetry that essentially defines the Standard Model. Roughly, the three factors of the gauge symmetry give rise to the three fundamental interactions. The fields fall into different representations of the various symmetry groups of the Standard Model (see table). Upon writing the most general Lagrangian, one finds that the dynamics depend on 19 parameters, whose numerical values are established by experiment. The parameters are summarized in the table above (note: with the Higgs mass is at 125 GeV, the Higgs self-coupling strength $\lambda \sim 1/8$).

Quantum chromodynamics sector

Main article: Quantum chromodynamics

The quantum chromodynamics (QCD) sector defines the interactions between quarks and gluons, with SU(3) symmetry, generated by T^a. Since leptons do not interact with gluons, they are not affected by this sector. The Dirac Lagrangian of the quarks coupled to the gluon fields is given by

$$\mathcal{L}_{QCD} = i\overline{U}(\partial_\mu - ig_s G_\mu^a T^a)\gamma^\mu U + i\overline{D}(\partial_\mu - ig_s G_\mu^a T^a)\gamma^\mu D.$$

G_μ^a is the SU(3) gauge field containing the gluons, γ^μ are the Dirac matrices, D and U are the Dirac spinors associated with up and down-type quarks, and g_s is the strong coupling constant.

Electroweak sector

Main article: Electroweak interaction

The electroweak sector is a Yang–Mills gauge theory with the simple symmetry group U(1)×SU(2)L,

$$\mathcal{L}_{EW} = \sum_\psi \bar{\psi}\gamma^\mu \left(i\partial_\mu - g'\frac{1}{2}Y_W B_\mu - g\frac{1}{2}\vec{\tau}_L \vec{W}_\mu \right) \psi$$

where $B\mu$ is the U(1) gauge field; YW is the weak hypercharge—the generator of the U(1) group; $\vec{W}_\mu$ is the three-component SU(2) gauge field; $\vec{\tau}_L$ are the Pauli matrices—infinitesimal generators of the SU(2) group. The subscript L indicates that they only act on left fermions; g' and g are coupling constants.

Higgs sector

Main article: Higgs mechanism

In the Standard Model, the Higgs field is a complex scalar of the group SU(2)L:

$$\varphi = \frac{1}{\sqrt{2}} \begin{pmatrix} \varphi^+ \\ \varphi^0 \end{pmatrix},$$

where the indices + and 0 indicate the electric charge (Q) of the components. The weak isospin (YW) of both components is 1.

Before symmetry breaking, the Higgs Lagrangian is:

$$\mathcal{L}_H = \varphi^\dagger \left(\partial^\mu - \frac{i}{2} \left(g'Y_W B^\mu + g\vec{\tau}\vec{W}^\mu \right) \right) \left(\partial_\mu + \frac{i}{2} \left(g'Y_W B_\mu + g\vec{\tau}\vec{W}_\mu \right) \right) \varphi - \frac{\lambda^2}{4} \left(\varphi^\dagger \varphi - v^2 \right)^2,$$

which can also be written as:

$$\mathcal{L}_H = \left| \left(\partial_\mu + \frac{i}{2} \left(g'Y_W B_\mu + g\vec{\tau}\vec{W}_\mu \right) \right) \varphi \right|^2 - \frac{\lambda^2}{4} \left(\varphi^\dagger \varphi - v^2 \right)^2.$$

37.5 Fundamental forces

Main article: Fundamental interaction

The Standard Model classified all four fundamental forces in nature. In the Standard Model, a force is described as an exchange of bosons between the objects affected, such as a photon for the electromagnetic force and a gluon for the strong interaction. Those particles are called force carriers.[27]

37.6   Tests and predictions

The Standard Model (SM) predicted the existence of the W and Z bosons, gluon, and the top and charm quarks before these particles were observed. Their predicted properties were experimentally confirmed with good precision. To give an idea of the success of the SM, the following table compares the measured masses of the W and Z bosons with the masses predicted by the SM:

The SM also makes several predictions about the decay of Z bosons, which have been experimentally confirmed by the Large Electron-Positron Collider at CERN.

In May 2012 BaBar Collaboration reported that their recently analyzed data may suggest possible flaws in the Standard Model of particle physics.[29][30] These data show that a particular type of particle decay called "B to D-star-tau-nu" happens more often than the Standard Model says it should. In this type of decay, a particle called the B-bar meson decays into a D meson, an antineutrino and a tau-lepton. While the level of certainty of the excess (3.4 sigma) is not enough to claim a break from the Standard Model, the results are a potential sign of something amiss and are likely to impact existing theories, including those attempting to deduce the properties of Higgs bosons.[31]

On December 13, 2012, physicists reported the constancy, over space and time, of a basic physical constant of nature that supports the *standard model of physics*. The scientists, studying methanol molecules in a distant galaxy, found the change ($\Delta\mu/\mu$) in the proton-to-electron mass ratio μ to be equal to "$(0.0 \pm 1.0) \times 10^{-7}$ at redshift z = 0.89" and consistent with "a null result".[32][33]

37.7   Challenges

See also: Physics beyond the Standard Model

Self-consistency of the Standard Model (currently formulated as a non-abelian gauge theory quantized through path-integrals) has not been mathematically proven. While regularized versions useful for approximate computations (for example lattice gauge theory) exist, it is not known whether they converge (in the sense of S-matrix elements) in the limit that the regulator is removed. A key question related to the consistency is the Yang–Mills existence and mass gap problem.

Experiments indicate that neutrinos have mass, which the classic Standard Model did not allow.[34] To accommodate this finding, the classic Standard Model can be modified to include neutrino mass.

If one insists on using only Standard Model particles, this can be achieved by adding a non-renormalizable interaction of leptons with the Higgs boson.[35] On a fundamental level, such an interaction emerges in the seesaw mechanism where heavy right-handed neutrinos are added to the theory. This is natural in the left-right symmetric extension of the Standard Model[36][37] and in certain grand unified theories.[38] As long as new physics appears below or around 10^{14} GeV, the neutrino masses can be of the right order of magnitude.

Theoretical and experimental research has attempted to extend the Standard Model into a Unified field theory or a Theory of everything, a complete theory explaining all physical phenomena including constants. Inadequacies of the Standard Model that motivate such research include:

- The model does not explain gravitation, although physical confirmation of a theoretical particle known as a graviton would account for it to a degree. Though it addresses strong and electroweak interactions, the Standard Model does not consistently explain the canonical theory of gravitation, general relativity, in terms of quantum field theory. The reason for this is, among other things, that quantum field theories of gravity generally break down before reaching the Planck scale. As a consequence, we have no reliable theory for the very early universe.

- Some physicists consider it to be *ad hoc* and inelegant, requiring 19 numerical constants whose values are unrelated and arbitrary. Although the Standard Model, as it now stands, can explain why neutrinos have masses, the specifics of neutrino mass are still unclear. It is believed that explaining neutrino mass will require an additional 7 or 8 constants, which are also arbitrary parameters.

- The Higgs mechanism gives rise to the hierarchy problem if some new physics (coupled to the Higgs) is present at high energy scales. In these cases, in order for the weak scale to be much smaller than the Planck scale, severe fine tuning of the parameters is required; there are, however, other scenarios that include quantum gravity in which such fine tuning can be avoided.[39] There are also issues of Quantum triviality, which suggests that it may not be possible to create a consistent quantum field theory involving elementary scalar particles.

- The model is inconsistent with the emerging "Standard Model of cosmology." More common contentions include the absence of an explanation in the Standard Model for the observed amount of cold dark matter (CDM) and its contributions to dark energy, which are many orders of magnitude too large. It is also difficult to accommodate the observed predominance of matter over antimatter (matter/antimatter asymmetry). The isotropy and homogeneity of the visible universe over large distances seems to require a mechanism like cosmic inflation, which would also constitute an extension of the Standard Model.

- The existence of ultra-high-energy cosmic rays are difficult to explain under the Standard Model.

Currently, no proposed Theory of Everything has been widely accepted or verified.

37.8 See also

- Fundamental interaction:

  - Quantum electrodynamics

  - Strong interaction: Color charge, Quantum chromodynamics, Quark model

  - Weak interaction: Electroweak theory, Fermi theory of beta decay, Weak hypercharge, Weak isospin

- Gauge theory: Nontechnical introduction to gauge theory

- Generation

- Higgs mechanism: Higgs boson, Higgsless model

- J. C. Ward

- J. J. Sakurai Prize for Theoretical Particle Physics

- Lagrangian

- Open questions: BTeV experiment, CP violation, Neutrino masses, Quark matter, Quantum triviality

- Penguin diagram

- Quantum field theory

- Standard Model: Mathematical formulation of, Physics beyond the Standard Model

37.9 Notes and references

[1] Technically, there are nine such color–anticolor combinations. However, there is one color-symmetric combination that can be constructed out of a linear superposition of the nine combinations, reducing the count to eight.

37.10   References

[1] R. Oerter (2006). *The Theory of Almost Everything: The Standard Model, the Unsung Triumph of Modern Physics* (Kindle ed.). Penguin Group. p. 2. ISBN 0-13-236678-9.

[2] In fact, there are mathematical issues regarding quantum field theories still under debate (see e.g. Landau pole), but the predictions extracted from the Standard Model by current methods applicable to current experiments are all self-consistent. For a further discussion see e.g. Chapter 25 of R. Mann (2010). *An Introduction to Particle Physics and the Standard Model.* CRC Press. ISBN 978-1-4200-8298-2.

[3] Sean Carroll, Ph.D., Cal Tech, 2007, The Teaching Company, *Dark Matter, Dark Energy: The Dark Side of the Universe*, Guidebook Part 2 page 59, Accessed Oct. 7, 2013, "...Standard Model of Particle Physics: The modern theory of elementary particles and their interactions ... It does not, strictly speaking, include gravity, although it's often convenient to include gravitons among the known particles of nature..."

[4] S.L. Glasho w (1961). "Partial-symmetries of weak interactions". *Nuclear Physics* **22** (4): 579–588. Bibcode:1961NucPh..22 ·· doi:10.1016/0029-5582(61)90469-2.

[5] S. Weinberg (1967). "A Model of Leptons". *Physical Review Letters* **19** (21): 1264–1266. Bibcode:1967PhRvL..19.1264W. doi:10.1103/PhysRevLett.19.1264.

[6] A. Salam (1968). N. Svartholm, ed. *Elementary Particle Physics: Relativistic Groups and Analyticity.* Eighth Nobel Symposium. Stockholm: Almquvist and Wiksell. p. 367.

[7] F. Englert, R. Brout (1964). "Broken Symmetry and the Mass of Gauge Vector Mesons". *Physical Review Letters* **13** (9): 321–323. Bibcode:1964PhRvL..13..321E. doi:10.1103/PhysRevLett.13.321.

[8] P.W. Higgs (1964). "Broken Symmetries and the Masses of Gauge Bosons". *Physical Review Letters* **13** (16): 508–509. Bibcode:1964PhRvL..13..508H. doi:10.1103/PhysRevLett.13.508.

[9] G.S. Guralnik, C.R. Hagen, T.W.B. Kibble (1964). "Global Conservation Laws and Massless Particles". *Physical Review Letters* **13** (20): 585–587. Bibcode:1964PhRvL..13..585G. doi:10.1103/PhysRevLett.13.585.

[10] F.J. Hasert; et al. (1973). "Searc h for elasti c muon-neutrino electron scattering". *Physics Letters B* **46** (1): 121. Bibcode: doi:10.1016/0370-2693(73)90494-2.

[11] F.J. Hasert; et al. (1973). "Observation of neutrino-like interactions without muon or electron in the Gargamelle neutrino experiment". *Physics Letters B* **46** (1): 138. Bibcode:1973PhLB...46..138H. doi:10.1016/0370-2693(73)90499-1.

[12] F.J. Hasert; et al. (1974). "Observation of neutrino-like interactions without muon or electron in the Gargamelle neutrino experiment". *Nuclear Physics B* **73** (1): 1. Bibcode:1974NuPhB..73....1H. doi:10.1016/0550-3213(74)90038-8.

[13] D. Haidt (4 October 2004). "The discovery of the weak neutral currents". *CERN Courier.* Retrieved 8 May 2008.

[14] "Details can be worked out if the situation is simple enough for us to make an approximation, which is almost never, but often we can understand more or less what is happening." from *The Feynman Lectures on Physics*, Vol 1. pp. 2–7

[15] G.S. Guralnik (2009). "The History of the Guralnik, Hagen and Kibble development of the Theory of Spontaneous Symmetry Breaking and Gauge Particles". *International Journal of Modern Physics A* **24** (14): 2601–2627. arXiv:0907.3466. Bibcode:2009IJMPA..24.2601G. doi:10.1142/S0217751X09045431.

[16] B.W. Lee, C. Quigg, H.B. Thacker (1977). "Weak interactions at very high energies: The role of the Higgs-boson mass". *Physical Review D* **16** (5): 1519–1531. Bibcode:1977PhRvD..16.1519L. doi:10.1103/PhysRevD.16.1519.

[17] "Huge $10 billion collider resumes hunt for 'God particle'". CNN. 11 November 2009. Retrieved 2010-05-04.

[18] M. Strassler (10 July 2012). "Higgs Discovery: Is it a Higgs?". Retrieved 2013-08-06.

[19] "CERN experiments observe particle consistent with long-sought Higgs boson". CERN. 4 July 2012. Retrieved 2012-07-04.

[20] "Observation of a New Particle with a Mass of 125 GeV". CERN. 4 July 2012. Retrieved 2012-07-05.

[21] "ATLAS Experiment". ATLAS. 1 January 2006. Retrieved 2012-07-05.

[22] "Confirmed: CERN discovers new particle likely to be the Higgs boson". *YouTube*. Russia Today. 4 July 2012. Retrieved 2013-08-06.

[23] D. Overbye (4 July 2012). "A New Particle Could Be Physics' Holy Grail". *New York Times*. Retrieved 2012-07-04.

[24] "New results indicate that new particle is a Higgs boson". CERN. 14 March 2013. Retrieved 2013-08-06.

[25] S. Braibant, G. Giacomelli, M. Spurio (2009). *Particles and Fundamental Interactions: An Introduction to Particle Physics*. Springer. pp. 313–314. ISBN 978-94-007-2463-1.

[26] Kayser, Boris (2009). "Are neutrinos their own antiparticles?" (PDF). *Journal of Physics: Conference Series* (IOP Publishing) **173** (1): 012013.

[27] http://home.web.cern.ch/about/physics/standard-model Official CERN website

[28] http://www.pha.jhu.edu/~{}dfehling/particle.gif

[29] "BABAR Data in Tension with the Standard Model". SLAC. 31 May 2012. Retrieved 2013-08-06.

[30] BaBar Collaboration (2012). "Evidence for an excess of $B \rightarrow D^{(*)} \tau^- \nu\tau$ decays". *Physical Review Letters* **109** (10): 101802. arXiv:1205.5442. Bibcode:2012PhRvL.109j1802L. doi:10.1103/PhysRevLett.109.101802.

[31] "BaBar data hint at cracks in the Standard Model". *e! Science News*. 18 June 2012. Retrieved 2013-08-06.

[32] J. Bagdonaite; et al. (2012). "A Stringent Limit on a Drifting Proton-to-Electron Mass Ratio from Alcohol in the Early Universe". *Science* **339** (6115): 46. Bibcode:2013Sci...339...46B. doi:10.1126/science.1224898.

[33] C. Moskowitz (13 December 2012). "Phew! Universe's Constant Has Stayed Constant". Space.com. Retrieved 2012-12-14.

[34] "Particle chameleon caught in the act of changing". CERN. 31 May 2010. Retrieved 2012-07-05.

[35] S. Weinber g (1979). "Baryon and Lepton Nonconserving Processes". *Physical Review Letters* **43** (21): 1566. Bibcode: doi:10.1103/PhysRevLett.43.1566.

[36] P. Minkowski (1977). "$\mu \rightarrow e \gamma$ at a Rate of One Out of 10_9Muon Decays?". *Physics Letters B* **67** (4): 421. Bibcode:1977PhLB.. doi:10.1016/0370-2693(77)90435-X.

[37] R. N. Mohapatra, G. Senjanovic (1980). "Neutrino Mass and Spontaneous Parity Nonconservation". *Physical Review Letters* **44** (14): 912–915. Bibcode:1980PhRvL..44..912M. doi:10.1103/PhysRevLett.44.912.

[38] M. Gell-Mann, P. Ramond and R. Slansky (1979). F. van Nieuwenhuizen and D. Z. Freedman, ed. *Supergravity*. North Holland. pp. 315–321. ISBN 0-444-85438-X.

[39] Salvio, Strumi a (2014-03-17). "Agravity". *JHEP 1406 (2014) 080*. arXiv:1403.4226. Bibcode:2014JHEP...06..080S. do

37.11 Further reading

- R. Oerter (2006). *The Theory of Almost Everything: The Standard Model, the Unsung Triumph of Modern Physics*. Plume.

- B.A. Schumm (2004). *Deep Down Things: The Breathtaking Beauty of Particle Physics*. Johns Hopkins University Press. ISBN 0-8018-7971-X.

- "The Standard Model of Particle Physics Interactive Graphic".

Introductory textbooks

- I. Aitchison, A. Hey (2003). *Gauge Theories in Particle Physics: A Practical Introduction*. Institute of Physics. ISBN 978-0-585-44550-2.

- W. Greiner, B. Müller (2000). *Gauge Theory of Weak Interactions*. Springer. ISBN 3-540-67672-4.

- G.D. Coughlan, J.E. Dodd, B.M. Gripaios (2006). *The Ideas of Particle Physics: An Introduction for Scientists.* Cambridge University Press.

- D.J. Griffiths (1987). *Introduction to Elementary Particles.* John Wiley & Sons. ISBN 0-471-60386-4.

- G.L. Kane (1987). *Modern Elementary Particle Physics.* Perseus Books. ISBN 0-201-11749-5.

Advanced textbooks

- T.P. Cheng, L.F. Li (2006). *Gauge theory of elementary particle physics.* Oxford University Press. ISBN 0-19-851961-3. Highlights the gauge theory aspects of the Standard Model.

- J.F. Donoghue, E. Golowich, B.R. Holstein (1994). *Dynamics of the Standard Model.* Cambridge University Press. ISBN 978-0-521-47652-2. Highlights dynamical and phenomenological aspects of the Standard Model.

- L. O'Raifeartaigh (1988). *Group structure of gauge theories.* Cambridge University Press. ISBN 0-521-34785-8.

- Nagashima Y. Elementary Particle Physics: Foundations of the Standard Model, Volume 2. (Wiley 2013) 920 рапуы

- Schwartz, M.D. Quantum Field Theory and the Standard Model (Cambridge University Press 2013) 952 pages

- Langacker P. The standard model and beyond. (CRC Press, 2010) 670 pages Highlights group-theoretical aspects of the Standard Model.

Journal articles

- E.S. Abers, B.W. Lee (1973). "Gauge theories". *Physics Reports* **9**: 1–141. Bibcode:1973PhR.....9....1A. doi: 1573(73)90027-6.

- M. Baak; et al. (2012). "The Electroweak Fit of the Standard Model after the Discovery of a New Boson at the LHC". *The European Physical Journal C* **72** (11). arXiv:1209.2716. Bibcode:2012EPJC...72.2205B. doi:10.1140/epjc/s10052-012-2205-9.

- Y. Hayato; et al. (1999). "Search for Proton Decay through $p \to \nu K^+$ in a Large Water Cherenkov Detector". *Physical Review Letters* **83** (8): 1529. arXiv:hep-ex/9904020. Bibcode:1999PhRvL..83.1529H. doi:10.1103/

- S.F. Novaes (2000). "Standard Model: An Introduction". arXiv:hep-ph/0001283 [hep-ph].

- D.P. Roy (1999). "Basic Constituents of Matter and their Interactions — A Progress Report". arXiv:hep-ph/9912523 [hep-ph].

- F. Wilczek (2004). "The Universe Is A Strange Place". *Nuclear Physics B - Proceedings Supplements* **134**: 3. arXiv:astro-ph/0401347. Bibcode:2004NuPhS.134....3W. doi:10.1016/j.nuclphysbps.2004.08.001.

37.12   External links

- "The Standard Model explained in Detail by CERN's John Ellis" omega tau podcast.

- "The Standard Model" The Standard Model on the CERN web site explains how the basic building blocks of matter interact, governed by four fundamental forces.

- "Standard Model" on YouTube

Chapter 38

Standard Model (mathematical formulation)

For a less mathematical description, see Standard Model.

This article describes the mathematics of the **Standard Model** of particle physics, a gauge quantum field theory con-

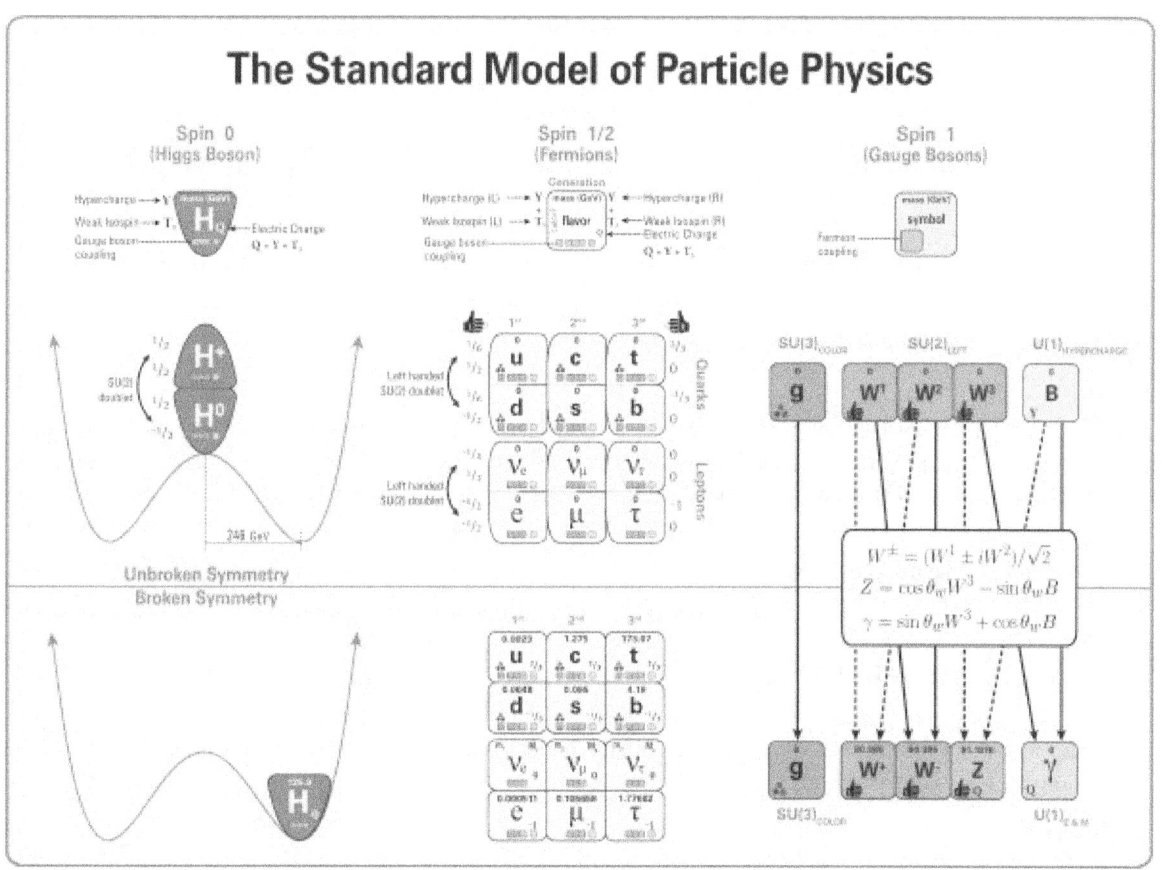

Standard Model of Particle Physics. The diagram shows the elementary particles of the Standard Model (the Higgs boson, the three generations of quarks and leptons, and the gauge bosons), including their names, masses, spins, charges, chiralities, and interactions with the strong, weak and electromagnetic forces. It also depicts the crucial role of the Higgs boson in electroweak symmetry breaking, and shows how the properties of the various particles differ in the (high-energy) symmetric phase (top) and the (low-energy) broken-symmetry phase (bottom).

taining the internal symmetries of the unitary product group SU(3) × SU(2) × U(1). The theory is commonly viewed as containing the fundamental set of particles – the leptons, quarks, gauge bosons and the Higgs particle.

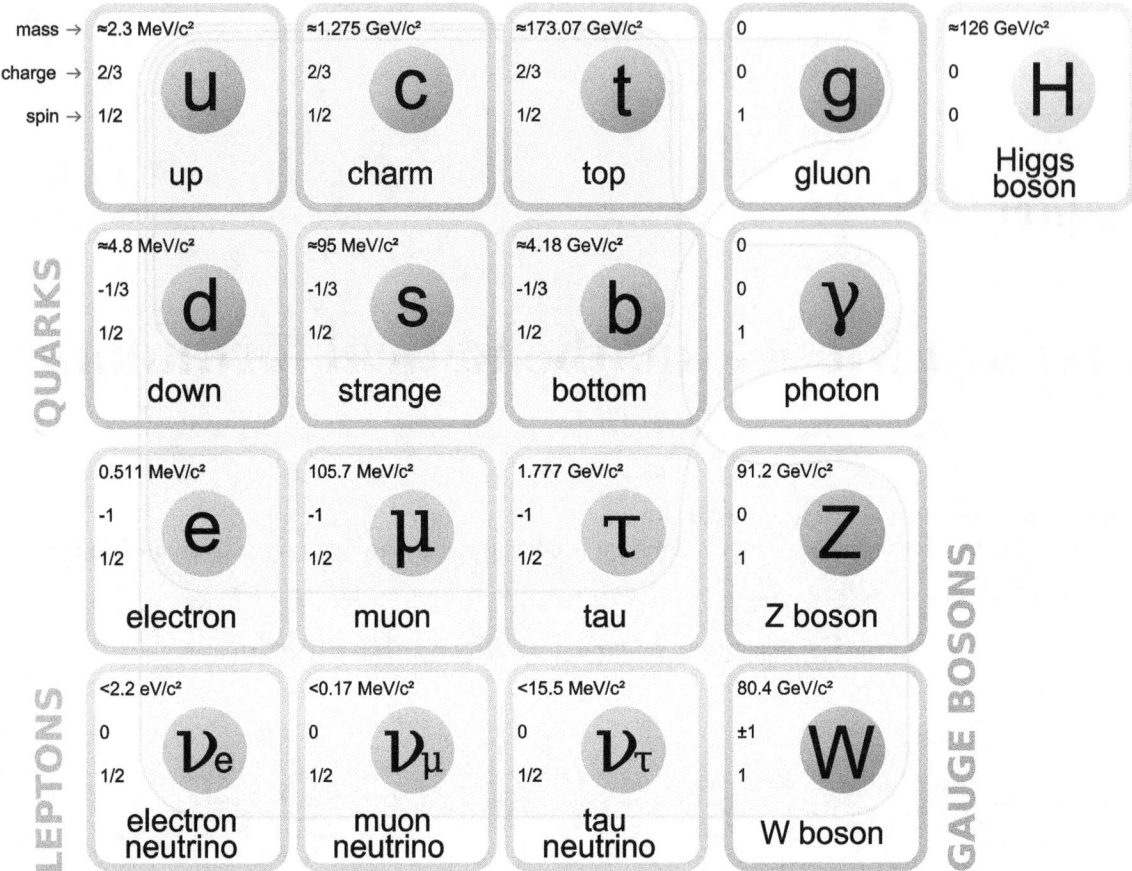

The Standard Model of Particle Physics: More Schematic Depiction

The Standard Model is renormalizable and mathematically self-consistent,[1] however despite having huge and continued successes in providing experimental predictions it does leave some unexplained phenomena. In particular, although the physics of special relativity is incorporated, general relativity is not, and the Standard Model will fail at energies or distances where the graviton is expected to emerge. Therefore in a modern field theory context, it is seen as an effective field theory.

This article requires some background in physics and mathematics, but is designed as both an introduction and a reference.

38.1   Quantum field theory

The standard model is a quantum field theory, meaning its fundamental objects are *quantum fields* which are defined at all points in spacetime. These fields are

- the fermion field, ψ, which accounts for "matter particles";
- the electroweak boson fields W_1, W_2, W_3 , and B;
- the gluon field, G_a; and
- the Higgs field, φ.

That these are *quantum* rather than *classical* fields has the mathematical consequence that they are operator-valued. In particular, values of the fields generally do not commute. As operators, they act upon the quantum state (ket vector).

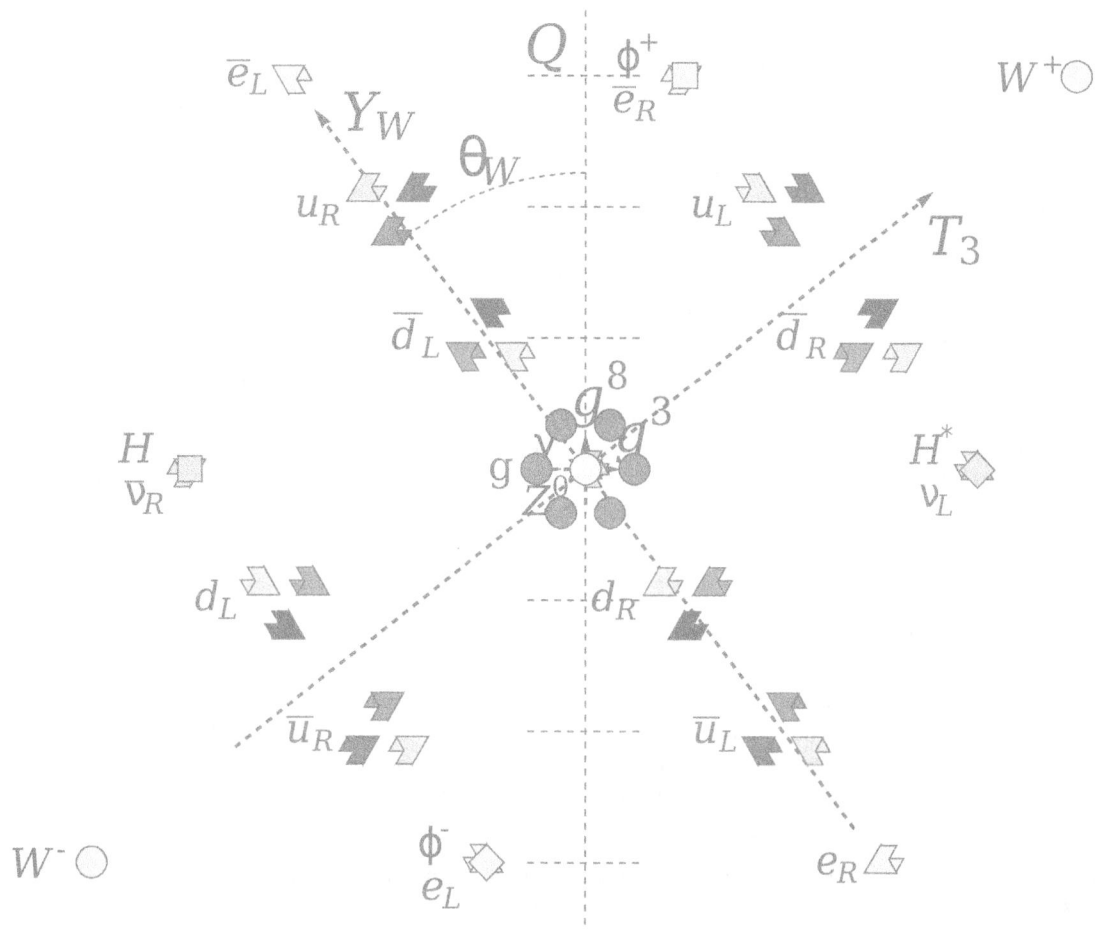

The pattern of weak isospin T_3, *weak hypercharge YW, and color charge of all known elementary particles, rotated by the weak mixing angle to show electric charge Q, roughly along the vertical. The neutral Higgs field (gray square) breaks the electroweak symmetry and interacts with other particles to give them mass.*

The dynamics of the quantum state and the fundamental fields are determined by the Lagrangian density $\mathcal{L}$ (usually for short just called the Lagrangian). This plays a role similar to that of the Schrödinger equation in non-relativistic quantum mechanics, but a Lagrangian is not an equation – rather, it is a polynomial function of the fields and their derivatives, and used with the principle of least action. While it would be possible to derive a system of differential equations governing the fields from the Langrangian, it is more common to use other techniques to compute with quantum field theories.

The standard model is furthermore a gauge theory, which means there are degrees of freedom in the mathematical formalism which do not correspond to changes in the physical state. The gauge group of the standard model is $SU(3) \times SU(2) \times U(1)$, where $U(1)$ acts on B and φ, $SU(2)$ acts on W and φ, and $SU(3)$ acts on G. The fermion field ψ also transforms under these symmetries, although all of them leave some parts of it unchanged.

38.1.1   The role of the quantum fields

In classical mechanics, the state of a system can usually be captured by a small set of variables, and the dynamics of the system is thus determined by the time evolution of these variables. In classical field theory, the *field* is part of the state of the system, so in order to describe it completely one effectively introduces separate variables for every point in spacetime (even though there are many restrictions on how the values of the field "variables" may vary from point to point, for example in the form of field equations involving partial derivatives of the fields).

In quantum mechanics, the classical variables are turned into operators, but these do not capture the state of the system, which is instead encoded into a wavefunction ψ or more abstract ket vector. If ψ is an eigenstate with respect to an operator P, then $P\psi = \lambda\psi$ for the corresponding eigenvalue λ, and hence letting an operator P act on ψ is analogous to multiplying ψ by the value of the classical variable to which P corresponds. By extension, a classical formula where all variables have been replaced by the corresponding operators will behave like an operator which, when it acts upon the state of the system, multiplies it by the analogue of the quantity that the classical formula would compute. The formula as such does however not contain any information about the state of the system; it would evaluate to the same operator regardless of what state the system is in.

Quantum fields relate to quantum mechanics as classical fields do to classical mechanics, i.e., there is a separate operator for every point in spacetime, and these operators do not carry any information about the state of the system; they are merely used to exhibit some aspect of the state, at the point to which they belong. In particular, the quantum fields are *not* wavefunctions, even though the equations which govern their time evolution may be deceptively similar to those of the corresponding wavefunction in a semiclassical formulation. There is no variation in strength of the fields between different points in spacetime; the variation that happens is rather one of phase factors.

38.1.2   Vectors, scalars, and spinors

Mathematically it may look as though all of the fields are vector-valued (in addition to being operator-valued), since they all have several components, can be multiplied by matrices, etc., but physicists assign a more specific physical meaning to the word: a **vector** is something which transforms like a four-vector under Lorentz transformations, and a **scalar** is something which is invariant under Lorentz transformations. The B, W_j, and G_a fields are all vectors in this sense, so the corresponding particles are said to be vector bosons. The Higgs field φ is a scalar.

The fermion field ψ does transform under Lorentz transformations, but not like a vector should; rotations will only turn it by half the angle a proper vector should. Therefore these constitute a third kind of quantity, which is known as a spinor.

It is common to make use of abstract index notation for the vector fields, in which case the vector fields all come with a Lorentzian index μ, like so: B^μ, W_j^μ, and G_a^μ. If abstract index notation is used also for spinors then these will carry a spinorial index and the Dirac gamma will carry one Lorentzian and two spinorian indices, but it is more common to regard spinors as column matrices and the Dirac gamma $\gamma\mu$ as a matrix which additionally carries a Lorentzian index. The Feynman slash notation can be used to turn a vector field into a linear operator on spinors, like so: $\slashed{B} = \gamma^\mu B_\mu$; this may involve raising and lowering indices.

38.2   Alternative presentations of the fields

As is common in quantum theory, there is more than one way to look at things. At first the basic fields given above may not seem to correspond well with the "fundamental particles" in the chart above, but there are several alternative presentations which, in particular contexts, may be more appropriate than those that are given above.

38.2.1   Fermions

Rather than having one fermion field ψ, it can be split up into separate components for each type of particle. This mirrors the historical evolution of quantum field theory, since the electron component ψ_e (describing the electron and its antiparticle the positron) is then the original ψ field of quantum electrodynamics, which was later accompanied by $\psi\mu$

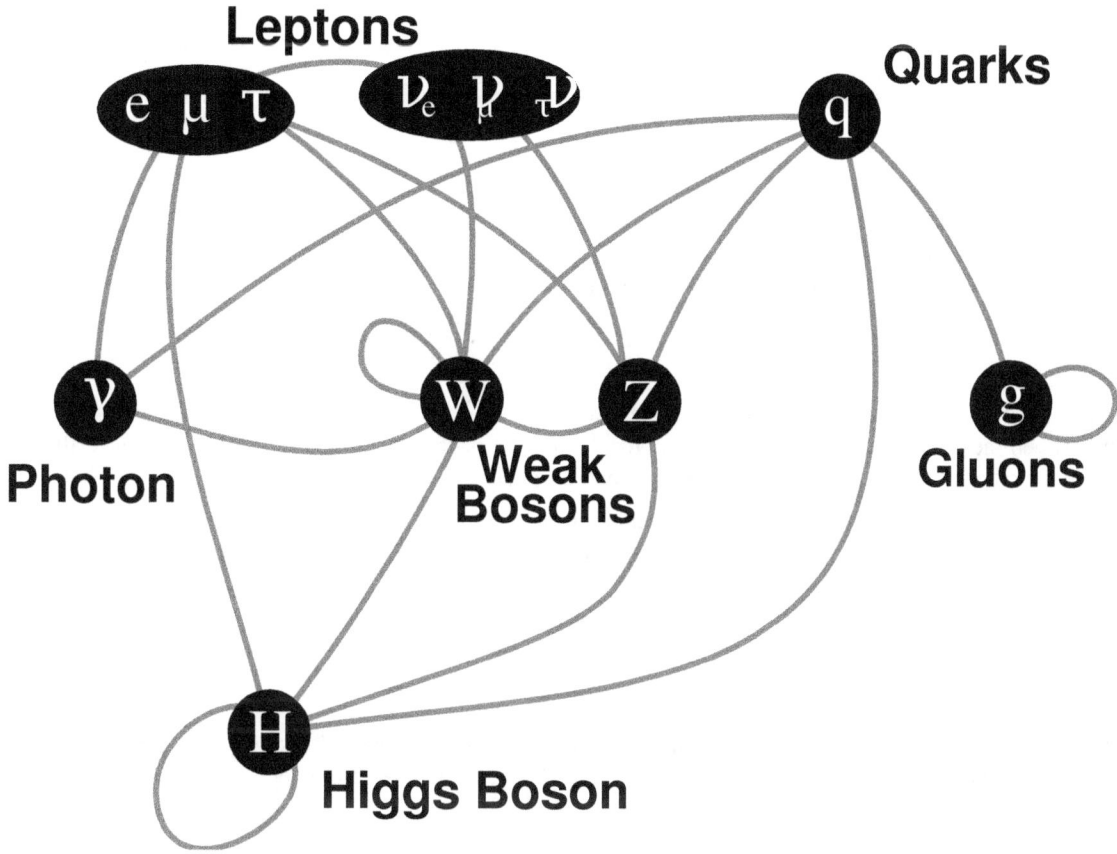

Connections denoting which particles interact with each other.

and ψτ fields for the muon and tauon respectively (and their antiparticles). Electroweak theory added $\psi_{\nu_e}, \psi_{\nu_\mu}$, and ψ_{ν_τ} for the corresponding neutrinos, and the quarks add still further components. In order to be four-spinors like the electron and other lepton components, there must be one quark component for every combination of flavour and colour, bringing the total to 24 (3 for charged leptons, 3 for neutrinos, and 2·3·3 = 18 for quarks).

An important definition is the barred fermion field $\bar{\psi}$ is defined to be $\psi^\dagger \gamma^0$, where † denotes the Hermitian adjoint and γ^0 is the zeroth gamma matrix. If ψ is thought of as an $n \times 1$ matrix then $\bar{\psi}$ should be thought of as a $1 \times n$ matrix.

A chiral theory

An independent decomposition of ψ is that into chirality components:

"Left" chirality: $\psi^L = \frac{1}{2}(1 - \gamma_5)\psi$

"Right" chirality: $\psi^R = \frac{1}{2}(1 + \gamma_5)\psi$

where γ_5 is the fifth gamma matrix. This is very important in the Standard Model because *left and right chirality components are treated differently by the gauge interactions.*

In particular, under weak isospin SU(2) transformations the left-handed particles are weak-isospin doublets, whereas the right-handed are singlets – i.e. the weak isospin of ψR is zero. Put more simply, the weak interaction could rotate e.g. a left-handed electron into a left-handed neutrino (with emission of a W⁻), but could not do so with the same right-handed particles. As an aside, the right-handed neutrino originally did not exist in the standard model – but the discovery of neutrino oscillation implies that neutrinos must have mass, and since chirality can change during the propagation of a massive particle, right-handed neutrinos must exist in reality. This does not however change the (experimentally-proven) chiral nature of the weak interaction.

Furthermore, U(1) acts differently on ψ_e^L than on ψ_e^R (because they have different weak hypercharges).

Mass and interaction eigenstates

A distinction can thus be made between, for example, the mass and interaction eigenstates of the neutrino. The former is the state which propagates in free space, whereas the latter is the *different* state that participates in interactions. Which is the "fundamental" particle? For the neutrino, it is conventional to define the "flavour" (ν

e, ν

μ, or ν

τ) by the interaction eigenstate, whereas for the quarks we define the flavour (up, down, etc.) by the mass state. We can switch between these states using the CKM matrix for the quarks, or the PMNS matrix for the neutrinos (the charged leptons on the other hand are eigenstates of both mass and flavour).

As an aside, if a complex phase term exists within either of these matrices, it will give rise to direct CP violation, which could explain the dominance of matter over antimatter in our current universe. This has been proven for the CKM matrix, and is expected for the PMNS matrix.

Positive and negative energies

Finally, the quantum fields are sometimes decomposed into "positive" and "negative" energy parts: $\psi = \psi^+ + \psi^-$. This is not so common when a quantum field theory has been set up, but often features prominently in the process of quantizing a field theory.

38.2.2   Bosons

Due to the Higgs mechanism, the electroweak boson fields W_1, W_2, W_3 , and B "mix" to create the states which are physically observable. To retain gauge invariance, the underlying fields must be massless, but the observable states can *gain masses* in the process. These states are:

The massive neutral (Z) boson:

$$Z = \cos\theta_W W_3 - \sin\theta_W B$$

The massless neutral boson:

$$A = \sin\theta_W W_3 + \cos\theta_W B$$

The massive charged W bosons:

$$W^\pm = \frac{1}{\sqrt{2}}\left(W_1 \mp iW_2\right)$$

where θW is the Weinberg angle.

The A field is the photon, which corresponds classically to the well-known electromagnetic four-potential – i.e. the electric and magnetic fields. The Z field actually contributes in every process the photon does, but due to its large mass, the contribution is usually negligible.

38.3 Perturbative QFT and the interaction picture

Much of the qualitative descriptions of the standard model in terms of "particles" and "forces" comes from the perturbative quantum field theory view of the model. In this, the Langrangian is decomposed as $\mathcal{L} = \mathcal{L}_0 + \mathcal{L}_I$ into separate *free field* and *interaction* Langrangians. The free fields care for particles in isolation, whereas processes involving several particles arise through interactions. The idea is that the state vector should only change when particles interact, meaning a free particle is one whose quantum state is constant. This corresponds to the interaction picture in quantum mechanics.

In the more common Schrödinger picture, even the states of free particles change over time: typically the phase changes at a rate which depends on their energy. In the alternative Heisenberg picture, state vectors are kept constant, at the price of having the operators (in particular the observables) be time-dependent. The interaction picture constitutes an intermediate between the two, where some time dependence is placed in the operators (the quantum fields) and some in the state vector. In QFT, the former is called the free field part of the model, and the latter is called the interaction part. The free field model can be solved exactly, and then the solutions to the full model can be expressed as perturbations of the free field solutions, for example using the Dyson series.

It should be observed that the decomposition into free fields and interactions is in principle arbitrary. For example renormalization in QED modifies the mass of the free field electron to match that of a physical electron (with an electromagnetic field), and will in doing so add a term to the free field Lagrangian which must be cancelled by a counterterm in the interaction Lagrangian, that then shows up as a two-line vertex in the Feynman diagrams. This is also how the Higgs field is thought to give particles mass: the part of the interaction term which corresponds to the (nonzero) vacuum expectation value of the Higgs field is moved from the interaction to the free field Lagrangian, where it looks just like a mass term having nothing to do with Higgs.

38.3.1 Free fields

Under the usual free/interaction decomposition, which is suitable for low energies, the free fields obey the following equations:

- The fermion field ψ satisfies the Dirac equation; $(i\hbar\partial\!\!\!/ - m_f c)\psi_f = 0$ for each type f of fermion.

- The photon field A satisfies the wave equation $\partial_\mu \partial^\mu A^\nu = 0$.

- The Higgs field φ satisfies the Klein–Gordon equation.

- The weak interaction fields Z, $W^\pm$ also satisfy the Klein–Gordon equation.

These equations can be solved exactly. One usually does so by considering first solutions that are periodic with some period L along each spatial axis; later taking the limit: $L \to \infty$ will lift this periodicity restriction.

In the periodic case, the solution for a field F (any of the above) can be expressed as a Fourier series of the form

$$F(x) = \beta \sum_{\mathbf{p}} \sum_{r} E_{\mathbf{p}}^{-\frac{1}{2}} \left(a_r(\mathbf{p}) u_r(\mathbf{p}) e^{-\frac{ipx}{\hbar}} + b_r^\dagger(\mathbf{p}) v_r(\mathbf{p}) e^{\frac{ipx}{\hbar}} \right)$$

where:

- β is a normalization factor; for the fermion field ψ_f it is $\sqrt{m_f c^2 / V}$, where $V = L^3$ is the volume of the fundamental cell considered; for the photon field A^μ it is $\hbar c / \sqrt{2V}$.

- The sum over $\mathbf{p}$ is over all momenta consistent with the period L, i.e., over all vectors $\frac{2\pi\hbar}{L}(n_1, n_2, n_3)$ where n_1, n_2, n_3 are integers.

- The sum over r covers other degrees of freedom specific for the field, such as polarization or spin; it usually comes out as a sum from 1 to 2 or from 1 to 3.

- $E_\mathbf{p}$ is the relativistic energy for a momentum $\mathbf{p}$ quantum of the field, $= \sqrt{m^2c^4 + c^2\mathbf{p}^2}$ when the rest mass is m.

- $ar(\mathbf{p})$ and $b_r^\dagger(\mathbf{p})$ are annihilation and creation respectively operators for "a-particles" and "b-particles" respectively of momentum $\mathbf{p}$; "b-particles" are the antiparticles of "a-particles". Different fields have different "a-" and "b-particles". For some fields, a and b are the same.

- $ur(\mathbf{p})$ and $vr(\mathbf{p})$ are non-operators which carry the vector or spinor aspects of the field (where relevant).

- $p = (E_\mathbf{p}/c, \mathbf{p})$ is the four-momentum for a quantum with momentum $\mathbf{p}$. $px = p_\mu x^\mu$ denotes an inner product of four-vectors.

In the limit $L \to \infty$, the sum would turn into an integral with help from the V hidden inside β. The numeric value of β also depends on the normalization chosen for $u_r(\mathbf{p})$ and $v_r(\mathbf{p})$.

Technically, $a_r^\dagger(\mathbf{p})$ is the Hermitian adjoint of the operator $ar(\mathbf{p})$ in the inner product space of ket vectors. The identification of $a_r^\dagger(\mathbf{p})$ and $ar(\mathbf{p})$ as creation and annihilation operators comes from comparing conserved quantities for a state before and after one of these have acted upon it. $a_r^\dagger(\mathbf{p})$ can for example be seen to add one particle, because it will add 1 to the eigenvalue of the a-particle number operator, and the momentum of that particle ought to be $\mathbf{p}$ since the eigenvalue of the vector-valued momentum operator increases by that much. For these derivations, one starts out with expressions for the operators in terms of the quantum fields. That the operators with † are creation operators and the one without annihilation operators is a convention, imposed by the sign of the commutation relations postulated for them.

An important step in preparation for calculating in perturbative quantum field theory is to separate the "operator" factors a and b above from their corresponding vector or spinor factors u and v. The vertices of Feynman graphs come from the way that u and v from different factors in the interaction Lagrangian fit together, whereas the edges come from the way that the as and bs must be moved around in order to put terms in the Dyson series on normal form.

38.3.2   Interaction terms and the path integral approach

The Lagrangian can also be derived without using creation and annihilation operators (the "canonical" formalism), by using a "path integral" approach, pioneered by Feynman building on the earlier work of Dirac. See e.g. Path integral formulation on Wikipedia or A. Zee's QFT in a nutshell. This is one possible way that the Feynman diagrams, which are pictorial representations of interaction terms, can be derived relatively easily. A quick derivation is indeed presented at the article on Feynman diagrams.

38.4   Lagrangian formalism

We can now give some more detail about the aforementioned free and interaction terms appearing in the Standard Model Lagrangian density. Any such term must be both gauge and reference-frame invariant, otherwise the laws of physics would depend on an arbitrary choice or the frame of an observer. Therefore the global Poincaré symmetry, consisting of translational symmetry, rotational symmetry and the inertial reference frame invariance central to the theory of special relativity must apply. The local SU(3) × SU(2) × U(1) gauge symmetry is the internal symmetry. The three factors of the gauge symmetry together give rise to the three fundamental interactions, after some appropriate relations have been defined, as we shall see.

A complete formulation of the Standard Model Lagrangian with all the terms written together can be found e.g. here.

38.4.1   Kinetic terms

A free particle can be represented by a mass term, and a *kinetic* term which relates to the "motion" of the fields.

Standard Model Interactions
(Forces Mediated by Gauge Bosons)

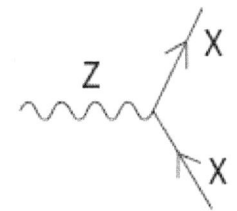

X is any fermion in
the Standard Model.

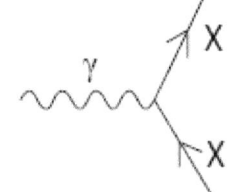

X is electrically charged.

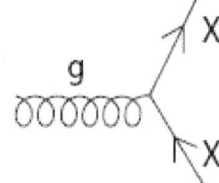

X is any quark.

U is a up-type quark;
D is a down-type quark.

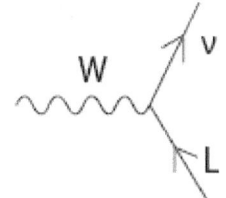

L is a lepton and ν is the
corresponding neutrino.

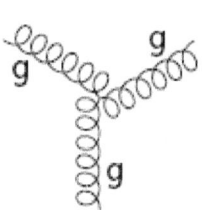

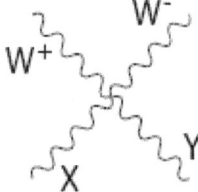

X is a photon or Z-boson.

X and Y are any two
electroweak bosons such
that charge is conserved.

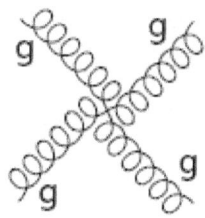

The above interactions show some basic interaction vertices – Feynman diagrams in the standard model are built from these vertices. Higgs boson interactions are however not shown, and neutrino oscillations are commonly added. The charge of the W bosons are dictated by the fermions they interact with.

Fermion fields

The kinetic term for a Dirac fermion is

$$i\bar{\psi}\gamma^{\mu}\partial_{\mu}\psi$$

where the notations are carried from earlier in the article. ψ can represent any, or all, Dirac fermions in the standard model. Generally, as below, this term is included within the couplings (creating an overall "dynamical" term).

Gauge fields

For the spin-1 fields, first define the field strength tensor

$$F^a_{\mu\nu} = \partial_\mu A^a_\nu - \partial_\nu A^a_\mu + g f^{abc} A^b_\mu A^c_\nu$$

for a given gauge field (here we use A), with gauge coupling constant g. The quantity f^{abc} is the structure constant of the particular gauge group, defined by the commutator

$$[t_a, t_b] = i f^{abc} t_c,$$

where t_i are the generators of the group. In an Abelian (commutative) group (such as the U(1) we use here), since the generators t_a all commute with each other, the structure constants vanish. Of course, this is not the case in general – the standard model includes the non-Abelian SU(2) and SU(3) groups (such groups lead to what is called a Yang–Mills gauge theory).

We need to introduce three gauge fields corresponding to each of the subgroups SU(3) × SU(2) × U(1).

- The gluon field tensor will be denoted by $G^a_{\mu\nu}$, where the index a labels elements of the **8** representation of colour SU(3). The strong coupling constant is conventionally labelled g_s (or simply g where there is no ambiguity). *The observations leading to the discovery of this part of the Standard Model are discussed in the article in quantum chromodynamics.*

- The notation $W^a_{\mu\nu}$ will be used for the gauge field tensor of SU(2) where a runs over the 3 generators of this group. The coupling can be denoted g_w or again simply g. The gauge field will be denoted by W^a_μ.

- The gauge field tensor for the U(1) of weak hypercharge will be denoted by Bμν, the coupling by g′, and the gauge field by Bμ.

The kinetic term can now be written simply as

$$\mathcal{L}_{\text{kin}} = -\frac{1}{4} B_{\mu\nu} B^{\mu\nu} - \frac{1}{2} \text{tr} W_{\mu\nu} W^{\mu\nu} - \frac{1}{2} \text{tr} G_{\mu\nu} G^{\mu\nu}$$

where the traces are over the SU(2) and SU(3) indices hidden in W and G respectively. The two-index objects are the field strengths derived from W and G the vector fields. There are also two extra hidden parameters: the theta angles for SU(2) and SU(3).

38.4.2 Coupling terms

The next step is to "couple" the gauge fields to the fermions, allowing for interactions.

Electroweak sector

Main article: Electroweak interaction

The electroweak sector interacts with the symmetry group U(1) × SU(2)L, where the subscript L indicates coupling only to left-handed fermions.

$$\mathcal{L}_{\text{EW}} = \sum_\psi \bar{\psi} \gamma^\mu \left(i\partial_\mu - g'\frac{1}{2} Y_{\text{W}} B_\mu - g\frac{1}{2} \boldsymbol{\tau} \mathbf{W}_\mu \right) \psi$$

Where Bμ is the U(1) gauge field; YW is the weak hypercharge (the generator of the U(1) group); $\mathbf{W}\mu$ is the three-component SU(2) gauge field; and the components of $\boldsymbol{\tau}$ are the Pauli matrices (infinitesimal generators of the SU(2) group) whose eigenvalues give the weak isospin. Note that we have to redefine a new U(1) symmetry of *weak hypercharge*, different from QED, in order to achieve the unification with the weak force. The electric charge Q, third component of weak isospin T_3 (also called T_z, I_3 or I_z) and weak hypercharge YW are related by

$$Q = T_3 + \tfrac{1}{2}Y_W,$$

or by the alternate convention $Q = T_3 + YW$. The first convention (used in this article) is equivalent to the earlier Gell-Mann–Nishijima formula. We can then define the conserved current for weak isospin as

$$\mathbf{j}_\mu = \frac{1}{2}\bar{\psi}_L \gamma_\mu \boldsymbol{\tau} \psi_L$$

and for weak hypercharge as

$$j_\mu^Y = 2(j_\mu^{em} - j_\mu^3)$$

where j_μ^{em} is the electric current and j_μ^3 the third weak isospin current. As explained above, *these currents mix* to create the physically observed bosons, which also leads to testable relations between the coupling constants.

To explain in a simpler way, we can see the effect of the electroweak interaction by picking out terms from the Lagrangian. We see that the SU(2) symmetry acts on each (left-handed) fermion doublet contained in ψ, for example

$$-\frac{g}{2}(\bar{\nu}_e\ \bar{e})\tau^+\gamma_\mu(W^-)^\mu\begin{pmatrix}\nu_e\\e\end{pmatrix} = -\frac{g}{2}\bar{\nu}_e\gamma_\mu(W^-)^\mu e$$

where the particles are understood to be left-handed, and where

$$\tau^+ \equiv \frac{1}{2}(\tau^1+i\tau^2) = \begin{pmatrix}0 & 1\\0 & 0\end{pmatrix}$$

This is an interaction corresponding to a "rotation in weak isospin space" or in other words, a *transformation between eL and veL via emission of a* W^- *boson*. The U(1) symmetry, on the other hand, is similar to electromagnetism, but acts on all "*weak hypercharged*" fermions (both left and right handed) via the neutral Z^0, as well as the *charged* fermions via the photon.

Quantum chromodynamics sector

Main article: Quantum chromodynamics

The quantum chromodynamics (QCD) sector defines the interactions between quarks and gluons, with SU(3) symmetry, generated by T_a. Since leptons do not interact with gluons, they are not affected by this sector. The Dirac Lagrangian of the quarks coupled to the gluon fields is given by

$$\mathcal{L}_{\text{QCD}} = i\overline{U}\left(\partial_\mu - ig_s G_\mu^a T^a\right)\gamma^\mu U + i\overline{D}\left(\partial_\mu - ig_s G_\mu^a T^a\right)\gamma^\mu D.$$

where D and U are the Dirac spinors associated with up- and down-type quarks, and other notations are continued from the previous section.

38.4.3   Mass terms and the Higgs mechanism

Mass terms

The mass term arising from the Dirac Lagrangian (for any fermion ψ) is $-m\bar{\psi}\psi$ which is *not* invariant under the electroweak symmetry. This can be seen by writing ψ in terms of left and right handed components (skipping the actual calculation):

$$-m\bar{\psi}\psi = -m(\bar{\psi}_L\psi_R + \bar{\psi}_R\psi_L)$$

i.e. contribution from $\bar{\psi}_L\psi_L$ and $\bar{\psi}_R\psi_R$ terms do not appear. We see that the mass-generating interaction is achieved by constant flipping of particle chirality. The spin-half particles have no right/left chirality pair with the same SU(2) representations and equal and opposite weak hypercharges, so assuming these gauge charges are conserved in the vacuum, none of the spin-half particles could ever swap chirality, and must remain massless. Additionally, we know experimentally that the W and Z bosons are massive, but a boson mass term contains the combination e.g. $A^\mu A\mu$, which clearly depends on the choice of gauge. Therefore, none of the standard model fermions *or* bosons can "begin" with mass, but must acquire it by some other mechanism.

The Higgs mechanism

Main article: Higgs mechanism

The solution to both these problems comes from the Higgs mechanism, which involves scalar fields (the number of which depend on the exact form of Higgs mechanism) which (to give the briefest possible description) are "absorbed" by the massive bosons as degrees of freedom, and which couple to the fermions via Yukawa coupling to create what looks like mass terms.

In the Standard Model, the Higgs field is a complex scalar of the group SU(2)L:

$$\phi = \frac{1}{\sqrt{2}}\begin{pmatrix}\phi^+\\\phi^0\end{pmatrix},$$

where the superscripts + and 0 indicate the electric charge (Q) of the components. The weak hypercharge (YW) of both components is 1.

The Higgs part of the Lagrangian is

$$\mathcal{L}_H = \left[\left(\partial_\mu - igW_\mu^a t^a - ig'Y_\phi B_\mu\right)\phi\right]^2 + \mu^2\phi^\dagger\phi - \lambda(\phi^\dagger\phi)^2,$$

where $\lambda > 0$ and $\mu^2 > 0$, so that the mechanism of spontaneous symmetry breaking can be used. There is a parameter here, at first hidden within the shape of the potential, that is very important. In a unitarity gauge one can set $\varphi^+ = 0$ and make φ^0 real. Then $\langle\phi^0\rangle = v$ is the non-vanishing vacuum expectation value of the Higgs field. v has units of mass, and it is the only parameter in the Standard Model which is not dimensionless. It is also much smaller than the Planck scale; it is approximately equal to the Higgs mass, and sets the scale for the mass of everything else. This is the only real fine-tuning to a small nonzero value in the Standard Model, and it is called the Hierarchy problem. Quadratic terms in Wμ and Bμ arise, which give masses to the W and Z bosons:

$$M_W = \tfrac{1}{2}v|g|$$
$$M_Z = \tfrac{1}{2}v\sqrt{g^2 + g'^2}$$

The Yukawa interaction terms are

$$\mathcal{L}_{YU} = U_L G_u U_R \phi^0 - \overline{D}_L G_u U_R \phi^- + \overline{U}_L G_d D_R \phi^+ + \overline{D}_L G_d D_R \phi^0 + hc$$

where $G_{u,d}$ are 3×3 matrices of Yukawa couplings, with the ij term giving the coupling of the generations i and j.

Neutrino masses

As previously mentioned, evidence shows neutrinos must have mass. But within the standard model, the right-handed neutrino does not exist, so even with a Yukawa coupling neutrinos remain massless. An obvious solution[2] is to simply *add a right-handed neutrino* νR resulting in a **Dirac mass** term as usual. This field however must be a sterile neutrino, since being right-handed it experimentally belongs to an isospin singlet ($T_3 = 0$) and also has charge $Q = 0$, implying $YW = 0$ (see above) i.e. it does not even participate in the weak interaction. Current experimental status is that evidence for observation of sterile neutrinos is not convincing.[3]

Another possibility to consider is that the neutrino satisfies the **Majorana equation**, which at first seems possible due to its zero electric charge. In this case the mass term is

$$-\frac{m}{2} \left(\overline{\nu}^C \nu + \overline{\nu} \nu^C \right)$$

where C denotes a charge conjugated (i.e. anti-) particle, and the terms are consistently all left (or all right) chirality (note that a left-chirality projection of an antiparticle is a right-handed field; care must be taken here due to different notations sometimes used). Here we are essentially flipping between LH neutrinos and RH anti-neutrinos (it is furthermore possible but *not* necessary that neutrinos are their own antiparticle, so these particles are the same). However for the left-chirality neutrinos, this term changes weak hypercharge by 2 units - not possible with the standard Higgs interation, requiring the Higgs field to be extended to include an extra triplet with weak hypercharge 2[4] - whereas for right-chirality neutrinos, no Higgs extensions are necessary. For both left and right chirality cases, Majorana terms violate lepton number, but possibly at a level beyond the current sensitivity of experiments to detect such violations.

It is possible to include **both** Dirac and Majorana mass terms in the same theory, which (in contrast to the Dirac-mass-only approach) can provide a "natural" explanation for the smallness of the observed neutrino masses, by linking the RH neutrinos to yet-unknown physics around the GUT scale[5] (see seesaw mechanism).

Since in any case new fields must be postulated to explain the experimental results, neutrinos are an obvious gateway to searching physics beyond the Standard Model.

38.5 Detailed Information

This section provides more detail on some aspects, and some reference material.

38.5.1 Field content in detail

The Standard Model has the following fields. These describe one *generation* of leptons and quarks, and there are three generations, so there are three copies of each field. By CPT symmetry, there is a set of right-handed fermions with the opposite quantum numbers. The column "**representation**" indicates under which representations of the gauge groups that each field transforms, in the order (SU(3), SU(2), U(1)). Symbols used are common but not universal; superscript C denotes an antiparticle; and for the U(1) group, the value of the weak hypercharge is listed. Note that there are twice as many left-handed lepton field components as left-handed antilepton field components in each generation, but an equal number of left-handed quark and antiquark fields.

38.5.2 Fermion content

This table is based in part on data gathered by the Particle Data Group.[6]

[1] These are not ordinary abelian charges, which can be added together, but are labels of group representations of Lie groups.

[2] Mass is really a coupling between a left-handed fermion and a right-handed fermion. For example, the mass of an electron is really a coupling between a left-handed electron and a right-handed electron, which is the antiparticle of a left-handed positron. Also neutrinos show large mixings in their mass coupling, so it's not accurate to talk about neutrino masses in the flavor basis or to suggest a left-handed electron antineutrino.

[3] The Standard Model assumes that neutrinos are massless. However, several contemporary experiments prove that neutrinos oscillate between their flavour states, which could not happen if all were massless. It is straightforward to extend the model to fit these data but there are many possibilities, so the mass eigenstates are still open. See neutrino mass.

[4] W.-M. Yao *et al.* (Particle Data Group) (2006). "Review of Particle Physics: Neutrino mass, mixing, and flavor change" (PDF). *Journal of Physics G* **33**: 1. arXiv:astro-ph/0601168. Bibcode:2006JPhG...33....1Y. doi:10.1088/0954-3899/33/1/001.

[5] The masses of baryons and hadrons and various cross-sections are the experimentally measured quantities. Since quarks can't be isolated because of QCD confinement, the quantity here is supposed to be the mass of the quark at the renormalization scale of the QCD scale.

38.5.3 Free parameters

Upon writing the most general Lagrangian without neutrinos, one finds that the dynamics depend on 19 parameters, whose numerical values are established by experiment. With neutrinos 7 more parameters are needed, 3 masses and 4 PMNS matrix parameters, for a total of 26 parameters.[7] The neutrino parameter values are still uncertain. The 19 certain parameters are summarized here.

The choice of free parameters is somewhat arbitrary. In the table above, gauge couplings are listed as free parameters, therefore with this choice Weinberg angle is not a free parameter - it is defined as $\tan\theta_W = \frac{g_1}{g_2}$. Likewise, fine structure constant of QED is $\alpha = \frac{1}{4\pi}\frac{(g_1 g_2)^2}{g_1^2+g_2^2}$.

Instead of fermion masses, dimensionless Yukawa couplings can be chosen as free parameters. For example, electron mass depends on the Yukawa coupling of electron to Higgs field, and its value is $m_e = G_e v$.

Instead of the Higgs mass, the Higgs self-coupling strength $\lambda \sim 1/8$ can be chosen as a free parameter.

38.5.4 Additional symmetries of the Standard Model

From the theoretical point of view, the Standard Model exhibits four additional global symmetries, not postulated at the outset of its construction, collectively denoted **accidental symmetries**, which are continuous U(1) global symmetries. The transformations leaving the Lagrangian invariant are:

$$\psi_q(x) \to e^{i\alpha/3}\psi_q$$

$$E_L \to e^{i\beta}E_L \text{ and } (e_R)^c \to e^{i\beta}(e_R)^c$$

$$M_L \to e^{i\beta}M_L \text{ and } (\mu_R)^c \to e^{i\beta}(\mu_R)^c$$

$$T_L \to e^{i\beta}T_L \text{ and } (\tau_R)^c \to e^{i\beta}(\tau_R)^c$$

The first transformation rule is shorthand meaning that all quark fields for all generations must be rotated by an identical phase simultaneously. The fields ML, TL and $(\mu_R)^c$, $(\tau_R)^c$ are the 2nd (muon) and 3rd (tau) generation analogs of EL and $(e_R)^c$ fields.

By Noether's theorem, each symmetry above has an associated conservation law: the conservation of baryon number, electron number, muon number, and tau number. Each quark is assigned a baryon number of $\frac{1}{3}$, while each antiquark is assigned a baryon number of $-\frac{1}{3}$. Conservation of baryon number implies that the number of quarks minus the number of antiquarks is a constant. Within experimental limits, no violation of this conservation law has been found.

Similarly, each electron and its associated neutrino is assigned an electron number of +1, while the anti-electron and the associated anti-neutrino carry a −1 electron number. Similarly, the muons and their neutrinos are assigned a muon number of +1 and the tau leptons are assigned a tau lepton number of +1. The Standard Model predicts that each of these three numbers should be conserved separately in a manner similar to the way baryon number is conserved. These numbers are collectively known as lepton family numbers (LF).

In addition to the accidental (but exact) symmetries described above, the Standard Model exhibits several **approximate symmetries**. These are the "SU(2) custodial symmetry" and the "SU(2) or SU(3) quark flavor symmetry."

38.5.5 The U(1) symmetry

For the leptons, the gauge group can be written $SU(2)_l \times U(1)L \times U(1)R$. The two U(1) factors can be combined into $U(1)Y \times U(1)_l$ where l is the lepton number. Gauging of the lepton number is ruled out by experiment, leaving only the possible gauge group $SU(2)L \times U(1)Y$. A similar argument in the quark sector also gives the same result for the electroweak theory.

38.5.6 The charged and neutral current couplings and Fermi theory

The charged currents $j^{\pm} = j^1 \pm ij^2$ are

$$j_{\mu}^+ = \overline{U}_{iL}\gamma_{\mu}D_{iL} + \overline{\nu}_{iL}\gamma_{\mu}l_{iL}.$$

These charged currents are precisely those that entered the Fermi theory of beta decay. The action contains the charge current piece

$$\mathcal{L}_{CC} = \frac{g}{\sqrt{2}}(j_{\mu}^+ W^{-\mu} + j_{\mu}^- W^{+\mu}).$$

For energy much less than the mass of the W-boson, the effective theory becomes the current–current interaction of the Fermi theory.

However, gauge invariance now requires that the component W^3 of the gauge field also be coupled to a current that lies in the triplet of SU(2). However, this mixes with the U(1), and another current in that sector is needed. These currents must be uncharged in order to conserve charge. So we require the **neutral currents**

$$j_{\mu}^3 = \frac{1}{2}(\overline{U}_{iL}\gamma_{\mu}U_{iL} - \overline{D}_{iL}\gamma_{\mu}D_{iL} + \overline{\nu}_{iL}\gamma_{\mu}\nu_{iL} - \overline{l}_{iL}\gamma_{\mu}l_{iL})$$

$$j_{\mu}^{em} = \frac{2}{3}\overline{U}_i\gamma_{\mu}U_i - \frac{1}{3}\overline{D}_i\gamma_{\mu}D_i - \overline{l}_i\gamma_{\mu}l_i.$$

The neutral current piece in the Lagrangian is then

$$\mathcal{L}_{NC} = ej_{\mu}^{em}A^{\mu} + \frac{g}{\cos\theta_W}(J_{\mu}^3 - \sin^2\theta_W J_{\mu}^{em})Z^{\mu}.$$

38.6 See also

- Overview of Standard Model of particle physics

- Fundamental interaction

- Noncommutative standard model

- Open questions: CP violation, Neutrino masses, Quark matter

- Physics beyond the Standard Model

- Strong interactions: Flavour, Quantum chromodynamics, Quark model

- Weak interactions: Electroweak interaction, Fermi's interaction

- Weinberg angle

- Symmetry in quantum mechanics

38.7   References and external links

[1] In fact, there are mathematical issues regarding quantum field theories still under debate (see e.g. Landau pole), but the predictions extracted from the Standard Model by current methods are all self-consistent. For a further discussion see e.g. R. Mann, chapter 25.

[2] https://fas.org/sgp/othergov/doe/lanl/pubs/00326607.pdf

[3] http://t2k-experiment.org/neutrinos/oscillations-today/

[4] https://fas.org/sgp/othergov/doe/lanl/pubs/00326607.pdf

[5] http://www.mpi-hd.mpg.de/personalhomes/schwetz/tueb-2.pdf

[6] W.-M. Yao *et al.* (Particle Data Group) (2006). "Review of Particle Physics: Quarks" (PDF). *Journal of Physics G* **33**: 1. arXiv:astro-ph/0601168. Bibcode:2006JPhG...33....1Y. doi:10.1088/0954-3899/33/1/001.

[7] Mark Thomson (5 September 2013). *Modern Particle Physics*. Cambridge University Press. pp. 499–500. ISBN 978-1-107-29254-3.

- *An introduction to quantum field theory*, by M.E. Peskin and D.V. Schroeder (HarperCollins, 1995) ISBN 0-201-50397-2.

- *Gauge theory of elementary particle physics*, by T.P. Cheng and L.F. Li (Oxford University Press, 1982) ISBN 0-19-851961-3.

- Standard Model Lagrangian with explicit Higgs terms (T.D. Gutierrez, ca 1999) (PDF, PostScript, and LaTeX version)

- *The quantum theory of fields* (vol 2), by S. Weinberg (Cambridge University Press, 1996) ISBN 0-521-55002-5.

- *Quantum Field Theory in a Nutshell* (Second Edition), by A. Zee (Princeton University Press, 2010) ISBN 978-1-4008-3532-4.

- *An Introduction to Particle Physics and the Standard Model*, by R. Mann (CRC Press, 2010) ISBN 978-1420082982

38.8 Text and image sources, contributors, and licenses

38.8.1 Text

- **Symmetry breaking** *Source:* https://en.wikipedia.org/wiki/Symmetry_breaking?oldid=699333374 *Contributors:* Michael Hardy, Kku, Charles Matthews, Phys, Aleron235, Giftlite, C8to, Thincat, FT2, Xezbeth, Gauge, TheParanoidOne, BD2412, Rjwilmsi, Nihiltres, Debivort, Hairy Dude, Splash, Dan Gluck, Thermochap, Ben MacDui, Parthasarathy.kr, R'n'B, DadaNeem, NigelHarris, Moose-32, Brews ohare, Torage, Stephen Poppitt, Addbot, DOI bot, Dranorter, Download, Ptbotgourou, AnomieBOT, RibotBOT, 🔢🔢, AllCluesKey, Enredanrestos, Citation bot 1, DrilBot, Mary at CERN, RjwilmsiBot, EmausBot, WikitanvirBot, Nyxhadanielle, Shivsagardharam, Qx2020, HMman, Mark viking, Isambard Kingdom and Anonymous: 21

- **Explicit symmetry breaking** *Source:* https://en.wikipedia.org/wiki/Explicit_symmetry_breaking?oldid=690133866 *Contributors:* Charles Matthews, Bevo, Keith Edkins, Lumidek, CALR, Malcolma, RL0919, SmackBot, QFT, Dan Gluck, Mbell, Sturm55, Guy Macon, Choihei, Cuzkatzimhut, Salvar, SieBot, Msrasnw, Dpmuk, Royalmate1, Addbot, Debresser, 🔢🔢, Henry Knight, Maschen, Markiewp, Helpful Pixie Bot, Camelnica, BG19bot, Bugkiller2015 and Anonymous: 15

- **Spontaneous symmetry breaking** *Source:* https://en.wikipedia.org/wiki/Spontaneous_symmetry_breaking?oldid=694158374 *Contributors:* AxelBoldt, Bryan Derksen, XJaM, Edward, Michael Hardy, Lexor, Charles Matthews, Timwi, Reddi, Phys, Bevo, Dusik, Nagelfar, Giftlite, Lethe, Alison, JeffBobFrank, Jcobb, Gotanda, Gadfium, DragonflySixtyseven, Lumidek, FT2, Hidaspal, Ascánder, Mal~enwiki, Bender235, Clement Cherlin, PhilHibbs, MPS, Shenme, Physicistjedi, Kocio, StuTheSheep, Linas, Jmhodges, Dennis Estenson II, Salix alba, Jehochman, BjKa, Chobot, YurikBot, Ugha, Bambaiah, Archelon, Zzuuzz, Reyk, Roques, RupertMillard, SmackBot, Maksim-e~enwiki, Complexica, Colonies Chris, Jmnbatista, Lagrangian, Akriasas, P199, JarahE, Hetar, Dan Gluck, JMK, Harej bot, Ezrakilty, Thijs!bot, Barticus88, Headbomb, Arcresu, Hillarryous, Dougher, Gökhan, JAnDbot, Yuksing, Attarparn, Jpod2, R'n'B, Natsirtguy, Lseixas, BernardZ, Cuzkatzimhut, Holme053, TXiKiBoT, Red Act, Michael H 34, Pamputt, Moose-32, SieBot, Wing gundam, Renatops, Denisarona, Mastertek, BlueDevil, MelonBot, Truthnlove, Addbot, Yakiv Gluck, Zahd, LaaknorBot, SpBot, OlEnglish, Luckas-bot, Yobot, Yotcmdr, Christopher Pritchard, Zimboz Montizawooba, Obersachsebot, False vacuum, Waleswatcher, Gsard, A. di M., 🔢🔢, CES1596, Freddy78, Pmokeefe, RobinK, Mary at CERN, Marie Poise, Slightsmile, Quondum, Shovkovy, Maschen, Boris Breuer, Vatsal19, Helpful Pixie Bot, Bibcode Bot, Ahhaha, Kalmiopsiskid, Fraulein451, Dexbot, Lugia2453, CMTdrew, Mparisi90, LudicrousTripe and Anonymous: 86

- **Age of the universe** *Source:* https://en.wikipedia.org/wiki/Age_of_the_universe?oldid=698591977 *Contributors:* SJK, Roadrunner, Boud, Dominus, Ixfd64, Ciphergoth, Ec5618, Universimmedia, Reddi, Furrykef, David Shay, BenRG, Northgrove, Zandperl, Altenmann, Gandalf61, Rholton, Rursus, Xanzzibar, Mattflaschen, David Gerard, Ancheta Wis, Giftlite, Art Carlson, Timpo, Herbee, Duncharris, Zeimusu, Karol Langner, Bodnotbod, TreyHarris, Canterbury Tail, Flex, Njh@bandsman.co.uk, Discospinster, Rich Farmbrough, KillerChihuahua, Vsmith, ArnoldReinhold, Martpol, Bender235, Mgedmin, Nodem, RJHall, El C, Art LaPella, Causa sui, BrokenSegue, Mtruch, I9Q79oL78KiL0QTFHgyc, Joe Jarvis, Donny11, Domster, Jcrocker, Alansohn, Mr Adequate, Matsw, Plumbago, AzaToth, Scott5114, Benna, Evil Monkey, Gpvos, Ndteegarden, Kazvorpal, Oleg Alexandrov, Simetrical, Woohookitty, FeanorStar7, Ylem, GeorgeTSLC, Pdn~enwiki, GregorB, Macaddct1984, Joke137, GSlicer, Sin-man, Ashmoo, MatthewDBA, Mendaliv, Sjö, Drbogdan, Rjwilmsi, Bill37212, Mike Peel, Yamamoto Ichiro, DVdm, Wavelength, Vuvar1, Jimp, Phantomsteve, RussBot, JabberWok, SpuriousQ, Archelon, Gaius Cornelius, Varnav, Wimt, Anomalocaris, Alcides, Wiki alf, Madcoverboy, Grafen, JPMcGrath, BOT-Superzerocool, Arcman, SamuelRiv, Light current, Emijrp, Deville, Arthur Rubin, JuJube, GraemeL, Johnpseudo, Ilmari Karonen, Luk, SmackBot, Unschool, Mehranwahid, Ashill, Melchoir, Speight, WilyD, Lengis, Canthusus, Onsly, Gilliam, Skizzik, Saros136, Bluebot, MalafayaBot, Scwlong, Petlif, Vanished User 0001, Xyzzyplugh, Jmlk17, Weirdy, Lhf, Evlekis, 🔢🔢🔢, J 1982, JoshuaZ, Gert2, Goodnightmush, RandomCritic, JHunterJ, Hypnosifl, Caiaffa, Autonova, Beefyt, Astrobayes, Shoeofdeath, JForget, Friendly Neighbour, CRGreathouse, Woudloper, MaxEnt, Mundaneman, Ttiotsw, Hispalois, Doug Weller, Fcn, Dchristle, Abtract, Xantharius, Thijs!bot, Cimbalom, Headbomb, Mrjinx, Second Quantization, Tkteun, Dawnseeker2000, AntiVandalBot, Voortle, Rico402, MER-C, Nthep, Ipoellet, TAnthony, Magioladitis, Bongwarrior, VoABot II, Ryanmcg2006, Ling.Nut, Animum, Charliet, Torchiest, WLU, ARC Gritt, Nikpapag, CommonsDelinker, Uncle Dick, Maurice Carbonaro, TomS TDotO, St.daniel, McSly, Kukec, Supuhstar, Mrg3105, NewEnglandYankee, Wesino, Ontarioboy, STBotD, Jamesontai, Ja 62, Wikieditor06, Deor, VolkovBot, Jmrowland, AlnoktaBOT, Jacroe, Lechatjaune, Anonymous Dissident, Inept, JayC, Joho00, Sevenchange, LeaveSleaves, PDFbot, Inseeisyou, Sciencegroupindia, Telecineguy, SheffieldSteel, SwordSmurf, James McBride, SieBot, Tiddly Tom, VVVBot, Sunrise, KryssTal, Nefariousski, Chrisrus, ClueBot, PipepBot, The Thing That Should Not Be, Desoto10, Shark96z, Drmies, VQuakr, Agge1000, ChandlerMapBot, CohesionBot, PixelBot, Estirabot, Sun Creator, Kaiba, Royalmate1, Panos84, 7, Roberto Mura, Editor2020, GKantaris, TimothyRias, Arianewiki1, Maky, AP Shinobi, Sturman4, Gwark, Rreagan007, SilvonenBot, Aunt Entropy, Wyatt915, Addbot, DOI bot, TheNeutroniumAlchemist, TutterMouse, CanadianLinuxUser, LaaknorBot, Quaristice, Bigzteve, Tide rolls, OlEnglish, Legobot, Yobot, Fraggle81, Amirobot, Aldebaran66, AnomieBOT, Neptune5000, Ufim, Mann jess, Mertzf, Citation bot, Nasnema, Quarkde, Gap9551, Dhtml12345, Sae1962, Steve Quinn, DivineAlpha, Citation bot 1, Newt Scamander, DrilBot, Pinethicket, Jonesey95, Tom.Reding, MastiBot, Pristino, IVAN3MAN, Kgrad, Comet Tuttle, MrX, Joshnankivel, Aoidh, Tbhotch, The Utahraptor, RjwilmsiBot, TheArguer, Ripchip Bot, Newty23125, DASHBot, Unstoppable777, EmausBot, Montgolfière, Tommy2010, Superman2410, Solomonfromfinland, Josve05a, JosueM, Arbnos, Russell.harper, Tiiliskivi, Surajt88, ChuispastonBot, Pandeist, Jeremyswest, Mjbmrbot, ClueBot NG, Happa, Tabletrack, Heyheyheyhohoho, O.Koslowski, Ryan Vesey, Helpful Pixie Bot, Ryan Zierke, Bibcode Bot, Quarkgluonsoup, Metricopolus, Musicalcrossbow, Bialy Goethe, Cadiomals, Drift chambers, Firstname1826, Sai.kiran0901, Dipsy7777, Craigc29, YFdyh-bot, Garamond Lethe, Trinitresque, Wjs64, Wwfls1, Tony Mach, Jamesx12345, Christiandad, Spencer.mccormick, John.Toth.uk..essex, The Sackinator, PirtleShell, Factschecking, Top420stud ent, JinQing123, Concord hioz, Monkbot, Chaguin1986, Neeraj Bhakta, Spideratseds, Mexicanprincetonguy, Tetra quark, Isambard Kingdom, Youknowwhatimsayin and Anonymous: 331

- **Chronology of the universe** *Source:* https://en.wikipedia.org/wiki/Chronology_of_the_universe?oldid=699277673 *Contributors:* AxelBoldt, Bryan Derksen, The Anome, Roadrunner, Heron, Modemac, Edward, Patrick, Boud, PhilipMW, Earth, Looxix~enwiki, Mkweise, Ahoerstemeier, Jebba, Darkwind, EdH, Timwi, A1r, Reddi, Doradus, Zoicon5, Maximus Rex, Topbanana, Lord Emsworth, TravelingDude, Chrisjj, Flockmeal, BenRG, Robbot, Astronautics~enwiki, Jotomicron, Gandalf61, Raeky, Tobias Bergemann, Ancheta Wis, Giftlite, Lethe, Bkonrad, BrendanRyan, Eequor, Edcolins, OldakQuill, Coldacid, Opera hat, Piotrus, CSTAR, Tothebarricades.tk, Icairns, JDoolin, Lumidek, Darksun, Njh@bandsman.co.uk, D6, FT2, Pjacobi, ArnoldReinhold, Dave souza, Dbachmann, El C, Art LaPella, Drhex, CDN99, Bobo192,

Army1987, Evolauxia, Maurreen, I9Q79oL78KiL0QTFHgyc, La goutte de pluie, Holdek, Pearle, Demi, *Kat*, RJFJR, Geraldshields11, Vuo, Gene Nygaard, Mindmatrix, FeanorStar7, Jeff3000, Duncan.france, Colin Watson, Joke137, Alan Canon, Christopher Thomas, RichardWeiss, Surnólë, Zalasur, Sjö, Drbogdan, Rjwilmsi, Zbxgscqf, Captain Disdain, Mike Peel, SeanMack, Krash, Yamamoto Ichiro, Nihiltres, Harmil, Who, SouthernNights, Pathoschild, Mathrick, Masnevets, Wjfox2005, YurikBot, Spacepotato, Mushin, Vuvar1, Jimp, TheDoober, Ericorbit, Stephenb, NawlinWiki, Siddiqui, Chrisbrl88, Raven4x4x, Pudist, Dbfirs, Bota47, Superiority, Leptictidium, Lenis, Enormousdude, Aremisasling, Arthur Rubin, Tevildo, GraemeL, Peyna, CWenger, ScoutNZ, Kevin, Geoffrey.landis, Albert ip, RG2, Hirudo, Paul Erik, GrinBot~enwiki, Serendipodous, KnightRider~enwiki, SmackBot, RDBury, MrDemeanour, Mehranwahid, Ashill, WildElf, AnOddName, David G Brault, Gilliam, BirdValiant, Saros136, JRSP, Bluebot, Stevenwagner, MrDrBob, Mergatroidal, Colonies Chris, Scwlong, V1adis1av, Hve, Chlewbot, Andy120290, Jgoulden, Huon, Emre D., John D. Croft, Pwjb, Yevgeny Kats, CIS, SashatoBot, Lambiam, Swatjester, Writtenonsand, Count Caspian, Ckatz, Ex nihil, JHunterJ, Stwalkerster, SQGibbon, Hypnosifl, JeffW, Vanished user, Pegasus1138, Chetvorno, Fairyhairycarpetfluff, CWY2190, MaxEnt, Ramitmahajan, LouisBB, Gogo Dodo, Huysman, Michael C Price, Abtract, Thijs!bot, Fournax, Quarkchord, Sopranosmob781, Headbomb, Najro, Marek69, James086, Second Quantization, Peter Gulutzan, AntiVandalBot, Obiwankenobi, Prolog, DarkAudit, Jdfekete, Shlomi Hillel, JAnDbot, MER-C, Xeno, Rothorpe, Ophion, Dward Fardhard, RogierBrussee, VoABot II, Nyq, Coredumperror, Catgut, Cgingold, Tenacious221, DinoBot, MartinBot, Anaxial, CommonsDelinker, Leyo, J.delanoy, Love Krittaya, ARTE, Clemm, Madhava 1947, STBotD, Fuenfundachtzig, GrahamHardy, Idioma-bot, Sheliak, Black Kite, VolkovBot, TreasuryTag, Jmrowland, Cosmic Latte, Andrius.v, Clarince63, Seraphim, UnitedStatesian, Madhero88, Hanjabba, Richwil, James McBride, Hibou21, MCTales, Alcmaeonid, NHRHS2010, Thw1309, Weezymagic, Llehctimj, Momo san, Oxymoron83, Globaleducator, Steven Crossin, Sean.hoyland, Namiluko, Denisarona, Escape Orbit, Martarius, ClueBot, Snigbrook, Taffboyz, Gawaxay, Wanderer57, Wysprgr2005, Cp111, Nondescripts, TheOldJacobite, CounterVandalismBot, Thomas Kist, ChandlerMapBot, Rprpr, DragonBot, Excirial, Gnome de plume, CrazyChemGuy, Eeekster, Scog, Soccerguy7735, Panos84, Subash.chandran007, DumZiBoT, XLinkBot, Tarlneustaedter, Scottman162, Jmacwiki, Aunt Entropy, CosmologyProfessor, Maldek, Addbot, Some jerk on the Internet, Medich1985, Spiderbo13, Ersik, E.pajer, Markkujn, 84user, F Notebook, Tide rolls, ⁇⁇, Whatismetric, Yinweichen, Yobot, TaBOT-zerem, THEN WHO WAS PHONE?, Robert Treat, AnomieBOT, Materialscientist, Citation bot, LilHelpa, Obersachsebot, HenryCorp, Wikikone, ProtectionTaggingBot, Omnipaedista, The Wiki Octopus, Smallman12q, Paine Ellsworth, Kikuyu3, Sae1962, Oashi, Citation bot 1, Dogaru Florin, Pinethicket, Tom.Reding, Naturehead, Seryo93, Saayiit, Extra999, Canuck100, JLincoln, Orca54, DeltaAce, Michaelbernal, Carlborden, DexDor, Tesseract2, EmausBot, GoingBatty, Rowan casey, Solomonfromfinland, Hhhippo, John Cline, Cogiati, Susfele, Medeis, Thedropsoffire, Mikeyfresh1556, Makecat, Wagino 20100516, Kirothereaper, RockMagnetist, Pandeist, AUN4, ClueBot NG, Astrocog, Jack Greenmaven, Gilderien, Piast93, Bped1985, James childs, Snotbot, Frietjes, Tinyplanetmusic, Cpluhar, Finchy0, Apopp8333, Bibcode Bot, Lowercase sigmabot, Rarelight, Cadiomals, Harizotoh9, Andi2011, Duxwing, Amphibio, ShizlGzngar, Justincheng12345-bot, Soulbust, Avneref, Wjs64, Mixert, 069952497a, SomeFreakOnTheInternet, AlbertoEspancho, Ivander-Clarent, Jwratner1, Kogge, SpeedEvil, Concord hioz, Monkbot, Sofia Koutsouveli, Wikipedian 2, Officer Sid X, Julietdeltalima, OmoiEgaite, Tetra quark, Xytreyum, Isambard Kingdom, Therewasnoothernameguys1, KSFT, Esadri21, Chemistry1111 and Anonymous: 351

- **Cosmic Calendar** *Source:* https://en.wikipedia.org/wiki/Cosmic_Calendar?oldid=699541982 *Contributors:* Skysmith, Denni, Btljs, Beland, Discospinster, RJHall, Alansohn, Woohookitty, Tripodics, Chiphead, MikailBorg, Drbogdan, Rjwilmsi, Vegaswikian, DVdm, Taurrandir, Ericorbit, Yamara, Efbrazil, Jecowa, Cmglee, SmackBot, Brian Patrie, Hmains, Paul H., Mgiganteus1, JohnSmart, Jonathan Tweet, Carpentc, Peter Gulutzan, AgentPeppermint, Dr. Submillimeter, CaptainIlama, Acroterion, Edward321, Ivazir, Antonio Lopez, Denisarona, Rawilson52, Niceguyedc, LupusCanidae, XLinkBot, Kasper2006, Addbot, Micromaster, CanadianLinuxUser, Zuidpoort, Tide rolls, DaveChild, Systemizer, Fraggle81, South Bay, Tom87020, Secdio, Aashaa, Stvltvs, Ilovemichaelcrichton, Codon3, Tom.Reding, Turkeymaster77, Jsha1, Pbrower2a, Onel5969, Ripchip Bot, DiogenesTCP, EmausBot, WikitanvirBot, ZéroBot, Yiosie2356, Mentibot, ClueBot NG, Mikewhite.bromham, Widr, ChrisGualtieri, 331dot, Lemaroto, Mohamed-Ahmed-FG, Mr. Online Wiki, Concord hioz, Alvandria, Wikipedian 2, Tetra quark, Kevt2002, AHercog, BHProf, Bomby7777 and Anonymous: 51

- **Planck epoch** *Source:* https://en.wikipedia.org/wiki/Planck_epoch?oldid=698825471 *Contributors:* XJaM, Patrick, Schneelocke, Doradus, Dragons flight, BenRG, Gandalf61, Discospinster, Hidaspal, ArnoldReinhold, Ascánder, Closeapple, Army1987, I9Q79oL78KiL0QTFHgyc, Alansohn, H2g2bob, Kazvorpal, Fred J, Alienus, Tizio, Rkeene0517, Jinma, YurikBot, Jimp, Yamara, Dbfirs, JonathanD, SigmaEpsilon, SG-Dentarthurdent, KnightRider~enwiki, SmackBot, DHN-bot~enwiki, Rcbutcher, Scwlong, Tsca.bot, Hgilbert, Yevgeny Kats, JHunterJ, Rdanhof, Michael C Price, Thijs!bot, Epbr123, Peter Gulutzan, Seaphoto, Tlabshier, Darklilac, Bongwarrior, JefeMixtli, TXiKiBoT, Joopercoopers, EvanCarroll, UnitedStatesian, Eubulides, SieBot, Henry Delforn (old), J8079s, Sammycef, NuclearWarfare, BOTarate, Pgallert, Chanakal, Addbot, Physicman123, West.andrew.g, Jarble, Luckas-bot, AnomieBOT, Bookaneer, Bluee Mountain, Abhall98, SassoBot, Salbers, DrArthurRubinPHD, Citation bot 1, Dogaru Florin, Bird in flight, Heurisko, Slawekb, ZéroBot, John Cline, ClueBot NG, Helpful Pixie Bot, Makecat-bot, Jwratner1, Tetra quark, Isambard Kingdom, SocraticOath, Arielweil, Sweepy and Anonymous: 44

- **Supersymmetry breaking** *Source:* https://en.wikipedia.org/wiki/Supersymmetry_breaking?oldid=660670697 *Contributors:* Phys, Lumidek, Sam Korn, Bluemoose, Conscious, Dialectric, SmackBot, Blehfu, Addbot, FrescoBot, Jwratner1, MarkovianStumble and Anonymous: 5

- **Spacetime** *Source:* https://en.wikipedia.org/wiki/Spacetime?oldid=699964313 *Contributors:* Paul Drye, The Cunctator, Dreamyshade, Bryan Derksen, Malcolm Farmer, Josh Grosse, XJaM, Karl Palmen, Stevertigo, Patrick, Infrogmation, Smelialichu, Michael Hardy, Wshun, Pit~enwiki, Dcljr, Karada, Mcarling, Looxix~enwiki, William M. Connolley, Snoyes, Kingturtle, Glenn, Loren Rosen, HolIgor, Adam Bishop, Dcoetzee, Reddi, Jay, E23~enwiki, Omegatron, Fvw, Robbot, Kristof vt, Goethean, Ashley Y, Sverdrup, Blainster, DHN, Papadopc, Tobias Bergemann, Finlander, Matt Gies, Giftlite, ByteCoder, Wolfkeeper, Herbee, Tom Radulovich, Everyking, Snowdog, Michael Devore, Niteowlneils, Yekrats, Eequor, Utcursch, Beland, Karol Langner, Wikimol, JimWae, Karl-Henner, Adashiel, ELApro, Chris Howard, Juan Ponderas, Discospinster, Rich Farmbrough, Cacycle, Ascánder, Dolda2000, Bender235, Ben Standeven, El C, Rgdboer, Lankiveil, Shoujun, Teorth, Che090572, Rbj, Tobacman, I9Q79oL78KiL0QTFHgyc, Como, Obradovic Goran, Free Bear, Keenan Pepper, Sourcer66~enwiki, Riana, Geoff-codes, Rey-Brujo, Arag0rn, DonQuixote, Eddie Dealtry, DominicC13, H2g2bob, Loxley~enwiki, Camw, StradivariusTV, TheNightFly, Pkeck, ^demon, Doran, Jeff3000, Mpatel, GregorB, Palica, Graham87, Deltabeignet, Li-sung, Mkn1234, MekaD, Rjwilmsi, KYPark, Kinu, Vary, MarSch, FayssalF, Lebha, Mathbot, Alexjohnc3, Jrtayloriv, Exelban, Pete.Hurd, Tardis, Chobot, Tene, DVdm, VolatileChemical, YurikBot, Wavelength, Splintercellguy, Wolfmankurd, CanadianCaesar, Yamara, NawlinWiki, Mipadi, Trovatore, Schlafly, JocK, Crasshopper, Tony1, T, Zythe, Gadget850, Sahands, Light current, Zzuuzz, StuRat, KGasso, JoanneB, Heathhunnicutt, Anclation~enwiki, RG2, Teply, Mejor Los Indios, Qero, Eigenlambda, Sardanaphalus, SmackBot, RDBury, Formativ, Maksim-e~enwiki, Forteller~enwiki, RaulMiller, Ashill, Kurochka, Lestrade, InverseHypercube, KnowledgeOfSelf, C.Fred, AndreasJS, Jaytan, Alex earlier account, JeffieAlex, Yamaguchi⁇⁇, Gilliam, Nick-Garvey, JMiall, Oli Filth, TheScurvyEye, Silly rabbit, Complexica, Dabigkid, Jerome Charles Potts, Nbarth, Sbharris, Bryan Truitt, Can't

Oleg Alexandrov, Kelly Martin, Linas, BoLingua, Duncan.france, Christopher Thomas, BD2412, Rjwilmsi, Koavf, Strait, R.e.b., Jehochman, FlaBot, Goudzovski, Markdroberts, Gareth E. Kegg, Chobot, Algebraist, YurikBot, Bambaiah, Wester, Darsie, AVM, Bhny, Stephenb, Długosz, Dna-webmaster, Tetracube, Caco de vidro, Finell, Triple333, SmackBot, Maksim-e~enwiki, ZerodEgo, Chris the speller, Sbharris, Jmnbatista, Lambiam, JorisvS, Ckatz, Meco, Newone, Benabik, MarsRover, Myasuda, Xxanthippe, Michael C Price, Quibik, Ldussan, Difty, Thijs!bot, Epbr123, Headbomb, West Brom 4ever, Mattfiller, D.H, RogierBrussee, VoABot II, Bakken, Jpod2, RickyCayley, JohnWilliams, Hekerui, Rif Winfield, MartinBot, Haydarhan, Gillleke, Cuzkatzimhut, VolkovBot, Off-shell, LokiClock, TXiKiBoT, Calwiki, Moose-32, Ptrslv72, Coffee, Gerakibot, Likebox, JacquesPHI, Henry Delforn (old), Pac72, Mr. Stradivarius, LoserJoke, ClueBot, General Epitaph, Wwheaton, Drmies, Auntof6, Brews ohare, M.O.X, Crowsnest, Nettings, XLinkBot, Scvblwxq, Addbot, Eric Drexler, SpBot, Bob K31416, Barak Sh, Tide rolls, Yoavd, Luckas-bot, Yobot, Ptbotgourou, Fraggle81, Galaxydraem, AnomieBOT, Ciphers, Rubinbot, ArthurBot, LilHelpa, Xqbot, TheAMmollusc, Capricorn42, DSisyphBot, RibotBOT, Waleswatcher, Davdde, Benzen, FrescoBot, BenzolBot, XeBot, Citation bot 1, Benji1986, O.anatinus, RedBot, MastiBot, Aknochel, Beth Ann Lindstrom, Felix0411, Meier99, Mary at CERN, WildBot, EmausBot, Wikitanvirbot, Japs 88, LHC Tommy, Slawekb, JSquish, Quondum, L Kensington, Cerlbar, Zueignung, BabbaQ, CBuiltother, PhysicsAboveAll, Giuseppe Vitiello, Jj1236, Parthdu, Curb Chain, Bibcode Bot, Tirebiter78, Ownedroad9, ChrisGualtieri, Dexbot, Abits52, Konbini, Curious-Mind01, Ajsal.ea, Itchmean, Cjean42, Crigeos, Crbeals, Jwratner1, Atotalstranger, Jzampardi, KasparBot and Anonymous: 146

- **QCD vacuum** *Source:* https://en.wikipedia.org/wiki/QCD_vacuum?oldid=685880330 *Contributors:* William Avery, Michael Hardy, SebastianHelm, Charles Matthews, Phys, Wmahan, CALR, Pak21, David Schaich, MuDavid, Pavel Vozenilek, Smalljim, Cmdrjameson, Fwb22, Guy Harris, Andrew Gray, Axl, JeeAge, Ceyockey, Linas, Rjwilmsi, Mushin, Bambaiah, Kirill Lokshin, Salsb, SmackBot, QFT, TriTertButoxy, JarahE, CmdrObot, Usgnus, Myasuda, Difluoroethene, Headbomb, OrenBochman, QuantumEngineer, R'n'B, Cuzkatzimhut, Gonamar, Neparis, Likebox, Paolo.dL, Xidong, ClueBot, CristianCantoro, Sun Creator, Brews ohare, SchreiberBike, Addbot, DOI bot, Mjamja, Tassedethe, Lightbot, AnomieBOT, Omnipaedista, Citation bot 1, Tom.Reding, Thinking of England, Trappist the monk, Puzl bustr, ZéroBot, Josve05a, Maschen, QuantumSquirrel, Hindustanilanguage, Bibcode Bot, Trompedo, Katherine Pendleton, Stamptrader and Anonymous: 25

- **Goldstone boson** *Source:* https://en.wikipedia.org/wiki/Goldstone_boson?oldid=699352787 *Contributors:* Bryan Derksen, Michael Hardy, Hike395, Charles Matthews, Andrewman327, Phys, BenRG, Jni, Wile E. Heresiarch, MilkMiruku, Centrx, Giftlite, JeffBobFrank, Pharotic, Icairns, Brianhe, Pt, MPS, Jag123, Delius, Lysdexia, Linas, Mandarax, Qwertyus, Kbdank71, Rjwilmsi, Zbxgscqf, Chobot, SCZenz, Chaiken, Tom Lougheed, Yevgeny Kats, Amtiss, JorisvS, WhiteHatLurker, JarahE, Stephen B Streater, Drfizz, WISo, Michael C Price, Oreo Priest, Uisqebaugh, Jpod2, Maliz, R'n'B, Nono64, HEL, Cuzkatzimhut, Antixt, Moose-32, Tiddly Tom, Likebox, AnonyScientist, MystBot, Addbot, Zahd, Yobot, AdamSiska, AnomieBOT, Citation bot, Xqbot, DSisyphBot, Omnipaedista, Gsard, Locobot, JanderClamber, Citation bot 1, Merongb10, Gil987, Mary at CERN, JSquish, Suslindisambiguator, Wabbott9, Bibcode Bot, ChrisGualtieri, EuroCarGT, Monkbot, Ibnbaja and Anonymous: 23

- **1964 PRL symmetry breaking papers** *Source:* https://en.wikipedia.org/wiki/1964_PRL_symmetry_breaking_papers?oldid=698374715 *Contributors:* Nealmcb, AnonMoos, Auric, Giftlite, Kaldari, Rich Farmbrough, FT2, David Schaich, Count Iblis, GregorB, Rjwilmsi, Tim!, Teply, Jmnbatista, CmdrObot, Cydebot, KnightMove, Headbomb, Nick Number, SamIAmNot, Vanished user ty12kl89jq10, Guillaume2303, Duncan.Hull, Lamro, Falcon8765, Moose-32, Ptrslv72, Paradoctor, Randy Kryn, NuclearWarfare, XLinkBot, Good Olfactory, OlEnglish, Yobot, Anypodetos, AnomieBOT, Citation bot, Omnipaedista, Fortdj33, Meier99, Mary at CERN, Magmalex, RjwilmsiBot, DASHBot, GoingBatty, KHamsun, LHC Tommy, Brendanpbehan, Helpful Pixie Bot, Bibcode Bot, EuroAgurbash, Dexbot, Hmainsbot1, SFK2, Stamptrader, Karthikprabhu22 and Anonymous: 25

- **Symmetry (physics)** *Source:* https://en.wikipedia.org/wiki/Symmetry_(physics)?oldid=696484335 *Contributors:* The Anome, Heron, Stevertigo, Patrick, Michael Hardy, Bloodshedder, Rorro, Giftlite, BenFrantzDale, Netoholic, NetBot, I9Q79oL78KiL0QTFHgyc, Physicistjedi, Danski14, Mattpickman, Reaverdrop, Oleg Alexandrov, Woohookitty, StradivariusTV, Pol098, Commander Keane, Mpatel, BD2412, Eyu100, Mathbot, Nihiltres, Jrtayloriv, X42bn6, Bhny, Archelon, Paul D. Anderson, Sbyrnes321, SmackBot, Incnis Mrsi, Complexica, Mets501, Quodfui, JRSpriggs, Myasuda, AndrewHowse, Cydebot, Michael C Price, Christian75, Divey, YK Times, PhilKnight, Email4mobile, Homunq, CodeCat, Janus Shadowsong, YohanN7, Paradoctor, Fratrep, ClueBot, Ottre, Bob108, Brews ohare, SchreiberBike, Thamuzino, Rror, Bradv, Addbot, Fyrael, Stylus881, PV=nRT, Luckas-bot, Yobot, Hotbody, Manganite, Rubinbot, Point-set topologist, Locobot, A. di M., FrescoBot, HRoestBot, MastiBot, 8af4bf06611c, ZéroBot, Maschen, Shivsagardharam, Vkpd11, Muskid, Dexbot, Hemlisp, Augustus Leonhardus Cartesius, Rabbitflyer, KHEname, BioticPixels, Matli, Ryanexler and Anonymous: 40

- **Lie group** *Source:* https://en.wikipedia.org/wiki/Lie_group?oldid=696102994 *Contributors:* AxelBoldt, Zundark, Jos h Grosse, XJaM, Miguel~enwiki,Stevertigo, Xavic69, Michae l Hardy, TakuyaMurata, GTBacchus, Looxix~enwiki, Barak~enwiki, Charles Matthews, Dysprosia, Jitse Niesen,Zoicon5, Davi d Shay, Itai, Phys, Jos h Cherry, Saaska, Tobias Bergemann, Weialawaga~enwiki, Tosha, Giftlite, JamesMLane, BenFrantzDale,Lethe, Fropuff, Wgmccallum, Jason Quinn, Bobblewik, DefLog~enwiki, Lockeownzj00, Beland , Pmanderson, Abdull, Dablaze, MuDavid,Paul August, ChrisJ, Bender235, Tompw, Rgdboer, Kwamikagami, Shanes, Cherlin, Msh 210, PAR, Ale x Varghese, Oleg Alexandrov, Zn-trip, Joriki, Linas, Dzordzm, Isnow, SDC, AnmaFinotera, Frankie1969, Graham 87, Porcher, Rjwilmsi, NatusRoma, MarSch, Salix alba,HappyCamper, R.e.b., VKokielov, BMF81, Masnevets, Chobot, Algebraist, Wavelength, Hillman, RussBot, Michae l Slone, KSmrq, Arch-elon, Buster79, Arkapravo, Smaines, Orthografer, Ekeb, Kier07, Pred, RodVance, JDspeeder1, SmackBot, Incnis Mrsi, Tom Lougheed,FlashSheridan, Davewild, Mhss, Kmarinas86, Bluebot, Badger014, Sill y rabbit, DHN-bot~enwiki, Bears16, Akriasas, KeithB, Lambiam,Ninte, Siva1979, John, Ulner, Jim.belk, Michae l Kinyon, Inquisitus, Mathchem271828, Rschwieb, Krasnoludek, Yggdrasil014, CRGreathouse,CBM, Logical2u, Myasuda, Kupirijo, MotherFunctor, The real dan, Dr.enh, Xantharius, Thijs!bot, Headbomb, JustAGal, RichardVeryard,RobHar, Salgueiro~enwiki, Dougher, Len Raymond, JAnDbot, De flective, Unifey~enwiki, Homeworlds, Magioladitis, Bongwarrior, Cmelby,WhatamIdoing, Sullivan.t.j, Davi d Eppstein, The Real Marauder, Benjamin.friedrich, Davi d J Wilson, Jesper Carlstrom, Maproom, To-myDuby, Rocket71048576, Pidara, Fylwind, Dorftrottel, Lseixas, Borat fan, Cuzkatzimhut, Trevorgoodchild, JohnBlackburne, Ndbrian1, James.r.a.gray, Hesam7, Geometry guy, Jmath666, Eubulides, Brian Huff man, GenuineOlegend, Drorata, Arcfrk, Smylei, Oscarbaltazar,YohanN7, JackSchmidt, S2000magician, Beastinwith, Mr. Stradivarius, Deciwill, Sidiropo, Leontios, Heckledpie, Cacadril, SchreiberBike,Marc van Leeuwen, MystBot, Addbot, Topology Expert, LaaknorBot, Ozob, Tanath, Tide rolls, Luckas-bot, Yobot, Ht686rg90, Niout,Amirobot, AnomieBOT, Citation bot, ArthurBot, Br77rino, Kaoru Itou, FrescoBot, Anterior1, Sławomir Biały, RedBot, Tinfoilcat, Emaus-Bot, KbReZiE 12, Darkfi ght, Slawekb, Suslindisambiguator, Maschen, Zueignung, Anita 5192, ClueBot NG, Mgvongoeden, Kasirbot, HelpfulPixi e Bot, Daviddwd, BG19bot, CitationCleanerBot, Fraisière, NotWith, MathKnight-at-TAU, Suhagja, Brirush, CsDix, Sol1, Blackbombchu,Pwm86, Mathphysman, Abitslow, Cbartondock, Victoryhuy, KasparBot, Egdunne, Referencing, Chemistry1111 and Anonymous: 111

- **Discrete symmetry** *Source:* https://en.wikipedia.org/wiki/Discrete_symmetry?oldid=678186111 *Contributors:* Edward, Michael Hardy, Phys, Lumidek, Jag123, Gary, SmackBot, David Eppstein, Addbot, Jeff Muscato, Dc987, Yellow octopus, Reak spoughly, Azzifeldman and Anonymous: 2

- **T-symmetry** *Source:* https://en.wikipedia.org/wiki/T-symmetry?oldid=687699177 *Contributors:* Zundark, Roadrunner, Stevertigo, Patrick, JohnOwens, Michael Hardy, Tim Starling, Albertplanck, Chinju, Karada, SebastianHelm, Stevenj, Angela, Ehn, Phys, Mp~enwiki, SJRubenstein, Giftlite, Lee J Haywood, Pcarbonn, Carandol~enwiki, AmarChandra, Hidaspal, Pjacobi, Martpol, ReallyNiceGuy, Dataphile, Roy-Boy, I9Q79oL78KiL0QTFHgyc, Cherlin, Bucephalus, Jheald, Reaverdrop, Mpatel, Marudubshinki, Strait, Lionelbrits, Moskvax, Wavelength, Mushin, Bambaiah, Yamara, Ihope127, SCZenz, 2over0, StuRat, Reyk, Tim R, SmackBot, QFT, Daqu, Richard L. Peterson, Zarniwoot, Hypnosifl, Mets501, Dan Gluck, JRSpriggs, CmdrObot, Galo1969X, Thijs!bot, Mbell, Headbomb, Pjvpjv, Infophile, Timeron~enwiki, Laksman~enwiki, JayJung, Magioladitis, Thasaidon, Grimlock, Bjheiden, Slash, Maurice Carbonaro, VolkovBot, Red Act, Someguy1221, Eubulides, SieBot, Likebox, Mr. Stradivarius, PhysicsGrad2013, Dave.bradi, EverettYou, Agor153, Forbes72, Addbot, LaaknorBot, Legobot, Luckas-bot, Yobot, Manganite, Citation bot, Companicus, Xqbot, Minibikini, Rurigok, FrescoBot, Tom.Reding, Dude1818, Al8217, Justpoppingintosayhi, EmausBot, Ebrambot, Pandeist, Bibcode Bot, Shuikouhw, Adamb76, Mark viking, Kfitzell29, Wish vishal and Anonymous: 45

- **Parity (physics)** *Source:* https://en.wikipedia.org/wiki/Parity_(physics)?oldid=691540112 *Contributors:* Patrick, TakuyaMurata, Charles Matthews,Phys, SoLando, Tobias Bergemann, Giftlite, Xerxes314, Beland, Karol Langner, Lumidek, CALR, Pak21, Nvj, Cmdrjameson, Eruantalon,Sergi o Macías, Wtmitchell, Knowledge Seeker, Count Iblis, Oleg Alexandrov, Joriki, Marudubshinki, Ae 77, Nihiltres, Thecurran, Wave-length, Bambaiah, Archelon, Pseudomonas, Kabirramola, E2mb0t~enwiki, Elkman, GrinBot~enwiki, SmackBot, Incnis Mrsi, Tom Lougheed,Leifisme, QFT, Wiki me, Akriasas, WhiteHatLurker, Erwin, JarahE, JRSpriggs, Raghunathan, Usgnus, Cydebot, Michae l C Price, Thijs!bot,Barticus88, Mbell, Headbomb, Pjvpjv, Dougher, Magioladitis, Thasaidon, Dirac66, HEL, Tarotcards, Idioma-bot, Gerri t C. Groenenboom,Cuzkatzimhut, Red Act, Pamputt, Antixt, SieBot, BotMultichill, Wing gundam, Paolo.dL, Anchor Link Bot, ClueBot, Sun Creator, DumZiBoT,Lazyrussian, Rror, TravisAF, Addbot, Luckas-bot, Yobot, Tonyrex, PianoDan, Citation bot, ArthurBot, Omnipaedista, Sahehco, Theaucitron,Sławomir Biały, Crai g Pemberton, Merongb10, RedBot, TobeBot, Heurisko, Linguisticgeek, Queller69, RjwilmsiBot, EmausBot, Albear-And,ZéroBot, Quondum, Kmva, ClueBot NG, Greedohun, Tamila Shalumova, Helpful Pixi e Bot, Bibcode Bot, Vkpd11, Slumdog2011, Goodbear3,MuonRay, Abitslow, JellyPatotie, Are you freaking kidding me and Anonymous: 64

- **Glide reflection** *Source:* https://en.wikipedia.org/wiki/Glide_reflection?oldid=679497811 *Contributors:* Patrick, Michae l Hardy, Glenn, Charles Matthews , Henrygb , Tosha , Tomruen , Wgw 4, Alexb @cut -the -knot .com , Aholtman , Mathbot , CiaPan , Roboto de Ajvol , SmackBot , Smith 609,Ezrakilty , Cydebot , Thijs!bot, .anacondabot , Davi d Eppstein , Katalaveno , Aboluay , ClueBot , Addbot , Ronhjones , Lesze k Jańczuk , Luckas -bot,AnomieBOT , MathsPoetry , N4m3, Akerans , Solomon 7968 , Trevayne 08, Brad 7777 , Kelvinsong , Dexbot and Anonymous: 7

- **C-symmetry** *Source:* https://en.wikipedia.org/wiki/C-symmetry?oldid=548565045 *Contributors:* Tim Starling, Albertplanck, SebastianHelm, Charles Matthews, The Anomebot, Phys, Jmabel, Xerxes314, Hidaspal, Pjacobi, Bender235, Linas, Mike Peel, Mathbot, YurikBot, Mushin, RussBot, Huatulco~enwiki, Spike Wilbury, SmackBot, Tom Lougheed, Gyrobo, QFT, Erwin, Alan.ca, Thijs!bot, Headbomb, Davidhorman, Alphachimpbot, Magioladitis, Thasaidon, Grimlock, Maliz, A4bot, Venny85, AlleborgoBot, WikHead, Addbot, Mpfiz, Zorrobot, Legobot, Luckas-bot, Citation bot, Maxis ftw, Xqbot, Trebauchet1986, Thinking of England, RobinK, ZéroBot, Ego White Tray, QuantumSquirrel, Steve Bz, Gabobaby and Anonymous: 14

- **Symmetry in quantum mechanics** *Source:* https://en.wikipedia.org/wiki/Symmetry_in_quantum_mechanics?oldid=694968437 *Contributors:* Bearcat, BD2412, Wavelength, Colonies Chris, Cesium 133, Headbomb, Camrn86, YohanN7, AnomieBOT, FrescoBot, Lonaowna, Quondum, Maschen, Stefan Neumeier, AdventurousSquirrel, Khazar2, Dexbot, Mark viking, Mathphysman, Sumeruhazra, Henderson Duff and Anonymous: 3

- **Noether's theorem** *Source:* https://en.wikipedia.org/wiki/Noether'{}s_theorem?oldid=693352274 *Contributors:* Zundark, The Anome, Tarquin, Heron, Stevertigo, Michael Hardy, Albertplanck, Looxix~enwiki, Stevan White, AugPi, HolIgor, Charles Matthews, Dysprosia, Jitse Niesen, Phys, Tobias Bergemann, David Gerard, Weialawaga~enwiki, Ancheta Wis, Giftlite, Sj, Harp, Tseller, Lupin, Dratman, Jcobb, DefLog~enwiki, HorsePunchKid, Karol Langner, AmarChandra, Lumidek, Zowie, CALR, Nathan Penton, Luqui, Dbachmann, MuDavid, Bender235, Jnestorius, Ntmatter, Teorth, .:Ajvol:., Brim, Rmz, Babajobu, Osmodiar, 图图图, Kusma, Oleg Alexandrov, Linas, David Haslam, Ae-a, Pfalstad, Driftwoodzebulin, Marudubshinki, BD2412, Rjwilmsi, MarSch, R.e.b., Ems57fcva, Rangek, Gseryakov, Alfred Centauri, Chobot, DVdm, WriterHound, Nahmad~enwiki, RussBot, Archelon, PaulGarner, Tong~enwiki, Długosz, Jpowell, Light current, Enormousdude, 2over0, Reyk, CharlesHBennett, Fourohfour, Paul D. Anderson, SmackBot, BeteNoir, Tom Lougheed, InverseHypercube, Melchoir, JC-Santos, RDBrown, Silly rabbit, Complexica, Modest Genius, Chlewbot, Berland, QFT, Jwy, Yevgeny Kats, Lambiam, Kuru, Atoll, Pathosbot, JRSpriggs, Chetvorno, ZICO, CBM, Vyznev Xnebara, Ksoileau, Myasuda, Cydebot, WillowW, Michael C Price, Thijs!bot, Headbomb, Oreo Priest, JAnDbot, .anacondabot, Magioladitis, David Eppstein, Church of emacs, Andejons, Quantling, Lisagosselin, Missphysics, Cuzkatzimhut, AlnoktaBOT, Barbacana, Red Act, Imadeitmyself, Arcfrk, Veshapa, SteakNShake, Thehotelambush, Lisatwo, StewartMH, ClueBot, Mallodi, Twolves14, Jjauregui, Brews ohare, Bbbeard, 1ForTheMoney, AnonyScientist, Pointillist blur, Ziofil~enwiki, Addbot, Substar, Download, Favonian, Deamon138, Luckas-bot, TaBOT-zerem, Niout, Helena srilowa, Materialscientist, Citation bot, Qiushi, Xqbot, Omnipaedista, RibotBOT, Gsard, GliderMaven, FrescoBot, Sae1962, Craig Pemberton, Citation bot 1, Night Jaguar, RjwilmsiBot, Wikeithpedia, Vincent Semeria, Slawekb, Quondum, Zephyrus Tavvier, Maschen, Zfeinst, RockMagnetist, ClueBot NG, Starshipenterprise, Wgolf, Helpful Pixie Bot, Mlhalpern, Bibcode Bot, Petermahlzahn, F=q(E+v^B), Mark viking, Jb1944, Airwoz, Epic Wink, Kfitzell29, Theoretical wormhole, Ayegbayo and Anonymous: 128

- **Lorentz covariance** *Source:* https://en.wikipedia.org/wiki/Lorentz_covariance?oldid=697340387 *Contributors:* Roadrunner , Stevertigo , Patrick ,Michae l Hardy , Charles Matthews , Reddi , Doradus , Phys , Drxenocide , Tobias Bergemann , Greyengine 5, Herbee , Fropu ff, Yath , Lumidek ,Kate , Chris Howard , Masudr , Hidaspal , Teorth , Oleg Alexandrov , Ruud Koot , Mpatel , Rjwilmsi , YurikBot , Bhny , Archelon , Grafen , Schla fl y,Modify , GrinBot ~enwiki , Incnis Mrsi , Gilliam , AlexDitto , Cybercobra , Canadianshoper , JRSpriggs , CmdrObot , Saintrain , Thijs !bot, Omooney ,Headbomb , D.H, Escarbot , Tim Shuba , Pkoppenb , WolfmanSF , Maurice Carbonaro , Lantonov , Thurth , TXiKiBoT , Kawakameha , Antixt ,Neparis , YohanN 7, Henry Delforn (old) , Curtdbz , Ajoykt , Djr32, SchreiberBike , MystBot , Addbot , Michele .allegra , DOI bot , Aboctok , SpBot ,Luckas -bot , Yobot , AnomieBOT , Citation bot , J04 n, Pandamonia , FrescoBot , Nunc aut numquam , Citation bot 1, Haael , UncertaintyPrinci -ples , Fizz 2010 , Quondum , AManWithNoPlan , SDLEECY , Raidr, Kevin Gorman, Bibcode Bot, Mr.viktor.stepanov, Dilaton, Andyhowlett,Monkbot and Anonymous: 43

- **File:Mecanismo_de_Higgs_PH.png** *Source:* https://upload.wikimedia.org/wikipedia/commons/4/44/Mecanismo_de_Higgs_PH.png *License:* CC-BY-SA-3.0 *Contributors;* ? *Original artist:* ?

- **File:Mexican_hat_potential_polar.svg** *Source:* https://upload.wikimedia.org/wikipedia/commons/7/7b/Mexican_hat_potential_polar.svg *License:* Public domain *Contributors:* Own work by uploader, with gnuplot *Original artist:* RupertMillard

- **File:Mollifier_Illustration.svg** *Source:* https://upload.wikimedia.org/wikipedia/commons/3/37/Mollifier_Illustration.svg *License:* CC0 *Contributors:* <a href='//commons.wikimedia.org/wiki/File:Mollifier_illustration.png' class='image'><img alt='Mollifier illustration.png' src='https://upload.wikimedia.org/wikipedia/commons/thumb/5/53/Mollifier_illustration.png/100px-Mollifier_illustration.png' width='100' height='32' srcset='https://upload.wikimedia.org/wikipedia/commons/thumb/5/53/Mollifier_illustration.png/150px-Mollifier_illustration.png 1.5x, https://upload.wikimedia.org/wikipedia/commons/thumb/5/53/Mollifier_illustration.png/200px-Mollifier_illustration.png 2x' data-file-width='500' data-file-height='158' /></a> *Original artist:* Fred the Oyster

- **File:MontreGousset001.jpg** *Source:* https://upload.wikimedia.org/wikipedia/commons/4/45/MontreGousset001.jpg *License:* CC-BY-SA-3.0 *Contributors:* Self-published work by ZA *Original artist:* Isabelle Grosjean ZA

- **File:Noether.jpg** *Source:* https://upload.wikimedia.org/wikipedia/commons/e/e5/Noether.jpg *License:* Public domain *Contributors:* EmmyNoether (1882-1935) *Original artist:* Unknown<a href='//www.wikidata.org/wiki/Q4233718' title='wikidata:Q4233718'><img alt='wikidata:Q4233718' src='https://upload.wikimedia.org/wikipedia/commons/thumb/f/ff/Wikidata- logo.svg/20px- Wikidata- logo.svg.png' width='20' height='11'srcset='https://upload.wikimedia.org/wikipedia/commons/thumb/f/ff/Wikidata- logo.svg/30px- Wikidata- logo.svg.png 1.5x, https://upload.wikimedia.org/wikipedia/commons/thumb/f/ff/Wikidata- logo.svg/40px- Wikidata- logo.svg.png 2x' data-file-width='1050' data-file-height='590' /></a>

- **File:Nuvola_apps_edu_mathematics_blue- p.svg** *Source:* https://upload.wikimedia.org/wikipedia/commons/3/3e/Nuvola_apps_edu_blue-p.svg *License:* GPL *Contributors:* Derivative work from Image:Nuvola apps edu mathematics.png and Image:Nuvola apps edu mathematics-p.svg *Original artist:* David Vignoni (original icon); Flamurai (SVG convertion); bayo (color)

- **File:Nuvola_apps_katomic.png** *Source:* https://upload.wikimedia.org/wikipedia/commons/7/73/Nuvola_apps_katomic.png *License:* LGPL *Contributors:* http://icon-king.com *Original artist:* David Vignoni / ICON KING

- **File:Office-book.svg** *Source:* https://upload.wikimedia.org/wikipedia/commons/a/a8/Office-book.svg *License:* Public domain *Contributors:* This and myself. *Original artist:* Chris Down/Tango project

- **File:Orthogonality_and_rotation.svg** *Source:* https://upload.wikimedia.org/wikipedia/commons/5/5e/Orthogonality_and_rotation.svg *License:* CC0 *Contributors:* Own work *Original artist:* Maschen

- **File:POV-Ray-Dodecahedron.svg** *Source:* https://upload.wikimedia.org/wikipedia/commons/a/a4/Dodecahedron.svg *License:* CC-BY-SA-3.0 *Contributors:* Vectorisation of Image:Dodecahedron.jpg *Original artist:* User:DTR

- **File:Parametric_continuity_C0.svg** *Source:* https://upload.wikimedia.org/wikipedia/commons/e/e8/Parametric_continuity_C0.svg *License:* CC BY-SA 4.0 *Contributors:* 100px *Original artist:* Fred the Oyster

- **File:Parametric_continuity_vector.svg** *Source:* https://upload.wikimedia.org/wikipedia/en/f/f3/Parametric_continuity_vector.svg *License:* PD *Contributors:*
I made a vector version of File:Parametric_continuity_c1.gif, using Inkscape.
Original artist:
Factsofphotos (talk)

- **File:Parity_1drep.png** *Source:* https://upload.wikimedia.org/wikipedia/commons/f/fc/Parity_1drep.png *License:* Public domain *Contributors:* ? *Original artist:* ?

- **File:Parity_clocks_-_P-conservation.svg** *Source:* https://upload.wikimedia.org/wikipedia/commons/6/65/Parity_clocks_-_P-conservation.svg *License:* CC BY-SA 3.0 *Contributors:* Own work *Original artist:* SiBr4

- **File:Parity_clocks_-_P-violation.svg** *Source:* https://upload.wikimedia.org/wikipedia/commons/1/1c/Parity_clocks_-_P-violation.svg *License:* CC BY-SA 3.0 *Contributors:* Own work *Original artist:* SiBr4

- **File:Pfeil_SO.svg** *Source:* https://upload.wikimedia.org/wikipedia/commons/a/a1/Pfeil_SO.svg *License:* Public domain *Contributors:* made by me (Inkscape or Corel-Draw or Flash) *Original artist:* user:Mjchael

- **File:Phase-diag2.svg** *Source:* https://upload.wikimedia.org/wikipedia/commons/3/34/Phase-diag2.svg *License:* CC-BY-SA-3.0 *Contributors:* SVG conversion from raster image Image:Phase-diag.png; some additions from Image:Phase diagram.png *Original artist:* me

- **File:Phase_change_-_en.svg** *Source:* https://upload.wikimedia.org/wikipedia/commons/0/0b/Phase_change_-_en.svg *License:* Public domain *Contributors:* Own work *Original artist:* F l a n k e r, penubag

- **File:Portal-puzzle.svg** *Source:* https://upload.wikimedia.org/wikipedia/en/f/fd/Portal-puzzle.svg *License:* Public domain *Contributors:* ? *Original artist:* ?

- **File:Question_book-new.svg** *Source:* https://upload.wikimedia.org/wikipedia/en/9/99/Question_book-new.svg *License:* Cc-by-sa-3.0 *Contributors:*
Created from scratch in Adobe Illustrator. Based on Image:Question book.png created by User:Equazcion *Original artist:*
Tkgd2007

- **File:Question_dropshade.png** *Source:* https://upload.wikimedia.org/wikipedia/commons/d/dd/Question_dropshade.png *License:* Public domain *Contributors:* Image created by JRM *Original artist:* JRM

- **File:Rapid_Oscillation.svg** *Source:* https://upload.wikimedia.org/wikipedia/commons/7/7d/Rapid_Oscillation.svg *License:* CC BY 3.0 *Contributors:* Transferred from en.wikipedia; transfered to Commons by User:Pbroks13 using CommonsHelper.
Original artist: --<a href='//en.wikipedia.org/wiki/user:Pbroks13' class='extiw' title='en:user:Pbroks13'>**pbroks13**</a><a href='//en.wikipedia.org/wiki/User_talk:Pbroks13' class='extiw' title='en:User talk:Pbroks13'>talk?</a> Original uploader was Pbroks13 at en.wikipedia

38.8.3 Content license